Grand Challenges in Biology and Biotechnology

Emerging Issues and Trends

Series Editor

Pabulo Henrique Rampelotto, Porto Alegre, Rio Grande do Sul, Brazil

This subseries of "Grand Challenges in Biology and Biotechnology" publishes volumes that take an in-depth look at selected emerging topics and trends in various areas of human, animal, plant, environmental and micro- biology and biotechnology. Each book addresses a current topic in a specialized manner, covering new technologies, new areas of research, or a new concept or idea that may be in its infancy today but could develop into an important mainstream topic in the future.

Sean Michael Scully • Johann Orlygsson
Editors

Thermophilic Anaerobes

Phylogeny, Physiology and Biotechnological Applications

Springer

Editors
Sean Michael Scully
Faculty of Natural Resource Sciences
University of Akureyri
Akureyri, Iceland

Johann Orlygsson
Faculty of Natural Resource Sciences
University of Akureyri
Akureyri, Iceland

ISSN 2367-1017 ISSN 2367-1025 (electronic)
Grand Challenges in Biology and Biotechnology
ISSN 2948-2461 ISSN 2948-247X (electronic)
Emerging Issues and Trends
ISBN 978-3-031-41719-1 ISBN 978-3-031-41720-7 (eBook)
https://doi.org/10.1007/978-3-031-41720-7

This Springer imprint is published by the registered company Springer Nature Switzerland AG
The registered company address is: Gewerbestrasse 11, 6330 Cham, Switzerland

Paper in this product is recyclable.

Contents

Part I
General Topics

Geothermal Habitats and Adaptations of Thermophilic Microbes

Bjorn Thor Adalsteinsson and Gudmundur Oli Hreggvidsson

Abstract In this chapter, the main habitats of thermophiles, their discovery, and ecology are discussed. The focus of the discussion is on natural habitats associated with geothermal activity, their geological origin, and characteristics of different geothermal surface manifestations, including mud pools, solfatara fields, alkaline hot springs, and warm springs. The ecological discussion is primarily focused on strategies that thermophiles utilize to obtain energy.

1 Introduction: Brief History of Scientific Exploration of the Upper Thermal Boundary of Life

From the early nineteenth century, microorganisms were known to inhabit high-temperature environments and the first thermophilic bacterial strains were isolated toward the end of the century (reviewed in Allen 1953 and Brock 2001). By 1920, scientific interest in thermophiles had dwindled and was largely confined to moderate thermophiles studied in the context of food microbiology. It was the pioneering work of Thomas D. Brock in the 1960s and onward that sparked a new wave of scientific interest in thermophiles that has continued to this date. In the early 1960s, Brock was involved in microbiology research, including studies on the sulfur-oxidizing bacterium *Thiothrix mucor* and on cyanobacteria, and visited Yellowstone National Park in search of possible habitats for these organisms. During these visits in 1964–1965, Brock noted that hot springs not only gave rise to diverse and dense microbial life in efflux channels where temperatures were moderately high but noticed evidence of microbial life at elevated temperatures, previously thought to be devoid of life. Specifically, he noticed pink filaments in the geothermal water at 82 °C that he strongly suspected were of biological nature (Brock 1995). He returned to Yellowstone in 1966 and attempted to obtain a culture of the organism by

B. T. Adalsteinsson · G. O. Hreggvidsson (✉)
Matís, Reykjavík, Iceland

University of Iceland, Reykjavík, Iceland
e-mail: gudmundo@matis.is

S. M. Scully, J. Orlygsson (eds.), *Thermophilic Anaerobes*, Emerging Issues and Trends, https://doi.org/10.1007/978-3-031-41720-7_1

inoculating a sample of the pink filaments into media and cultivating the enrichment at 70 °C under aerobic conditions. Instead of the desired pink organism, Brock was searching for, the culture became dense with yellow-pigmented cells. From the culture, the strain *Thermus aquaticus* YT-1 was first isolated, and subsequently, Brock and his colleagues isolated several other bacterial strains. They showed that the species grows optimally at 70 °C and has a maximal growth temperature of 79 °C (Brock and Freeze 1969). This work was the first to report an organism that grows at a temperature above 70 °C. The publication of these findings along with the deposition of strain YT-1 in the American Type Culture Collection was hugely influential, not only leading to an interest in the exploration of life at elevated temperatures but also to influential innovations in biotechnology. Taq polymerase, derived from *T. aquaticus,* became the standard enzyme for the polymerase chain reaction (PCR). Brock continued his studies in Yellowstone for several years. During that period, he showed that when microscope slides were immersed in geothermal hot springs at high temperatures, even at or around 100 °C, they became covered in cells that could be observed under a microscope (Brock 1967; Bott and Brock 1969). He also isolated the first representatives of thermoacidophilic aerobic archaea *Sulfolobus acidocaldarius* and *Thermoplasma acidophilum* (Brock et al. 1972; Darland et al. 1970).

In the 1980s, Karl Stetter and his colleague Wolfram Zillig became interested in microbial life at high temperatures (Stetter 2006) and their studies would lead to further leaps in knowledge about thermophiles, the discovery of the great diversity of anaerobes in geothermal sites, and the corresponding ecological and metabolic variety of these organisms. At that time, the aerobic archaeon, *Sulfolobus acidocaldarius,* was the most extreme thermophile known, with an optimal growth rate at 75 °C and an upper-temperature limit for growth at 85 °C. On a trip to Iceland in 1980, Stetter and Zillig sampled multiple boiling hot springs and, under the microscope, observed that the water teemed with what appeared to be microorganisms. Stetter noted that when the blue redox indicator resazurin was incubated in the hot-spring water, it immediately turned pink, an indication that the water was reducing—i.e., an anaerobic environment. This turned out to be a highly important observation since most strains that were later isolated from comparable environments are indeed strict anaerobes, and hence, their isolation requires careful handling under conditions devoid of oxygen. From samples collected in Kerlingarfjöll in central Iceland, Stetter and Zillig isolated the methanogen *Methanothermus fervidus* and a strictly anaerobic species of *Thermoproteales,* both of which grew at a maximum temperature of 97 °C (Zillig et al. 1981; Stetter et al. 1981)—far beyond the maximal growth temperature of *S. acidocaldarius.* Searching for still more extreme thermophiles, Stetter sampled submarine hydrothermal vents off the coast of Italy in 1981. From these samples, strictly anaerobic *Pyrodictium* strains were isolated with optimal growth at 105 °C and an upper limit for growth at 110 °C (Stetter 1982). Later, Stetter and colleagues isolated *Pyrolobus fumarii* from a black smoker in the Atlantic Ocean, which has an optimal growth temperature of 106 °C, an upper growth limit of 113 °C, and can survive in an autoclave for an hour at 121 °C (Blöchl et al. 1997). The term hyperthermophile has been coined for microorganisms with optimum growth temperature above 80 °C and most of the new species isolated by Stetter and coworkers, at that time and in the following decades were anaerobic hyperthermophilic Archaea. These new species were chemolithotrophic, chemolithoautotrophic, or organotrophic, which harnessed energy by anaerobic respiration. Extremely thermophilic bacteria ($T_{opt} > 80$ °C) were also discovered such as the aerobic marine hyperthermophile *Aquifex pyrophilus* of the phylum *Aquificota* and hyperthermophilic fermentative marine species *Thermotoga maritima* of the phylum *Thermotogae.*

Research efforts on thermophilic microbiology and ecology increased extensively in the early 1990s when large international research projects (thermophiles and extremophiles) in the field were funded by the European Union. This helped to establish in Europe important research groups in the field and to advance collaborative research in the microbiology of terrestrial and marine geothermal habitats.

Geothermal areas are largely reduced and anaerobic habitats, with various adaptations to energy sources, physicochemical conditions, and scarcity of oxygen. Due to the high novelty and the extremophilic adaptations to both temperatures and pH, early research focused largely on the microbiology of Archaea, but early work on the microbiology and ecology of anaerobic and fermentative bacteria was carried out by Jurgen Wiegel in Yellowstone Park in the USA and by Birgitte Ahring (Denmark) in Iceland. Fermentative, anaerobic bacteria are important consumers of organic matter in microbial mats and sediments, and a number of thermophilic adaptations belonging to novel genera such as *Thermoanaerobacterium*, *Thermoanaerobacter*, *Thermotoga,* and *Caldicellulosiruptor* have been discovered and described from these biotopes.

Early work on aerobic Bacteria was mainly carried out by the effort of KO. Stetter, and R. Huber in Germany; RAD. Williams, R. Sharp, and NDH. Raven in Britain; JK. Kristjansson, G. Alfredsson, S. Hjörleifsdottir, GO. Hreggvidsson in Iceland; Da. Costa and H. Santos in Portugal; T. Oshima in Japan; and HW. Morgan and RM. Daniel in New Zealand. This included work on the heterotrophic genera *Thermus*, *Rhodothermus,* and *Geobacillus* and the autotrophic hydrogen oxidizing genera *Hydrogenobacter* and *Aquifex.*

Colorful photosynthetic microbial mats are a conspicuous feature of alkaline hot spring effluents composed of photosynthetic bacteria. They are abundant in summer and near disappearing in winter. Pioneering work on the microbial ecology of these microbial mats was done by DM. Ward, R. Castenholz, and SR. Miller who isolated and described novel species belonging to the phyla *Cyanobacteria* and *Chloroflexi.* Consequently, the photosynthetic temperature boundary at 70–74 °C was established, above which photosynthetic bacteria are not found.

2 Underlying Geology and Global Distribution of Geothermal Areas

Geothermal areas are the primary natural habitats of thermophiles. They are diverse in terms of temperature, water abundance, pH levels, and availability of various molecules that microorganisms utilize for growth—a result of different underlying geology, which is briefly discussed here.

Geothermal areas are primarily found at plate-tectonic margins, in regions of active volcanism, where recent plutonism has occurred, and at intracontinental rifts (Fig. 1, Nukman and Moeck 2013; Bogie et al. 2005; Faulds et al. 2009), due to the presence of a heat source in the form of magma, pluton, or mantle close to earth's surface (Moeck 2014).

These geothermal sites harbor biotopes of only thermophilic microbes completely different from those of the surrounding area. They have sporadic distribution, and they are far apart on a global scale, which gives these confined ecosystems distinct island characteristics.

Geothermal features arise because of different sub-surface geological phenomena that provide a heat source. These heat sources result in temperatures exceeding 200 ° C within a depth of 3.000 m (Moeck and Beardsmore 2014). The features are formed as a result of heat being transferred from the heat source to the surface via convection—i.e., through the movement of fluids. The source of the fluids is generally meteoric water (Deon et al. 2012). As the fluids are exposed to the heat source, their composition changes when salts, minerals, acids, and other chemicals are dissolved. In their "journey" from heat source to the surface, the fluids may be further altered chemically as they pass through different geological layers. On breaking the surface, the hot fluids are manifested in geothermal features including fumaroles, geysers, hot springs, mud pools, or solfatara fields, depending on the water availability and resultant chemical composition, pH, and temperature of the fluids when they reach the surface. Further alterations then occur at the surface when encountering atmospheric oxygen due to abiotic oxidation, mainly of H_2S.

A shallow, active magma chamber provides an intense heat source that can give rise to geothermal features. Such geothermal systems can arise in association with active volcanism, for example, in Iceland, Java, the South American Andes, and Taiwan (Moeck and Beardsmore 2014). Surface features may arise directly above the heat source, atop a respective volcano, in the so-called upflow zone, which is generally characterized by high temperatures, water scarcity, and acidity. Fluids may also move horizontally after exposure to the heat source and surface at the roots of a volcano, in the so-called outflow zone. During the horizontal movement, the fluid cools down and approaches near neutral pH (Hochstein 1988). Magmatic geothermal systems can also arise due to the presence of an active magma chamber without active volcanism, for example, in the Taupo Volcanic Zone in New Zealand (Bogie et al. 2005; Moeck and Beardsmore 2014).

Further, geothermal features can be formed in areas where recent plutonism has occurred—i.e., where magma has risen through the crust without reaching the

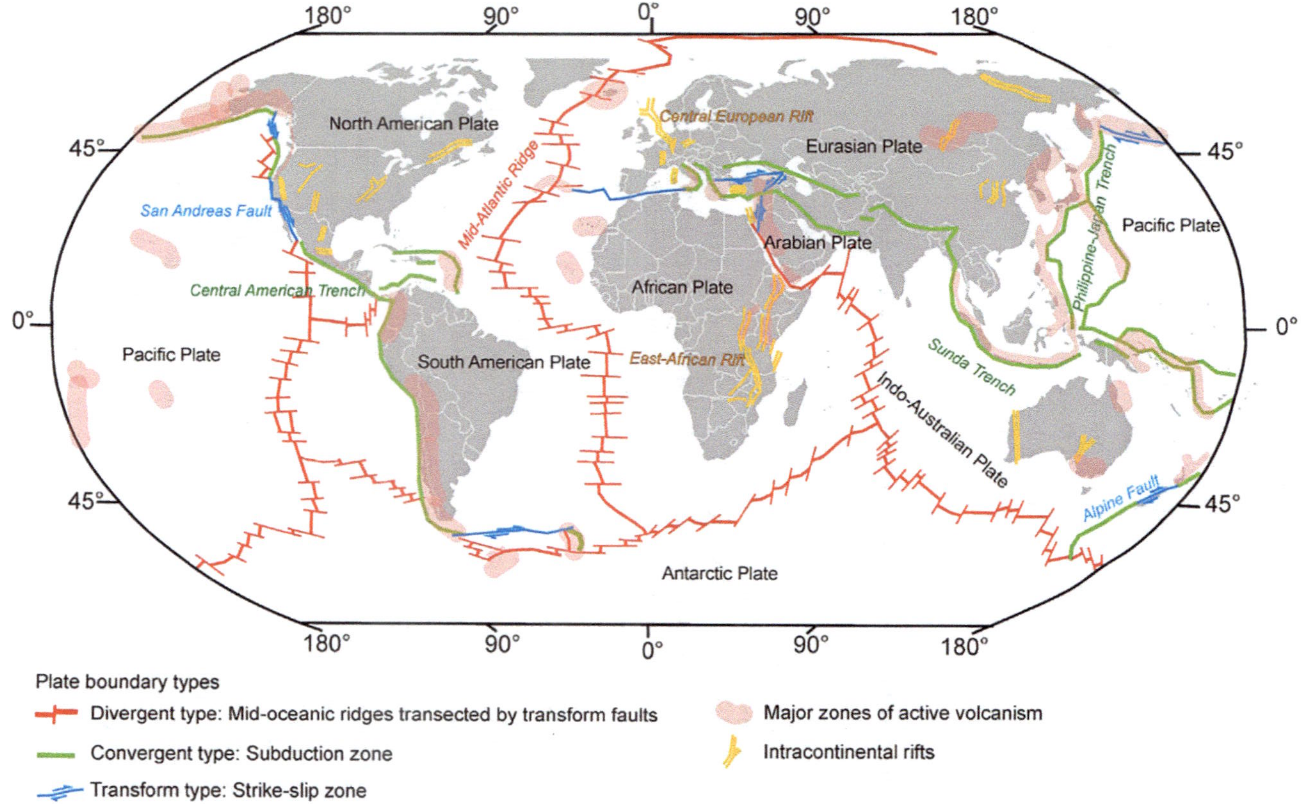

Fig. 1 Global distribution of plate boundaries, volcanism, and rifts. Geothermal surface features are found at plate-tectonic margins, in regions of active volcanism, where recent plutonism has occurred, and at intracontinental rifts. Adapted from Moeck (2014)

surface, is crystallizing, and radiates heat. They are generally formed at convergent continent–continent margins surrounded by mountains that provide a rich source of meteoric water to sustain convection (Moeck and Beardsmore 2014). Examples include the Geysers geothermal field in California and the Larderello geothermal system in Italy (Argus and Gordon 2001; Bertini et al. 2006).

Finally, geothermal features can be observed in areas where extension has caused crustal thinning such that the mantle is elevated to levels close to the surface, providing a heat source. Geothermal features form in this context where meteoric water can seep deep into the crust to interact with the heat source via faults or otherwise permeable layers. Geothermal systems of this type are found in western Turkey, East African rift, Upper Rhine graben in central Europe, and the Great Basin in the USA (Moeck and Beardsmore 2014).

2.1 High- and Low-Temperature Geothermal Areas

Geothermal areas with surface features can broadly be classified as high- or low-temperature fields.

In high-temperature fields, which usually coincide with active volcanic areas at high altitudes, temperatures exceed 200 °C at a depth of 1.000 m and groundwater levels are usually low. On the surface, they are characterized by the presence of steam, transformed and colorful soils, mud pools, and the release of gases, particularly N_2, CO_2, H_2S, and H_2. Below the surface, pH levels in these fields are circumneutral due to the presence of CO_2 and H_2S (p$Ka = 6.3$ and 7.2, respectively). However, as hydrogen sulfide reaches the surface, it is oxidized chemically due to exposure to atmospheric O_2 and biologically due to microbial respiration. This leads to the formation of H_2SO_4 ($H_2S + O_2 \rightarrow H_2SO_4$) that lowers the pH level at the surface and transforms surface rocks into mud. Most of the surface of the respective geothermal area is turned into an acidic solfatara field with white and yellow sulfur precipitations, generally with mud pools scattered about where some water is present. These areas are further characterized by instability, in that water levels may change dramatically over a short time span, and surface features may be "lost" and others may form at regular intervals.

In low-temperature fields, temperatures are lower than 150 °C at 1.000 m depth. On the surface, they are characterized by pools of liquid water at neutral or slightly alkaline pH. Water influx is relatively generous, as compared with high-temperature fields, giving rise to effluent streams with temperature gradients that sustain diverse microbial communities, often providing colorful layers to the otherwise largely white or off-white geothermal field. Sulfide levels are low, and with high efflux rates, H_2SO_4 does not accumulate. The water contains dissolved minerals and silica (SiO_2) when the fluids interact with silicate rocks and bicarbonate when they interact with carbonate rocks and gases of varying levels. The silica precipitates as waters cool down at the surface of the hot springs, forming silica sinters, and accordingly, the sinters form low, broad deposits that extend several meters from the respective hot spring. In contrast, calcium carbonate precipitates rapidly when CO_2 escapes from bicarbonate-rich fluids, causing supersaturation with respect to bicarbonate, and the formation of travertine. The rapid precipitation results in the formation of deposits near the edge of the respective hot spring, and as a result, the deposits form high-relief structures. In low-temperature fields, temperatures and water flow are generally stable.

2.2 *Surface Features in Terrestrial Geothermal Areas*

Solfatara fields are characterized by large surface areas covered in soft soil of varying hues of light brown—rock transformed by sulfuric acid (Fig. 2). Adding to the richness of color, the fields are generally scattered with tones of yellow due to

Fig. 2 Solfatara field. (*Top*): The hill Námafjall, east of Mývatn, Iceland, showing hues of *brown*, *yellow*, and *red* characteristic of solfatara fields. (*Bottom*): Close-up images from the same field, showing *red* (*left*) and *yellow* (*right*) deposits in more detail

Fig. 3 Mud pools. (*Top*): A mud pool in the Hverarönd geothermal area in the slopes of Námafjall, Mývatn, Iceland. The pool is about 5 m in diameter (rough estimate by the photographer). (*Bottom*): Closeup images of the thick bubbling fluids of two mud pools. The *bottom left* image is from the same pool as that on *top*, and the *bottom right* image is from a separate pool with thicker fluids. The images were taken after a period of heavy rainfall, which likely reduced the thickness of the pool fluids

the deposition of sulfur, tones of red due to the deposition of iron-containing compounds (e.g., hematite Fe_2O_3), and hues of dark gray due to the deposition of ferrous sulfide (FeS). The acidic soil harbors communities of strictly anaerobic archaea and at lower temperatures of *Thermoplasma* and *Picrophilus* archaea and *Thiomonas, Thiobacillus, Geobacillus,* and *Deinococcus* bacteria (Hreggvidsson et al. 2017).

Mud pools are formed in solfatara fields where the field's mud and liquid water mix in varying proportions (Fig. 3). The pools have no efflux channel, water is generally scarce, and depending on the water influx rate, the mud pot can be thick or thin. Often, bubbling is observed in the pool, a result of gases and steam passing to

Fig. 4 Fumaroles. Both images are from Hveraröndo, east of Mývatn in Iceland. Surrounding the fumarole in the top image is a mound has formed of about half a meter in height. The opening of the fumarole in the bottom image is very small, about 0.5–1 cm in diameter

the surface. Microorganisms frequently encountered in mud pools include *Stygiolobus* and *Sulfolobus* archaea, and *Hydrogenobaculum* bacteria (Hreggvidsson et al. 2017)

Fumaroles, often observed in solfatara fields, are small openings through which steam and volcanic gases (including CO_2, H_2S, and SO_2) are released (Fig. 4). The flow rate of the gases is generally quite high, and as it passes through the surrounding rocks and mud at the surface, a rather loud sound is often emitted, reminiscent of a violent storm

Boiling pits are small, shallow depressions in the surface with bubbling water. They can be found in high-temperature areas, in which case the water bubbles as a result of volcanic gases streaming through the water, and they can be found in low-temperature areas, in which case the bubbling is caused by boiling of the water

Alkaline hot springs (>50 °C) and warm springs (<50 °C) are common in low-temperature fields. They are pools of mineral water, circular in form, and of varying sizes, often 2–5 m^2. They generally have outlets where water from the pools flows out into the colder, surrounding environment, creating a temperature gradient that can sustain diverse life. In alkaline hot springs, pH levels are in the range of 7–10. These habitats often give rise to colorful microbial mats, with distinct "bands" of different colors corresponding to particular microorganisms. The boundaries reflect gradients in physicochemical parameters—pH, temperature, and fluid composition—and the ability or competence of respective microorganisms to thrive therein. Commonly observed colors and associated microorganisms include green from cyanobacteria; hues of orange and/or red from phototrophic *Chloroflexus* and *Roseiflexus*; and white, gray, gray with hues of blue, and black from hydrogen oxidizing *Sulfurihydrogenibium* and *Thermocrinis albus* (Hreggvidsson et al. 2017).

Sulfide-rich hot springs (65–85 °C) are rare but are occasionally found in low-temperature geothermal fields and in high-temperature fields where water is abundant. Like alkaline hot springs, they are water-rich, approximately circular in form, and often 2–5 m^2 in size. They differ in that the water contains high concentrations of sulfide. The springs have relatively high water flow rates and outlets, and therefore, sulfuric acid does not accumulate, and the pH level is around 5.5–6.5. They are often associated with microbial mats containing sulfide-utilizing species, in particular dominated by *Sulfurihydrogenobium* species (Hreggvidsson et al. 2017).

Steam vents are sometimes found in low-temperature fields, often in the absence of any other nearby surface features (Fig. 5). They form where steam from hot groundwater rises through porous layers such as a young lava field and rises to the surface. From a distance, they may therefore give the appearance that the lava itself is fuming. On closer inspection, one can identify discrete openings where the steam escapes, in which temperatures are in the range of 55–85 °C and pH of the surrounding soil is 7–8. Commonly, members of *Thermus, Chloroflexus, Actinobacteria,* and *Acidobacteria* are found in these environments (Hreggvidsson et al. 2017).

3 Other High-Temperature Environments, Natural and Anthropogenic

Various other thermal environments, natural and anthropogenic (man-made), sustain thermophilic microbial communities. They are, however, generally less extensively studied than those discussed above—in some cases a result of their rarity and/or inaccessibility—and will therefore only be briefly discussed here.

Intertidal and *submarine hot springs*: Geothermal fluids can in principle surface through earth's crust anywhere on the globe, given the presence of a geological heat source as discussed above—i.e., they can surface on dry land, on shore, or underwater. Intertidal hot springs are formed when geothermal fluids surface on an ocean's shore. They are quite unique environments in that temperatures fluctuate greatly with tidal movements. At high tide, the pools are covered in seawater, and temperatures match that of the respective ocean, except right at the hot-spring source. At low tide, the pool is quickly heated by the geothermal fluids. The temperature shift at low and

Fig. 5 Steam vents. *Top* image shows steam arising from multiple steam vents on and under the slopes of Jarðbaðshólar, east of Mývatn in Iceland. Note the larger plume in the foreground and the smaller plumes on the hill. The *bottom* image shows a steam vent opening. Note the lack of fluids and the transformed vegetation

high tide depends on the temperature of the geothermal fluids and that of the ocean water. These types of environments are for example found at various locations around the coast of Iceland (Bjornsdottir et al. 2021; Kale et al. 2013; Hobel et al. 2005), in Italy, New Zealand, Fiji (Burgess et al. 2007), and others. Submarine hot springs are formed when geothermal fluids discharge underwater. They can be located at depths from a few meters to a few kilometers and include vents off the coast of Milos Island, Greece (Sievert et al. 2000), off the northern coast of Iceland (Marteinsson et al. 2001), and deep-sea vents located at a great depth near Galápagos islands (Corliss et al. 1979). The hydrothermal fluids that emanate from the vents differ significantly in chemical composition and hence the associated microbial communities. In some systems, they contain high concentrations of sulfides (Kelley et al. 2002), while in others they are enriched in hydrogen and methane (Kelley et al. 2005). Heat at the source can be extremely high, but mixing with seawater causes a steep temperature gradient, thus sustaining communities of thermophiles, mesophiles, and psychrophiles.

Anthropogenic thermal environments: Various human activities have resulted in the formation of "non-natural" warm habitats that are conducive for the growth of thermophiles (Pask-Hughes and Williams 1975; Brock and Boyle 1973). Examples of such habitats include heat exchangers and pipelines in homes that carry hot water, e.g., in the context of delivering hot water from a boiler to a radiator; district heating systems that distribute hot water to entire towns or cities; thermophilic waste treatment plants; burning coal waste piles; and various industrial processes that involve heating, e.g., in the context of reducing microbial content in foods.

Transient natural thermal habitats can result from the self-heating of composts; of hay, straw, or other similar agricultural products that are stored in large quantities; or of manure. Though these types of environments may form in the absence of humans, they exist at a larger scale due to human activities.

4 Ecology in Geothermal Habitats

Temperature strongly affects the physiology of organisms and hence their ability to thrive in a given environment. Some multicellular organisms can regulate body heat and their cells are therefore partially protected from the surrounding heat, while in microorganisms, which are unicellular, cytoplasmic temperature directly follows that of the environment. Many cellular macromolecules like DNA and enzymes are quite vulnerable to loss of structure (and hence function) due to elevated heat since their structure is largely the result of weak chemical interactions (hydrogen bonds, etc.). Many thermophilic microorganisms have evolved with genomes that have high GC content and encode rigid enzymes/proteins to withstand thermal disruption (Feller 2010; Radestock and Gohlke 2011; Hu et al. 2022). These adaptations, however, render the organism's incapable of growth at lower temperatures. Each microorganism is adapted for growth at a particular temperature, the T_{opt}, where its growth is fastest. Growth rates generally reduce linearly some 15–25 °C from the T_{opt} to the lowest temperature that will sustain growth, the T_{min}. Above the T_{opt}, the growth rate is reduced more rapidly to a temperature, T_{max}, above which no growth is observed. There is no single accepted consensus for the defining growth temperature that would classify an organism as a thermophile. Brock proposed that this temperature should be 55–60 °C since habitats with temperatures below 55 °C are common in nature, while habitats with higher temperatures are rare. Further, he noted that no eukaryotes grow at temperatures beyond this limit, while certain bacteria and archaea thrive. According to this definition, the thermophiles are therefore exclusively prokaryotes. The terms moderate thermophile, thermophile, extreme thermophile, and hyperthermophile are now used in microbiology of thermophiles to describe different temperature adaptations. The demarcations are not clear, but the following criteria have been proposed for defining more accurately both thermophiles and hyperthermophiles, the former having $T_{max} \geq 65$ °C and the latter having $T_{opt} > 80$ °C (Hreggvidsson et al. 2017; Kristjansson and Stetter 1991). Taking into account the definition of Brock, a moderate thermophile would then have T_{max} higher than 55 °C and lower than 65 °C.

Thermophiles have diverse chemotropic catabolic processes for harnessing energy, both organotrophic and lithotrophic, and they use both autotrophic and organotrophic processes for supplying carbon to anabolic pathways. Autotrophic thermophiles include both photoautotrophs and chemolithoautotrophy. Photoautotrophs are, however, not found at temperatures beyond 70–74 °C, the photosynthetic boundary, while diverse chemolithoautotrophs that utilize various inorganic electron acceptors and donors thrive at more extreme temperatures. In addition to the primary production of organic chemicals that occurs in geothermal habitats, the organic material may be introduced into the habitat from outside sources—e.g., leaves or other plant material that are blown in a gust of wind into a hot spring. Overall, geothermal habitats are nevertheless generally oligotrophic. Sudden changes in temperature in hot springs can, however, lead to a rapid increase in nutritional availability. After a period of stable temperature, in which certain organisms can thrive at or close to their T_{opt}, with a corresponding accumulation of biomass, a sudden increase in temperature can result in their death and hence elevated nutritional level in the habitat that other more thermophilic species then utilize. Some chemolithoautotrophic thermophiles utilize oxygen as an electron acceptor in their metabolism. Oxygen concentration in geothermal fluids is, however, relatively low, as compared with concentrations found in lakes or seawater, since oxygen solubility in water decreases with increased temperature (Geng and Duan 2010).

Among archaea, the most extensively studied thermophiles are methanogens and a broad group of sulfur-metabolizing species. Methanogens obtain energy via an anaerobic respiratory pathway called methanogenesis that is uniquely found in archaea. They can be further classified as hydrogenotrophs, which comprise five orders, and methylotrophs that comprise a single order. Hydrogenotrophs obtain energy primarily by the reduction of CO_2 into CH_4 using H_2 as an electron donor, though a few other small organic molecules can act as electron donors as well. Methylotrophs are similarly capable of the reduction of CO_2 into CH_4, but are characterized by their ability to convert various methyl group-containing compounds—methanol, methyl amines, and methyl sulfides—and acetate into CH_4 (Costa and Leigh 2014). Examples of thermophilic methanogens are *Methanothermus fervidus* (T_{opt} = 83 °C, pH_{opt} = 6.5) that is found in anaerobic mud and soil and *Methanobacterium thermoautotrophicum* (T_{opt} = 65, pH_{opt} = 7.4) found in alkaline hot springs and sewage sludge (Stetter et al. 1981; Zeikus and Wolfe 1972).

Archaea use diverse aerobic and anaerobic metabolic pathways for energy conservation using sulfur-containing compounds as electron donors and acceptors (reviewed in Liu et al. 2012 and Hreggvidsson et al. 2017). The compounds include elemental sulfur S^0, sulfate, sulfite, thiosulfate, sulfide, and others. In aerobic sulfur oxidation, S^0 is the electron donor and oxygen is the electron acceptor. The process is, e.g., utilized by the thermoacidophilic *Sulfolobus acidocaldarius* (T_{opt} = 75 °C, pH_{opt} = 2.5) and *Acidianus infernus* (T_{opt} = 90 °C, pH_{opt} = 2) that inhabit acidic solfatara fields. Under anaerobic conditions, S^0 can be reduced for energy conservation by at least three mechanisms, all of which are commonly found in thermophiles that inhabit anaerobic geothermal soil. First, this can occur by autotrophic respiration using H_2 as an electron donor—e.g., in *A. infernus* and in *Thermoproteus tenax* (T_{opt} = 90 °C, pH_{opt} = 5) and *Pyrodictium occultum* (T_{opt} = 105 °C,

$pH_{opt} = 6.5$); second, by heterotrophic respiration with organic chemicals as electron donors—e.g., in *Thermoproteus tenax* (T_{opt} = 90 °C, pH_{opt} = 5); and third, by fermentation of organic chemicals—for example, *Pyrococcus furiosus* (T_{opt} = 100 °C, pH_{opt} = 6). Under anaerobic conditions, sulfate and sulfite can also be reduced with organic compounds or H_2 as electron donors—e.g., in *Archaeoglobus fulgidus* (T_{opt} = 83 °C, pH_{opt} = 7) and *A. profundus* (T_{opt} = 82 °C, pH_{opt} = 6). Other thermophilic archaea found in geothermal areas include the heterotrophs *Thermoplasma volcanium* (T_{opt} = 60 °C, pH_{opt} = 2) and *Sulfolobus acidocaldarius* (T_{opt} = 75 °C, pH_{opt} = 2.5) that can obtain energy through aerobic respiration of organic matter and *Pyrococcus furiosus* (T_{opt} = 100 °C, pH_{opt} = 6) that obtain energy through anaerobic fermentation of organic matter.

Some metabolic pathways for energy conservation in archaea are also found in bacteria, such as the Embden–Meyerhof and Entner–Doudoroff glycolytic pathways. The pathways are, however, partially different in the two domains, in that reactants are converted to products via different enzymes and hence through different intermediates (Bräsen et al. 2014). Methanogenesis is not found in bacteria, and, conversely, photo-autotrophy based on the electron transport chain is found in bacteria but not in archaea—e.g., in *Synechococcus lividus* (T_{opt} = 65 °C, pH_{opt} = 8) and *Chloroflexus aurantiacus* (T_{opt} = 56 °C, pH_{opt} = 8). Photoautotrophic thermophilic bacteria do, however, not survive at very high temperatures —the highest T_{max} observed is in the range of 70–75 °C. Other autotrophic thermophilic bacteria oxidize inorganic molecules (reviewed in Kristjansson et al. 2000), including hydrogen and hydrogen sulfide—e.g., aerobic species of the phylum *Aquificae,* the extreme thermophile, *Hydrogenobacter thermophilus* (T_{opt} = 72 °C, pH_{opt} = 6.8) and the hyperthermophile, *Thermocrinis ruber* (T_{opt} = 80 °C, pH_{opt} = 7–8.5), respectively, that inhabit alkaline hot springs. Other thermophilic autotrophic bacteria in the same habitat utilize sulfate as an electron acceptor and hydrogen as electron donor for energy conservation—e.g., *Thermodesulfobacterium thermophilum* (T_{opt} = 65 °C, pH_{opt} = 7.5).

Thermophilic aerobic chemoorganotrophic bacteria such as species belonging to the genera *Thermus, Geobacillus,* and *Rhodothermus* oxidize organic matter and thrive in circumneutral and alkaline environments with growth optima ≤80 ° C.

Anaerobic thermophilic and hyperthermophilic fermentative bacteria are isolated from microbial mats and geothermal anaerobic mud generally in the range of pH 5.5–7.5. Identified species are usually strict anaerobes, and fermentation is their predominant catabolic process. A great variety of these species belong to anaerobic genera of the phylum *Firmicutes,* e.g., *Clostridium, Thermoanaerobacter, Thermoanaerobacterium*, and the hyperthermophilic genus *Caldicellulosiruptor* (Scully and Orlygsson 2015; Willquist and van Niel 2012). Whereas *Firmicutes* has both aerobic and anaerobic members, *Thermotogae* has only the latter, glycolytic fermentative catabolism being the predominant catabolic process of the phylum. Temperature growth optima of the members range from 55° to 80 °C, and the optimum pH for growth ranges from 5.5 to 7.5. Different genera have been isolated from terrestrial and marine habitats, e.g., respectively, the genera *Fervidobacterium* and *Thermotoga*. Hyperthermophiles belonging to the phylum have only been identified in marine habitats, by species such as *Thermotoga maritima* ($T_{opt} = 80$ °C, $pH_{opt} = 6.5$) and *Thermotoga neapolitana* ($T_{opt} = 80$ °C, $pH_{opt} = 7$), the latter of which uses S^0 as an electron acceptor and have only been found in marine geothermal sites (Frock et al. 2010).

References

Allen MB (1953) The thermophilic aerobic spore-forming bacteria. Bacteriol Rev 17(2):125–173

Argus DF, Gordon RG (2001) Present tectonic motion across the coast ranges and San Andreas fault system in Central California. Geol Soc Am Bull 113(12):1580–1592. https://doi.org/10.1130/0016-7606(2001)113<1580:PTMATC>2.0.CO;2

Bertini G, Casini M, Gianelli G et al (2006) Geological structure of a long-living geothermal system, Larderello, Italy. Terra Nova 18(3):163–169. https://doi.org/10.1111/j.1365-3121.2006.00676.x

Bjornsdottir SH, Petursdottir SK, Gudmundsdottir EE et al (2021) *Rhodocaloribacter litoris* gen. nov., sp. nov., isolated from an intertidal hot spring. Int J Syst Evol Micr 71(10):005073. doi: https://doi.org/10.1099/ijsem.0.005073

Blöchl E, Rachel R, Burggraf S et al (1997) *Pyrolobus fumarii*, gen. and sp. nov., represents a novel group of archaea, extending the upper temperature limit for life to 113°C. Extremophiles 1(1): 14–21

Bogie I, Lawless JV, Rychagov S et al (2005) Magmatic-related hydrothermal systems: classification of the types of geothermal systems and their ore mineralization. In: Proceedings of geoconference in Russia. Kuril, pp. 51–73

Bott TL, Brock TD (1969) Bacterial growth rates above 90°C in Yellowstone hot springs. Science 164(3886):1411–1412

Bräsen C, Esser D, Rauch B et al (2014) Carbohydrate metabolism in Archaea: current insights into unusual enzymes and pathways and their regulation. Microbiol Mol Biol Rev 78(1):89–175. https://doi.org/10.1128/MMBR.00041-13

Brock TD (1967) Life at high temperatures: evolutionary, ecological, and biochemical significance of organisms living in hot springs is discussed. Science 158(3804):1012–1019

Brock TD (1995) The road to Yellowstone—and beyond. Annu Rev Microbiol 49(1):1–29

Brock TD (2001) The origins of research on thermophiles. In: Reysenbach A, Voytek M, Mancinelli R (eds) Thermophiles biodiversity, ecology, and evolution. Springer, Boston, pp 1–9

Brock TD, Boyle KL (1973) Presence of thermophilic bacteria in laundry and domestic hot-water heaters. Appl Microbiol 25(1):72–76. https://doi.org/10.1128/AEM.25.1.72-76.1973

Brock TD, Freeze H (1969) *Thermus aquaticus* gen. n. and sp. n., a nonsporulating extreme thermophile. J Bacteriol 98(1):289–297. doi:https://doi.org/10.1128/jb.98.1.289-297.1969

Brock TD, Brock KM, Belly RT et al (1972) *Sulfolobus*: a new genus of sulfur-oxidizing bacteria living at low pH and high temperature. Arch Mikrobiol 84(1):54–68. doi:https://doi.org/10.1007/BF00408082

Burgess EA, Wagner ID, Wiegel J (2007) Thermal environments and biodiversity. In: Gerday C, Glansdorff N (eds) Physiology and biochemistry of extremophiles. ASM Press, pp 11–29

Corliss JB, Dymond J, Gordon LI et al (1979) Submarine thermal springs on the galapagos rift. Science 203(4385):1073–1083. https://doi.org/10.1126/science.203.4385.1073

Costa KC, Leigh JA (2014) Metabolic versatility in methanogens. Curr Opin Biotechnol 29:70–75. https://doi.org/10.1016/j.copbio.2014.02.012

Darland G, Brock TD, Samsonoff W et al (1970) A thermophilic, acidophilic mycoplasma isolated from a coal refuse pile. Science 170(3965):1416–1418. https://doi.org/10.1126/science.170.3965.1416

Deon F, Moeck I, Sheytt T et al (2012) Preliminary assessment of the geothermal system of the Tiris volcanic area, East Java, Indonesia. In: 74th EAGE conference and exhibition incorporating EUROPEC 2012. European Association of Geoscientists & Engineers, p 293

Faulds JE, Bouchot V, Moeck I et al (2009) Structural controls on geothermal systems in western Turkey: a preliminary report. Geoth Res T 33(2009):375–382

Feller G (2010) Protein stability and enzyme activity at extreme biological temperatures. J Phys Condens Matter 22(32):323101. https://doi.org/10.1088/0953-8984/22/32/323101

Frock AD, Notey JS, Kelly RM (2010) The genus *Thermotoga*: recent developments. Environ Technol 31(10):1169–1181. https://doi.org/10.1080/09593330.2010.484076

Geng G, Duan ZH (2010) Prediction of oxygen solubility in pure water and brines up to high temperatures and pressures. Geochim Cosmochim Acta 74(19):5631–5640. https://doi.org/10.1016/j.gca.2010.06.034

Hobel CF, Marteinsson VT, Hreggvidsson GO et al (2005) Investigation of the microbial ecology of intertidal hot springs by using diversity analysis of 16S rRNA and chitinase genes. Appl Environ Microbiol 71(5):2771–2776. https://doi.org/10.1128/AEM.71.5.2771-2776.2005

Hochstein MP (1988) Assessment and modelling of geothermal reservoirs (small utilization schemes). Geothermics 17(1):15–49

Hreggvidsson GO, Petursdottir SK, Stefansson SK et al (2017) Divergence of species in the geothermal environment. In: Stan-Lotter H, Fendrihan S (eds) Adaption of microbial life to environmental extremes. Springer, Cham, pp 41–74

Hu EZ, Lan XR, Liu ZL et al (2022) A positive correlation between GC content and growth temperature in prokaryotes. BMC Genomics 23(1):110. https://doi.org/10.1186/s12864-022-08353-7

Kale V, Björnsdottir SH, Fridjonsson OH et al (2013). *Litorilinea aerophila* gen. nov., sp. nov., an aerobic member of the class Caldilineae, phylum Chloroflexi, isolated from an intertidal hot spring. Int J Syst Evol Microbiol 63:1149–1154. doi:https://doi.org/10.1099/ijs.0.044115-0

Kelley DS, Baross JA, Delaney JR (2002) Volcanoes, fluids, and life at mid-ocean ridge spreading centers. Annu Rev Earth Planet Sci 30(1):385–491. https://doi.org/10.1146/annurev.earth.30.091201.141331

Kelley DS, Karson JA, Fruh-Green GL, Yoerger DR, Shank TM, Butterfield DA et al (2005) A serpentinite-hosted ecosystem: the Lost City hydrothermal field. Science 307(5714):1428–1434

Kristjansson JK, Stetter KO (1991) Thermophilic bacteria. In: Kristjanson JK (ed) Thermophilic bacteria. CRC press, pp 1–18

Kristjansson JK, Hreggvidsson GO, Grant WD (2000) Taxonomy of extremophiles. In: Priest FG, Goodfellow M (eds) Applied microbial systematics. Chapman & Hall, London

Liu Y, Beer LL, Whitman WB (2012) Sulfur metabolism in archaea reveals novel processes. Environ Microbiol 14(10):2632–2644. https://doi.org/10.1111/j.1462-2920.2012.02783.x

Marteinsson VT, Kristjansson JK, Kristmannsdottir H et al (2001) Discovery and description of giant submarine smectite cones on the seafloor in Eyjafjordur, northern Iceland, and a novel thermal microbial habitat. Appl Environ Microbiol 67(2):827–833. https://doi.org/10.1128/AEM.67.2.827-833.2001

Moeck IS (2014) Catalog of geothermal play types based on geologic controls. Renew Sustain Energy Rev 37:867–882. https://doi.org/10.1016/j.rser.2014.05.032

Moeck I, Beardsmore G (2014) A new 'geothermal play type' catalog: streamlining exploration decision making. In: Proceedings of the thirty-ninth workshop on geothermal reservoir engineering, Stanford University, Stanford, CA, p 2014

Nukman M, Moeck I (2013) Structural controls on a geothermal system in the Tarutung Basin, north Central Sumatra. J Asian Earth Sci 74:86–96. https://doi.org/10.1016/j.jseaes.2013.06.012

Pask-Hughes R, Williams RAD (1975) Extremely thermophilic gram-negative bacteria from hot tap water. Microbiology 88(2):321–328. https://doi.org/10.1099/00221287-88-2-321

Radestock S, Gohlke H (2011) Protein rigidity and thermophilic adaptation. Proteins 79(4):1089–1108. https://doi.org/10.1002/prot.22946

Scully SM, Orlygsson J (2015) Recent advances in second generation ethanol production by thermophilic bacteria. Energies 8:1–30. https://doi.org/10.3390/en8010001

Sievert SM, Kuever J, Muyzer G (2000) Identification of 16S ribosomal DNA-defined bacterial populations at a shallow submarine hydrothermal vent near Milos Island (Greece). Appl Environ Microbiol 66(7):3102–3109. https://doi.org/10.1128/AEM.66.7.3102-3109.2000

Stetter KO (1982) Ultrathin mycelia-forming organisms from submarine volcanic areas having an optimum growth temperature of 105°C. Nature 300:258–260. https://doi.org/10.1038/300258a

Stetter KO (2006) History of discovery of the first hyperthermophiles. Extremophiles 10(5): 357–362. https://doi.org/10.1007/s00792-006-0012-7

Stetter KO, Thomm M, Winter J et al (1981) *Methanothermus fervidus*, sp. nov., a novel extremely thermophilic methanogen isolated from an Icelandic hot spring. ZBL Bact P: I. Abt. Originale C: Allgemeine, angewandte und ökologische Mikrobiologie 2(2):166–178. doi:https://doi.org/10.1016/S0721-9571(81)80038-5

Willquist K, van Niel EWJ (2012) Growth and hydrogen production characteristics of *Caldicellulosiruptor saccharolyticus* on chemically defined minimal media. Int J Hydrogen Energy 37:4925–4929. https://doi.org/10.1016/j.ijhydene.2011.12.055

Zeikus JG, Wolfe RS (1972) *Methanobacterium thermoautotrophicus* sp. n., an anaerobic, autotrophic, extreme thermophile. J Bacteriol 109(2):707–713. doi:https://doi.org/10.1128/jb.109.2.707-713.1972

Zillig W, Stetter KO, Schäfer W et al (1981) *Thermoproteales*: a novel type of extremely thermoacidophilic anaerobic archaebacteria isolated from Icelandic solfataras. ZBL Bact P: I. Abt Originale C: Allgemeine, angewandte und ökologische Mikrobiologie 2(3): 205–227. doi:https://doi.org/10.1016/S0721-9571(81)80001-4

Diversity of Thermophilic Prokaryotes

Oddur Vilhelmsson, M. Audur Sigurbjornsdottir, Gudny Vala Thorsteinsdottir, Martina Cascone, Davide Corso, Luca Tonietti, Flavia Migliaccio, Nunzia Nappi, Annarita Ricciardelli, Matteo Selci, Francesco Montemagno, Bernardo Barosa, Deborah Bastoni, Alessia Bastianoni, Angelina Cordone, and Donato Giovannelli

Abstract The term "thermophilic prokaryotes" covers an immense taxonomic and functional diversity of bacteria and archaea, spanning the length and breadth of the prokaryotic Tree of Life. Indeed, thermophiles are found within most major prokaryotic lineages and their functional diversity runs the gamut of biochemical and

---

O. Vilhelmsson (✉)
Natural Resource Sciences, University of Akureyri, Akureyri, Iceland

BioMedical Center, University of Iceland, Reykjavík, Iceland
e-mail: oddurv@unak.is

M. A. Sigurbjornsdottir
Natural Resource Sciences, University of Akureyri, Akureyri, Iceland

G. V. Thorsteinsdottir
Icelandic Institute of Natural History, Akureyri, Iceland

M. Cascone · D. Corso · F. Migliaccio · N. Nappi · A. Ricciardelli · M. Selci · F. Montemagno · B. Barosa · D. Bastoni · A. Bastianoni · A. Cordone
Department of Biology, University of Naples Federico II, Naples, Italy

Department of Science and Technology, Parthenope University of Naples, Naples, Italy

L. Tonietti
Department of Biology, University of Naples Federico II, Naples, Italy

Department of Science and Technology, Parthenope University of Naples, Naples, Italy

National Research Council – Institute of Marine Biological Resources and Biotechnologies - CNR-IRBIM, Ancona, Italy

D. Giovannelli
Department of Biology, University of Naples Federico II, Naples, Italy

Department of Science and Technology, Parthenope University of Naples, Naples, Italy

Department of Marine and Coastal Science, Rutgers University, New Brunswick, NJ, USA

Marine Chemistry and Geochemistry Department, Woods Hole Oceanographic Institution, Woods Hole, MA, USA

Earth-Life Science Institute, Tokyo Institute of Technology, Tokyo, Japan

Department of Science and Technology, University of Naples Parthenope, Naples, Italy

S. M. Scully, J. Orlygsson (eds.), *Thermophilic Anaerobes*, Emerging Issues and Trends, https://doi.org/10.1007/978-3-031-41720-7_2

physiological adaptations. Thus, examples can be found of thermophilic lithoautotrophs as well as chemoheterotrophs, obligate anaerobes and aerophiles, extreme halophiles, acidophiles and alkaliphiles, and more. Their ecology is likewise diverse, with thermophiles found in a variety of habitats ranging from hydrothermal vents to desert soil to industrial settings and wastewater treatment facilities. It goes without saying that such immense diversity cannot be reviewed comprehensively in a relatively short book chapter. We thus aim to present examples pulled from diverse taxa within the vast menagerie of prokaryotic thermophiles in order to give insights into the metabolic, taxonomic, and ecological diversity of thermophilic prokaryotes rather than attempting an exhaustive review.

Abbreviations

Comammox	Complete ammonia oxidation
MAG	Metagenome-assembled genome

1 Introduction

Thermophilic prokaryotes are found not only at various geothermally heated sites such as hot springs and hydrothermal vents, but also in a variety of other habitats where growth and metabolism at elevated temperatures may confer a selective advantage. Examples include hot desert soils, solar salterns, compost heaps, deep subsurface sites, such as oil reservoirs and deep mine shafts, and various industrial settings. It is therefore not surprising that thermophily has evolved independently within multiple and highly diverse clades and not necessarily limited to deep lineages of microbes. The occurrence of thermophily across multiple lineages can be seen as an evolutionary response to the environment and the ability of microorganisms to adjust and evolve. A recent review counted more than 1200 known thermophilic species of bacteria and archaea (DiGiacomo et al. 2022), making an exhaustive review of their diversity and phylogeny a difficult exercise of dubious applicability. This review is therefore intended to give snapshot examples pulled from the vast menagerie that is thermophile diversity. Notwithstanding the vast diversity of thermophiles, some generalizations can be made about their physiology and metabolism, especially as regards those thermophiles originating from geothermal sites, where the chemical composition of the environment is characteristically mineral rich, resulting in the microbial community to be largely supported by lithotrophic bacteria and archaea. Common metabolism in thermophilic bacteria and archaea is therefore sulfur and iron metabolism, along with gas metabolism such as H_2 and CO oxidation.

In this chapter, we will briefly discuss some of the major thermophilic and hyperthermophilic groups of Bacteria and Archaea, focusing on 14 phyla commonly

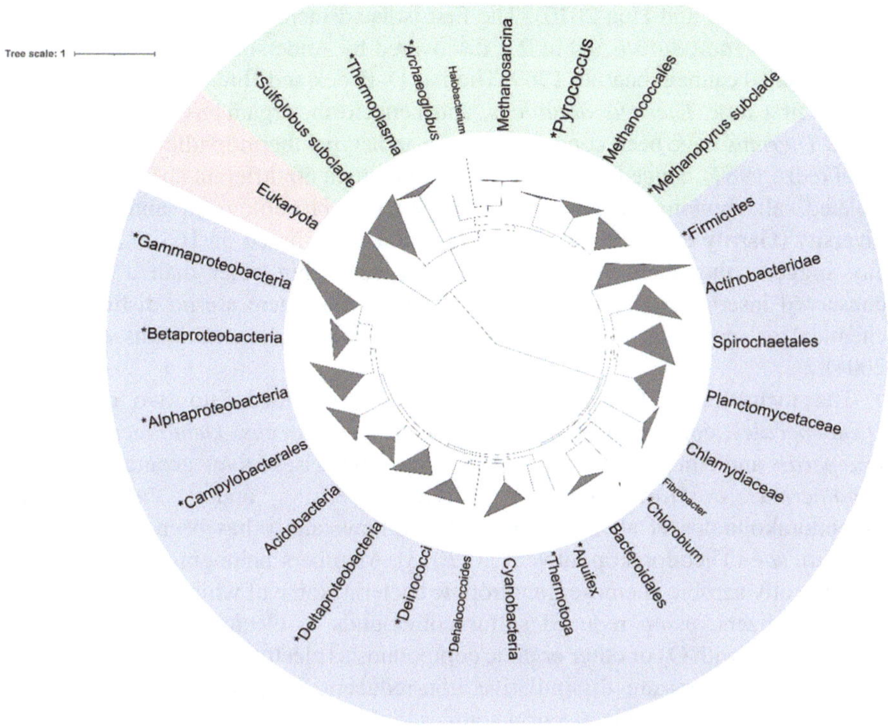

Fig. 1 Tree of life representing the phylogenetic diversity of thermophilic bacteria and archaea. The majority of thermophilic microbes discussed in the chapter belong to the phylogenetic groups shown with asterisk in the Tree of life [based on Ciccarelli et al. (2006) and vizualised with iTOL (Letunic and Bork 2007)]

occurring in diverse natural thermophilic environments. Phylogenetic tree of the different phyla discussed is shown in Fig. 1. We highlight their distribution, physiology, ecology, and potential biotechnological applications where known. Current valid taxonomy and nomenclature as presented on the List of Prokaryotic Names with Standing in Nomenclature (Parte et al. 2020) are followed in this chapter, which includes the major recent changes to the nomenclature of upper-level taxa validly published by Oren and Garrity (2021). To minimize confusion, some of the more commonly used older names are indicated where appropriate.

2 Deinococcota (Previously Deinococcus-Thermus)

The deeply branching Deinococcota (Oren and Garrity 2021) phylum, originally named as Deinococcus-Thermus (Weisburg et al. 1989), due to its resistance to extreme stressors including oxidation, radiation, desiccation, and high temperature, is known as one of the most extremophilic phyla of bacteria (Theodorakopoulos et al. 2013; Tian and Hua 2010). The first isolated member from this phylum was *Deinococcus radiodurans* strain R1, discovered by Anderson et al. (1956) from X-ray-irradiated canned meat. In 1969, Thomas D. Brock and Hudson Freeze described for the first time *Thermus aquaticus*, and henceforth, organisms belonging to the genus *Thermus* have been considered as the archetypal thermophilic bacteria (Brock and Freeze 1969). Since these discoveries, more than 60 different species have been isolated, all showing an enormous biochemical, physiological, and phenotypic diversity (Garrity et al. 2001b). Their close branching based on 16S rRNA phylogeny suggests that they share a common ancestor, but other than a number of conserved inserts or deletions specific to the groups, there are no distinctive biochemical or molecular signatures exclusive of the phylum (Griffiths and Gupta 2004).

The members of Deinococcota are currently divided into two orders, the *Deinococcales*, composed of the three genera *Deinococcus*, *Deinobacterium*, and *Truepera*, and the order *Thermales*, that comprises five genera: *Thermus*, *Meiothermus*, *Marinithermus*, *Oceanithermus*, and *Vulcanithermus* (Theodorakopoulos et al. 2013). Recently, a new family has been proposed, the *Truperaceae* (Theodorakopoulos et al. 2013). Members belonging to this phylum are generally aerobic chemo-organotrophic bacteria, some of which are dissimilative sulfur oxidizers, using reduced sulfur compounds as electron donors in energy conservation, with O_2 or other organic compounds as electron acceptor (Skirnisdottir et al. 2001), others are dissimilative iron-reducers, coupling for respiration the oxidation of H_2 or organic electron sources with the reduction of ferric iron (Fe^{3+}) as a terminal acceptor (Kieft et al. 1999). Although some members of the Deinococcota stain gram-positively, they are classified as gram-negative bacteria, due to their complex cell envelope that includes an outer membrane-like structure (Thompson and Murray 1981), and a peculiar peptidoglycan, in which ornithine represents the principal diamino acid in the cross-linked chains (Rainey and da Costa 2001). The members of the genus *Deinococcus* exhibit resistance to high ionizing and ultraviolet radiations, and desiccation (Albuquerque et al. 2005; Cox and Battista 2005; Slade and Radman 2011). In contrast, cultured representatives of the *Thermus* genus are either thermophilic or hyperthermophilic, while not having any unusual radiation nor stress resistance capabilities (Brock and Freeze 1969), and have a tolerance temperature range between 45 °C and 80 °C. Most of the organisms belonging to the group have a maximum growth temperature slightly under 80 °C (Brock and Freeze 1969; Chung et al. 2000), and just a few strains, belonging to *T. thermophilus,* can grow at 80 °C or above (Manaia et al. 1995).

This phylum comprises very different species that have been found and isolated from a number of different environments, such as air dust and air samples (Weon et al. 2007; Yoo et al. 2009), desert soils (Rainey et al. 2005), cold environments in Antarctica (Hirsch et al. 2004), hot springs or biofilms, radioactive sites (Siebert and Hirsch 1988), and Phoenix spacecraft surface (Stepanov et al. 2014).

The exploitation of the adaptation strategies of these organisms finds numerous applications in the biotechnological field, for an example the treatment of nuclear energy waste. Indeed, *D. radiodurans* can be directly used for the adsorption of uranium in radioactive wastewater and for the treatment of mixed radioactive wastes containing ionic mercury (Li et al. 2021). The most noteworthy discovery has been the isolation of the thermostable DNA polymerase (DNA pol) from *Thermus aquaticus* (Taq pol) which revolutionized the history of molecular biology through its use in the polymerase chain reaction (PCR) and paved the way for the development of modern biotechnology (Bessman et al. 1956; Chien et al. 1976).

2.1 Thermales

Bacteria belonging to *Thermales* order are generally rod-shaped, non-motile, gram-negative, aerobic or facultatively anaerobic thermophiles, generally isolated from hydrothermal areas with temperatures of 50–70 °C and neutral to alkaline pH (Rainey and da Costa 2001). The most representative species is *Thermus aquaticus* (Brock and Freeze 1969), which grows at temperatures ranging from 40 to 80 °C, with an optimum at 70 °C, and at a pH of 7.5–7.8.

The genus *Thermus* lists several thermophilic and hyperthermophilic species, mainly isolated from hydrothermal areas, such as Yellowstone National Park and Pacheteaus Calistoga in California (Brock and Freeze 1969), Japan (Oshima and Imahori 1974), Iceland (Pask-Hughes and Williams 1977), continental Portugal, the Island of São Miguel in the Azores (Manaia and da Costa 1991; Santos et al. 1989), and the Australian Artesian Basin (Denman et al. 1991). Isolates of the genus *Thermus* have also been obtained from abyssal geothermal areas in the mid-Atlantic Ridge and in the Guaymas Basin, Gulf of California, at depths of 3500 and 2000 m, respectively (Marteinsson 1999).

The most representative species of this genus is *Thermus aquaticus*, first isolated in 1969 from the Mushroom Spring in the Lower Geyser Basin of Yellowstone National Park by Thomas D. Brock and Hudson Freeze of Indiana University (Brock and Freeze 1969). *T. aquaticus* has been described as a gram-negative, non-sporulating, non-motile, rod-shaped bacterium, able to switch to a long filamentous shape at supraoptimal temperatures or in the stationary phase. Moreover, one of its main reported features is the production of peculiar large spherical bodies in older cultures, probably generated as temporary food and nucleotide storage (Brock and Freeze 1969). Another representative member of this genus is *Thermus thermophilus*, first isolated in 1968 from the thermal water of Mine Hot Spring, in Japan, and originally placed in the genus Flavobacterium (Oshima and Imahori 1971). *T. thermophilus* is used to grow at temperatures ranging from 47 to 85 °C, with an optimum between 65 and 72 °C, and at a pH range of 5.1 and 9.6, with an optimum pH value of around 7.5 (Oshima and Imahori 1974).

3 Nitrospirota (Previously Nitrospirae)

The first member of this phylum, *Nitrospira marina*, a chemolithotrophic nitrite-oxidizing bacterium isolated from water in the Gulf of Maine, was discovered by Watson in 1986. The phylum Nitrospirota, previously known as Nitrospirae, is a diverse group of Gram-negative, curved, vibrioid, or spiral-shaped bacteria (Garrity et al. 2001e). This phylum consists of three genera that have cultured representatives: *Nitrospira*, *Leptospirillum*, and *Thermodesulfovibrio* (Garrity et al. 2001e; Watson et al. 1986; Watson and Waterbury 1971). Phylogenetic surveys based on the 16S rRNA gene show how *Nitrospira* and *Leptospirillum* consistently cluster together, while *Thermodesulfovibrio* forms a separate branch.

The phylum comprises physiologically diverse bacteria, with a big representation of a chemolithoautotrophic, aerobic, nitrite-oxidizing bacteria (NOB) (*Nitrospira*), chemolithoautotrophic, aerobic, ferrous iron (FeIII) oxidizers (*Leptospirillum*), and anaerobic, thermophilic, chemo-organoheterotrophic, or hydrogenotrophic sulfate reducers (SRB) (*Thermodesulfovibrio*) (Daims 2014). Members of this phylum have been found in different natural and man-made ecosystems, like soil (Wang et al. 2019), fresh and groundwater (Ghimire-Kafle et al. 2023; Hovanec et al. 1998; Palomo et al. 2022), wastewater treatment plants (Daims 2014), marine sponge (Off et al. 2010), and geothermal springs with temperatures between 55 °C and 70 °C (Edwards et al. 2013; Lebedeva et al. 2011).

The genus *Nitrospira* represents the most abundant nitrite oxidizer, and it is almost ubiquitously present in oxic habitats, catalyzing an essential step of nitrification for biogeochemical nitrogen cycling. *Leptospirillum* tends to be a thermotolerant acidophilic genus, with optimal temperature growth ranging between 30 and 37 °C, and with only a few moderately thermophilic strains that can grow up to 40 °C (Vardanyan et al. 2023). The *Nitrospira* genus comprises only five formally identified members (Lücker et al. 2010), usually reported as moderately thermophilic, with growth temperature ranges between 28 and 44 °C (Ehrich et al. 1995; Lebedeva et al. 2008). Recently, the most thermophilic NOB known, *Nitrospira calida*, was isolated from Gorjachinsk Hot Spring in the Lake Baikal area (Russia), with a growth temperature optimum of 46–52 °C and an upper temperature for growth of 58 °C (Lebedeva et al. 2011). *Thermodesulfovibrio* is the most thermophilic genus of the phylum, with a temperature range for growth from 40 to 75 °C, and an upper pH limit for growth of 7.7–8.5 for all species, and a lower pH limit from 6.0 to 6.5 (Frank et al. 2016).

This phylum comprises three genera, all with relevant biotechnological use. *Leptospirillum* members are considered the most important microorganisms used in commercial bioleaching, and their iron oxidizing activity has been used in acid mine drainage, while *Thermodesulfovibrio* members are used to degrade organic compounds in anaerobic digesters. Moreover, *Nitrospira* species are used in wastewater treatment plants (WWTPs) to prevent eutrophication (Spasov et al. 2020; Vardanyan et al. 2023).

3.1 Nitrospirales

The genus *Nitrospira* represents a key nitrite oxidizer in nature, and it is almost ubiquitously present in oxic habitats, catalyzing the second and essential step of nitrification for biogeochemical nitrogen cycling. But their metabolism is not limited to nitrite oxidation or comammox, but they are also involved in other functionalities in the nitrogen cycle, either under anoxic or aerobic conditions. Members of this group are generally aerobic chemolithoautotrophic bacteria. They have been found in groundwater, geothermal springs, freshwater, soils, and wastewater treatment plants (WWTPs). But they could also colonize marine sponges, rhizospheres, and leaf surfaces of plants. According to many studies, *Nitrospira* generally have an optimum of growth between 30–35 °C and the optimum pH is in the range 8.0–8.3 (Mehrani et al. 2020), and the only true thermophile found in this genus so far is *Nitrospira calida*, with a growth temperature optimum of 46–52 °C and an upper temperature for growth of 58 °C (Lebedeva et al. 2011).

The genus *Thermodesulfovibrio* is classified as a group of strictly anaerobic, curved rod-shaped, thermophilic bacteria capable of reducing sulfate and other sulfur compounds and even perform oxidation of hydrogen and other organic compounds (Henry et al. 1994). Another ecologically significant physiological capability is the ability to undergo syntrophic degradation of organic compounds in the absence of sulfate. This group shows adaptations to the physico-chemical surroundings found in hot springs, as well as a diverse array of subterranean and terrestrial hot spring ecosystems, where they have been observed to thrive. The growth of *Thermodesulfovibrio* was observed at 40 °C and 70 °C with an optimum at 65 °C with a pH range between 6.8 and 7.0. (Haouari et al. 2008; Henry et al. 1994; Sekiguchi et al. 2008).

4 Chloroflexota (Previously Chloroflexi)

The phylum Chloroflexota (Oren and Garrity 2021), formerly known as Chloroflexi, contains a large number of thermophilic bacteria that differ widely both physiologically and ecologically. They are largely isolated from geothermal habitats, but also from compost heaps and industrial settings. Originally, the Chloroflexota were described as "Green non-sulfur bacteria," suggesting a very specific physiology. However, in recent years, with the increase of isolated strains and with metagenome-assembled genomes (MAGS) reconstructed from environmental samples, the phylum Chloroflexota has expanded to include vast metabolic diversity and abundant numbers in diverse subsurface ecosystems. Studies have been unraveling the true diversity of the phylum, which consists of a wide variety of anaerobic and aerobic bacteria, mesophilic and thermophilic, chemoheterotrophic, chemolithotrophic, or photolithotrophic. Nevertheless, the cell biology of members within the Chloroflexota remains poorly studied.

A commonly occurring physiological trait within the Chloroflexota is their filamentous cell structure with gliding motility (Garrity et al. 2001b) with only a few of them possessing flagella. The cell envelope of Chloroflexota is quite unique and has long been debated if they have a two-layer membrane (diderma) or only one layer (monoderm) since many of them lack the ability to form peptidoglycan (Kube et al. 2005), but it seems that species within the phylum might diverge in this sense as well.

4.1 Anaerolineae

Environmental DNA-based community analyses have revealed that members of the class Anaerolineae frequently occur in geothermal environments in widely disparate geographical locations. Examples include hot methane seeps in Iceland (Þorsteinsdóttir et al. 2020), hot spring microbial mats in Saudi Arabia (Yasir et al. 2020), hot spring water in the Himalayas (Sharma et al. 2020), sediments downstream from a sulfidic hot spring in Sumatra (Okumura et al. 2022), and seafloor hydrothermal vents (McGonigle et al. 2020).

Cultured thermophiles within this class all belong to the order Anaerolineales (Yamada et al. 2006) and are filamentous chemoheterotrophic anaerobes, oxidizing various carbohydrates, and proteinaceous compounds. They are likely to do so synergistically with hydrogenotrophic methanogens, as their growth is enhanced in their presence, indicating that they can use hydrogenotrophs as a hydrogen scavenging system (Yamada and Sekiguchi 2018). The family Anaerolineaceae contains thermophiles in at least four of its genera, *Anaerolinea*, *Bellilinea*, *Thermanaerothrix*, and *Thermomarinilinea*. These include *A. thermophila* isolated from hot spring sulfur-turf (Sekiguchi et al. 2003), *A. thermolimosa* and *B. caldifistulae* from anaerobic thermophilic granular sludge (Yamada et al. 2006, 2007), *Tt. daxenis* isolated from a deep hot aquifer (Gregoire et al. 2011), and the halophilic *Tm. lacunofontana* isolated from a shallow hydrothermal vent (Nunoura et al. 2013).

Most of the members of the Anaerolineae indicated by community analyses have not been cultured and attempts at reconstructing their genomes from metagenomic data are frequently complicated by binning problems (Singleton et al. 2021). Nevertheless, recent MAG-based studies (Ward et al. 2018; Braga et al. 2021; Liu et al. 2022; Rogers et al. 2023) have revealed this class to be considerably more metabolically diverse than previously believed, including sulfur oxidation, nitrogen cycling, and microaerophilic respiration (Rogers et al. 2023).

4.2 *Chloroflexia*

The class Chloroflexia comprises among others the well-known phototrophic mat-forming bacteria that often grow in great abundance in and around hot springs, frequently in association with Cyanobacteria (Pierson and Castenholz 1974; Giovannoni et al. 1987; Jørgensen and Nelson 1988). The chloroflexi proper, forming the family Chloroflexaceae, are gliding filamentous phototrophic thermophiles. Three species have been described, *Chloroflexus aurantiacus*, *C. aggregans*, and *C. islandicus* (Pierson and Castenholz 1974; Hanada et al. 1995; Gaisin et al. 2017). The family Roseoflexaceae contains additional thermophilic mat-formers, *Heliothrix oregonensis* (Pierson et al. 1985) and *Roseiflexus castenholzii* (Hanada et al. 2002). A cellulolytic filamentous thermophile, *Kallotenue papyrolyticum*, is found within the order Kallotenuales (Cole et al. 2013).

4.3 *Ktedonobacteria*

The class Ktedonobacteria harbors aerobic bacteria that stain gram-positive (Cavaletti et al. 2006). In DNA-based environmental studies, they are often encountered in environments such as geothermal fields, volcanic soils, caves, and gas vents (Hernández et al. 2020; Þorsteinsdóttir et al. 2020; Arif et al. 2021). Ktedonobacteria MAGs have been found to be abundant in young volcanic soil where they are thought to use reduced gases, such as H_2 and CO for growth (Hernández et al. 2020). Studies of their functional capabilities have shown diverse degradation abilities, indicating an ecological function comparable to that of actinomycetes and saprotrophic fungi. Genome sequencing has revealed large genome sizes, including megaplasmids, with several unknown functions, while screening has shown broad antimicrobial activities (Zheng et al. 2019). Cultured representatives of thermophiles within the class Ktedonobacteria are thus far limited to two species in one of its seven genera, *Thermosporothrix hazakensis* and *T. narukonensis* (Yabe et al. 2010, 2016).

4.4 *Other Chloroflexota Classes*

The class Ardenticatenia is yet represented only by a single thermophilic species, *Ardenticatena maritima*. This facultatively anaerobic chemoheterotroph reduces iron and nitrate under anaerobic conditions. It was isolated from a hydrothermal field in Japan (Kawaichi et al. 2013).

DNA-based studies on hot methane seeps in Iceland have indicated a dominating presence of uncultured members of the class *Dehalococcoidia* (Þorsteinsdóttir et al. 2020). This class contains obligate organohalide-respiring anaerobic bacteria

(Löffler et al. 2013). MAG-based studies have indicated that uncultured members of the class *Dehalococcoidia* may fix CO_2 via the Wood-Ljungdahl pathway, which has thus far not been observed among the culturable Chloroflexota (Hug et al. 2013).

A recently described class, Tepidiformia, is composed of only a single species, *Tepidiforma bonchosmolovskayae,* which was isolated from a hot spring in Russia. It is an aerobic, chemoheterotrophic thermophile that can grow chemolithoautotrophically with siderite ($FeCO_3$) as the electron donor and oxygen as electron acceptor (Kochetkova et al. 2020).

Thermoflexus hugenholtzii (Dodsworth et al. 2014) is a thermophilic filamentous bacterium within the class Thermoflexia. It is chemoheterotrophic, microaerophilic, and facultatively anaerobic. It is found in terrestrial geothermal systems, occurring prominently in geothermally heated sediments, also found in microbial mats in similar habitats.

5 Pseudomonadota

The phylum Pseudomonadota (Oren and Garrity 2021), previously named Proteobacteria, is an extremely diverse and cosmopolitan phylum. It harbors a number of different types of thermophiles, including anoxygenic phototrophs, heterotrophs, oligotrophs, and diverse chemolithotrophs.

5.1 *Acidithiobacillia*

Chemolithoautotrophic thermophiles are found within the genera *Acidithiobacillus* and *Thermithiobacillus*. These organisms are obligately aerobic sulfur oxidizers, using reduced inorganic sulfur such as thiosulfate, elemental sulfur, and hydrogen sulfide as electron donors (Boden and Hutt 2019a, b). While most members of these genera are mesophilic, moderate thermophiles are represented by *A. caldus* and *T. tepidarius*. An apparently obligate autotroph, *T. tepidarius* was isolated from a Roman bath and is adapted to a fluctuating environment in terms of oxygen partial pressure (Boden and Hutt 2019a). On the other hand, *A. caldus* is a strictly aerobic mixotroph (Boden and Hutt 2019b).

5.2 *Alphaproteobacteria*

Although the vast majority of the *Alphaproteobacteria* are considered mesophilic, generally not exhibiting growth at temperatures above 45 °C, this phylogenetically, ecologically, and metabolically highly diverse class nevertheless harbors an astonishing diversity of moderately thermophilic bacteria. Thermophilic members of this class range from obligate chemoheterotrophs to facultative phototrophs and mixotrophs. They have been isolated from habitats including hot springs, desert soils, and industrial environments.

Unusual cell morphologies, such as prosthecate cells, and complex life cycles are frequently observed among members of the *Caulobacteraceae* family. They are usually chemo-organotrophs and may be quite fastidious, often requiring supplemented peptides, amino acids, and B vitamins. Habitat range is wide, but they are most often found in oligotrophic habitats and are frequently associated with freshwater environments, including submerged surfaces to which they can attach and form a biofilm (Abraham et al. 2014). Some moderately thermophilic species are known within the genus *Phenylobacterium*, including *P. terrae* and *P. lituiforme*, isolated from Pakistani soil and a subsurface aquifer, respectively (Kanso and Patel 2004; Khan et al. 2018). Like other members of the genus, these species are heterotrophic. Atypically for the genus, which classically is described as comprising strict aerobes only, *P. lituiforme* is a facultative anaerobe, reflecting the oxygen-poor environment of the subsurface from where it was isolated.

In the order Parvarculales, the only thus far described species of the genus *Amphiplicatus*, *A. meriothermophilus*, is a prosthecate moderate thermophile and halophile isolated from a Chinese hot spring (Zhen-Li et al. 2014). This species can grow at an optimum temperature between 48 and 50 °C and optimum pH of 7.5.

Within the Rhizobiales, the family Hyphomicrobiaceae, although not particularly large, is morphologically and physiologically highly diverse (Oren and Xu 2014). Prosthecate and budding species are commonly found within the family. Most species are aerobic chemoheterotrophs, although some are facultatively anaerobic denitrifiers, and many are oligocarbophilic. Some are facultative photoheterotrophs, and a few facultative chemolithotrophs have been found within this family (Oren and Xu 2014). Thermophiles are found within the genera *Rhodoplanes* and *Dichotomicrobium*, including *D. thermohalophilum*, *R. azumiensis*, *R. tepidamas*, and *R. tepidicaeni* (DiGiacomo et al. 2022).

Moderately thermophilic species have also been described within the family *Methylobacteriaceae*, which is characterized by pigmented facultatively methylotrophic bacteria that can grow on one-carbon compounds, e.g., formate, formaldehyde, and methanol as sole carbon sources. The thermophiles are within the genus *Microvirga* and were isolated from plant root nodules and desert soil (Reeve et al. 2014; Amin et al. 2016). Other thermophiles within the *Rhizobiales* order include members of the genera *Chelativorans*, *Tepidamorphus*, *Tepdicaulis*, and *Mongoliimonas* (DiGiacomo et al. 2022).

In the Rhodobacterales, thermophily seems rather common in the metabolically diverse family Paracoccaceae (previously Rhodobacteraceae). It contains a variety of slight and moderate thermophiles, including members of the genera: *Albidovulum*, *Jhaorihella*, *Oceanicella*, *Paracoccus*, *Rubellimicrobium*, *Ruegeria*, *Rhodobacter*, *Rhodosalinus*, and *Tranquilimonas* (Xian et al. 2020; DiGiacomo et al. 2022).

The genus *Albidovulum* contains two validly described species, *A. inexpectatum* and *A. xiamenense*, both of which are thermophilic (Albuquerque et al. 2018). While capable of oxidizing thiosulfate to sulfate in the presence of organic substrates,

A. inexpectatum does not grow chemoautotrophically and is therefore characterized as a facultative mixotroph (Albuquerque et al. 2018).

Moderate thermophiles are also found among the sphingomonads, such as *Porphyrobacter tepidarius*, *Altererythrobacter lauratis*, *Sphingomonas astaxanthinifaciens*, and *Thermaurantiacus tibetensis* (Ming et al. 2021; DiGiacomo et al. 2022). Known for the budding mode of reproduction of several of its member species, the genus *Porphyrobacter* currently contains 10 described species (Parte et al. 2020), all of which are strictly aerobic chemo-organoheterotrophs. Members of this genus produce bacteriochlorophyll-*a* esterified with phytol and carotenoids (Hiraishi and Imhoff 2015). They can thus appear purple pigmented, although the presence of carotenoids often renders them orange or red (Tonon et al. 2014). The carotenoids are polar and include carotenoid sulfates and bacteriorubixanthin, as well as nostaxanthin in *P. tepidarius*, and form with bacteriochlorophyll-*a* the core light-harvesting complex and the photosynthetic reaction center. However, they seem to lack the peripheral light-harvesting center, as indeed do other bacteriochlorophyll-containing Alphaproteobacteria (Hiraishi and Imhoff 2015). Also, similarly to many other aerobic photosynthetic bacteria, *Porphyrobacter* species produce bacteriochlorophyll-*a* only in the dark.

The porphyrobacters are primarily associated with freshwater environments, including the neuston (the air-water interface) and some strains are known to degrade biphenyl and dibenzofuran (Hiraishi et al. 2002). The thermophilic species *P. tepidarius*, with an optimum temperature of about 40–48 °C, was isolated from cyanobacterial mats in a brackish hot spring in Japan (Hanada et al. 1997). Another similarly thermophilic (Topt ~45–50 °C) *Porphyrobacter* species, *P. cryptus*, was isolated in 2003 from a Portuguese hot spring (Rainey et al. 2003). Comparing the genomes of *Porphyrobacter* type strains, Xu et al. (2018) concluded that thermophily of *P. cryptus* was at least in part made possible by substitution of glycine and serine to alanine, increasing the frequency of alpha-helices and thus promoting thermostability of proteins.

Although closely related to *Porphyrobacter*, members of the genus *Altererythrobacter* produce carotenoids but no bacteriochlorophyll and do not photosynthesize (Tonon et al. 2014). Two thermophilic *Altererythrobacter* species, *A. palmitatis* and *A. lauratis,* have been described (Yuan et al. 2017) and were both isolated from Tibetan hot springs.

The only species thus far described in the genus *Thermaurantiacus* is the thermophilic species *T. tibetensis*, isolated from a microbial mat in a Tibetan hot spring (Ming et al. 2021). It is moderately thermophilic, with an optimal growth temperature of 45 °C and growth not observed above 55 °C. Carotenoids are present.

5.3 Betaproteobacteria

The order Burkholderiales contains bacteria that are highly diverse in terms of ecology, physiology, and metabolism, ranging from psychrophiles to thermophiles, and widely distributed in Nature. It contains obligate and facultative chemolithotrophs, nitrogen-fixers, and facultative anaerobes, as well as strict aerobes (Garrity et al. 2015). Accumulation of poly-hydroxybutyrate is a common feature among the Burkholderiales, consistent with organisms adjusted to an oligotrophic environment where storage of carbon in carbonosomes is likely to confer an advantage (Prieto et al. 2014). Among thermophiles in this order are members of the genera *Caldimonas*, *Tepidimonas*, and *Tepidicella*, as well as at least two species of *Schlegelella*, *S. aquatica,* and *S. thermodepolymerans* (Elbanna et al. 2003; Lütke-Eversloh et al. 2004; Chou et al. 2006). Both are moderately thermophilic, the former isolated from a Taiwanese hot spring, while the latter comprises strains isolated from sewage treatment sludges.

5.4 *Gammaproteobacteria*

The purple sulfur bacteria (PSB) are found within the order Chromatiales, primarily in the families Chromatiaceae and Ectothiorhodospiraceae. While most of the PSBs described so far are mesophilic, a few species in both families are thermophilic, notably *Thermochromatium tepidum* (Imhoff et al. 1998) and the extremely halotolerant *Halorhodospira halochloris* (Imhoff and Trüper 1977). Thermophilic PSBs are found in hot springs and warm soda lakes and salterns. The term "purple bacteria" refers to their reddish-purple color, which in turn is explained by the presence of bacteriochlorophyll and carotenoids. They are thus capable of anoxygenic photosynthesis and typically fix CO_2 via Rubisco and the Calvin-Benson cycle. Electron donors are usually reduced sulfur compounds, often stored in sulfur granules either intracellularly or, in the case of the Ectothiorhodospiraceae, on the cell surface. Non-phototrophic thermophiles are also found within the *Chromatiales*, such as the chemolithoautotrophic sulfur oxidizers *Thiofaba tepidiphila* and *Sulfurivermis fontis*, both isolated from Japanese hot springs (Mori and Suzuki 2008; Kojima et al. 2017).

5.5 *Deltaproteobacteria*

The class Deltaproteobacteria, recently proposed to be reclassified into the phyla *Desulfobacterota*, *Myxococcota*, and SAR324 (Langwig et al. 2022), is ubiquitous in marine sediments (Wang et al. 2012), soils (Delgado-Baquerizo et al. 2018), subterranean environments (Hug et al. 2016), wetlands (Liu et al. 2014), freshwater, and marine water columns (Swan et al. 2011; Sheik et al. 2015), playing an essential role in global biogeochemical cycling. Despite their importance, our understanding of these bacteria is biased toward cultured organisms. They include gram-negative bacteria, for the major part mesophilic anaerobes, many of which are interesting for their potential biotechnological applications. Notable metabolic features within this class are the sulfate respiration, using protein complexes sulfate adenylyltransferase (Sat), adenylyl sulfate reductase (Apr), and dissimilatory sulfite reductase (Dsr), most of which have already been characterized (Minz et al. 1999). Members of the class *Deltaproteobacteria* are also known for their ability to reduce elemental metals, such as iron, with species of the families *Geobacteraceae* and *Desulfuromonadaceae* able to use external surfaces as electron acceptors to complete respiration, producing an electrical current (Bond et al. 2002; Holmes et al. 2004). These organisms have a variety of other metabolic abilities, such as inorganic sulfur compounds disproportionation (mainly carried out by members of the orders *Desulfobacterales* and Desulfovibrionales and of one phylogenetically separate species, *Desulfomonile tiedjei* (Deweerd and Suflita 1990; Sorokin et al. 2008)), dissimilatory iron reduction (mainly within Desulfuromonadia) (Lovley and Phillips 1988; Lonergan et al. 1996), nitrogen fixation (Masuda et al. 2017; Tan et al. 2019), mercury methylation (Si et al. 2015), organohalide respiration (Liu and Häggblom 2018), and aliphatic and aromatic hydrocarbon degradation (Bergmann et al. 2011).

Among thermophilic *Deltaproteobacteria*, different strains have been characterized recently and seem to have very similar metabolisms. For instance, an anaerobic thermophilic chemolithoautotroph bacterium called *Desulfurirhabdus thermomarina* has been isolated from a shallow sea hydrothermal vent. It can grow at 50 °C by the disproportionation of sulfite and elemental sulfur (Allioux et al. 2020). Furthermore, among thermophilic *Deltaproteobacteria* there is *Dissulfurimicrobium hydrothermale* Sh68T, which is a motile, anaerobic, chemolithoautotrophic microorganism, isolated from a hydrothermal pond at Uzon caldera, Kamchatka, Russia. Similar to *D. thermomarina*, it is able to produce energy and grow by disproportionation of elemental sulfur, sulfite, and thiosulfate (Yvenou et al. 2021). Another thermophilic deltaproteobacterium able both to disproportionate sulfur compounds and accomplish dissimilatory sulfate reduction at high temperatures corresponds to *Dissulfuribacter thermophilus*, strain S69T (Slobodkin et al. 2013). This microorganism was isolated from a deep-sea hydrothermal vent chimney located on the Eastern Lau Spreading Center and Valu Fa Ridge, Pacific Ocean, at a depth of 1910 m using anoxic medium with elemental sulfur as the only energy source. Its temperature range for growth was 28–70 °C, with an optimum at 61 °C. A metabolically and phylogenetically similar deltaproteobacteria strain was also isolated from a shallow-sea hydrothermal vent where it participates in biogeochemical cycling of sulfur, most probably as a primary producer (Slobodkina et al. 2016).

A different strategy of bacterial predation is found in the members of the order *Myxococcales*, known for their predominantly aerobic, predatory lifestyle with the ability to produce a variety of secondary metabolites (Jurkevitch and Davidov 2007; Berleman and Kirby 2009). During the motile phase of their life, members of this order exhibit complex social behavior, swarm prey organisms, secrete enzymes and proteases to lyse the target cells (Hart and Zahler 1966), and are also able to sporulate. *Myxobacteria* have long been considered mesophiles (Reichenbach 1999). Recently, a thermotolerant strain GT-7, identified as belonging to the genus *Myxococcus* within the Cystobacterineae, was isolated from the soil. Its growth temperature range is between 42 and 44 °C (Gerth and Müller 2005). It is also known that the occurrence of the myxobacteria in hot spring microbial mats is relatively high and this is probably due to more stable microbial ecosystems existing in the mats, compared to the external environment (Iizuka et al. 2006).

Deltaproteobacteria are one of the most used groups of microorganisms for numerous biotechnological applications. For instance, a subtype of Class II type V CRISPR-Cas effectors from *Deltaproteobacteria* have been reported to produce extensive 5′-overhangs at cleaved DNA targets, which can make it usable for various applications (Selkova et al. 2020). Furthermore, *Deltaproteobacteria* can degrade Polycyclic Aromatic Hydrocarbons, enabling them significant environmental remediation functions in many areas, including groundwater pollution and oil spills (Viggi et al. 2017; Wang et al. 2021). Recent binding analysis of six previously uncharacterized proteins from the magnetotactic deltaproteobacterium *Desulfamplus magnetovallimortis* BW-1, identified two new magnetite-binding proteins which can be used as affinity tags for the immobilization of recombinant fusion proteins to magnetite (Pohl et al. 2021). Members of the *Deltaproteobacteria* class such as *Desulfovibrio magneticus* and *Geobacter sulfurreducens* are defined as "magnetotactic bacteria," because their ability to derive their magnetic orientation from magnetosomes, which are unique organelles containing nanometer-sized crystals of magnetic iron minerals. Nanowires enable bacteria to transfer electrons over micrometer distances to extracellular electron acceptors such as insoluble metal oxides or electrodes (Lovley and Malvankar 2015). Understanding how protein-based nanowires can conduct electrons is intriguing, as proteins are generally considered to be electrical insulators and are widely used in different biotechnological fields. Dissimilatory metal-reducing bacteria have been employed in microbial fuel cells and in bioremediation techniques, and bioengineered nanowires have been proposed for future use in nanobioelectronics (Boesen and Nielsen 2013).

6 Campylobacterota

The phylum Campylobacterota comprises a phylogenetically diverse bacterial group consisting of three classes: *Desulfurellia* (formerly corresponding to the order *Desulfurellales*) and *Campylobacteria* (formerly classified as *Epsilonproteobacteria*) (Waite et al. 2017). Although this phylum is widely known for its pathogenic members such as *Helicobacter pylori* and *Campylobacter jejuni* (Salama 2020; Yeh et al. 2021), a greater number of non-pathogenic mesophilic and thermophilic species have been discovered in diverse natural environments, including deep-sea hydrothermal fields (Shiotani et al. 2020), aquatic redox-stratified systems (Henkel et al. 2022), subterranean estuaries (Huang et al. 2021), terrestrial sulfidic caves (Bizic et al. 2023), terrestrial Mud Volcanoes (Mardanov et al. 2020), and oil fields (Nakagawa and Takaki 2009). These microorganisms are key players in element cycling (Lopez-Fernandez et al. 2018; Di Cesare et al. 2020; Seyler et al. 2021), and in deep-sea hydrothermal vents, they constitute the dominant members of the microbial community, particularly in sulfide chimney structures where they can represent up to 85% of the total microbial biomass (Nakagawa et al. 2006).

The taxonomically and metabolically diverse members of Campylobacterota are also responsible for chemosynthetic primary productivity of inorganic sources at sub seafloor level (Campbell et al. 2006; McNichol et al. 2018). Although the cultivation of thermophilic Campylobacterota members has increased with the refinement of cultivation conditions, the number of described thermophilic species still only represents 14% of the total number of validly published species within Campylobacterota (Shiotani et al. 2020). Consequently, there is insufficient information on their genomes and intra-specific diversity, leaving the classification of thermophiles unresolved. Within the Campylobacterota, thermophiles are found within the order *Nautiliales* and in the *incertae sedis* genera *Nitratiruptur*, while the majority of the other members in the *Campylobacterales* are represented by mesophiles with some exception showing thermotolerance. All the thermophiles in the phylum have been isolated from deep-sea hydrothermal vents where they are key primary producers sustaining the vent ecosystems (Campbell et al. 2006).

Cultivation, isolation, and characterization of these bacteria have enabled new discoveries about their evolution and diversification (Nakagawa et al. 2007), biogeography (Mino et al. 2017), and their potential of biotechnological applications related to global warming mitigation (Mino et al. 2018; Fukushi et al. 2020). For example, *Nitratiruptor labii* carries out the reduction of N_2O to N_2 (Fukushi et al. 2020), a reaction which could be exploited for nitrogen removal from wastewater. Future studies, including cultivation analysis of N_2O-reducing *Campylobacteria*, transcriptional analysis of nitrous oxide reductase genes, and evaluation of N_2O consumption thermodynamics and kinetics, may provide a broader view of mechanisms allowing to significant N_2O consumption rates.

Vent-derived hydrogen-converting enzymes detected in Campylobacterota have also been widely studied in recent years. Due to the wide range of thermal and chemical conditions featuring vent environments (Miyazaki et al. 2020), thermophilic Campylobacterota hydrogenases genes are of particular interest for the use in hydrogen production as a green energy source and energy generation in biofuel cells (Armstrong et al. 2009; Chenevier et al. 2013).

6.1 Nautiliales

Within the deeply branching group of the *Nautiliales*, sequences have been retrieved exclusively from hydrothermal systems, and cultured representatives of the family are thermophilic, autotrophic, and can reduce elemental sulfur with molecular hydrogen. Deep-sea hydrothermal environments can be regarded as one of the largest reservoirs of diverse environmental Campylobacterota on Earth, whose deeply branching groups correspond to *Nautiliales* and *Nitratiruptor*. The first hydrogen-oxidizing, sulfur-reducing, thermophilic chemolithoautotrophs

Nautiliales successful isolations were from *Alvinella pompejana* symbiont-associated biomass and tube samples (Campbell et al. 2001). *Alvinella pompejana* is a deep-sea hydrothermal vent polychaete, which has an episymbiont community integrated into its dorsal epithelium. Surveys of geochemical conditions within *A. pompejana* tubes revealed high temperature range (~20–80 °C), anoxia, surprisingly low or trace free hydrogen sulfide (from <0.2 to 46.53 μM), pH values between 5.3 and 6.4, high concentrations of potential electron acceptors such as sulfate, nitrate, and iron, and potentially lethal levels of heavy metals like zinc, nickel, vanadium, copper, lead, cadmium, cobalt, and silver (Di Meo-Savoie et al. 2004; Luther et al. 2001; Cary et al. 1998). The *Nautiliales* isolates from *Alvinella pompejana* were moderately thermophilic sulfur-reducing heterotrophs growing on formate as the energy and carbon source. In addition, two of the isolates were able to grow on sulfur using hydrogen as the electron donor. Optimal growth temperatures of the bacteria ranged from 41 to 45 °C (Campbell et al. 2001). Preliminary analysis of a metagenomic library of the *A. pompejana* symbionts supports the hypothesis that at least a portion of the symbiotic Campylobacterota detoxify sulfide by rendering it biologically unavailable through metal-transport and sulfide-oxidation processes so that *A. pompejana* can thrive in this extreme microhabitat (Barbara J. Campbell et al. 2006). Recently, several previously uncultivated, phylogenetically diverse Campylobacterota groups were isolated from various geologically and geographically distinct deep-sea hydrothermal fields, all showing a diverse range of physiological characteristics and utilization of electron donors, such as hydrogen and sulfur, and electron acceptors, such as sulfur and nitrogen. Additionally, they have been shown to fix inorganic carbon (Voordeckers et al. 2005; Nakagawa et al. 2005a, b) through two autotrophic pathways, the acetyl-coenzyme A pathway and the rTCA cycle.

Therefore, all the Campylobacterota studied so far are chemolithoautotrophs, and it is highly significant from the evolutionary point of view, because chemolithoautotrophy is thought to be the first type of metabolic pathway evolved on Earth (Russell and Hall 1997; Wächtershäuser 1990). Phylogenetic analysis of rTCA genes amplified directly from hydrothermal vent chimney samples, as well as enzymatic expression analyses of the beta subunit of the ATP citrate lyase (aclB), genetic analyses of the cultures of Candidatus *A. sulfidicus,* and the chemolithoautotrophic *Nautilia* sp. strain AmH (Wirsen et al. 2002; Campbell et al. 2003) have demonstrated the potential presence of the rTCA cycle for autotrophic carbon fixation in these environments (Wächtershäuser 1990; Campbell et al. 2003; Ken Takai et al. 2005). Oxygen is not essential for many of the isolated sulfur-reducing or sulfur-oxidizing Campylobacterota, especially for the deeply branching *Nautiliales*, which are obligate anaerobes that use various alternative electron acceptors, such as sulfite, elemental sulfur, and nitrate. Because the metabolic features and environmental adaptation of the studied Campylobacterota, including growth at high temperatures, anaerobic metabolism, and chemoautotrophy, are similar to the members of *Chlorobiaceae* and *Aquificales*, the evolution and significance of the Campylobacterota throughout Earth's history are interesting reasons to pursue in future research.

Nitratiruptor tergarcus is a thermophilic bacterium (growth temperature between 37 and 65 °C) which has been isolated from a deep-sea hydrothermal chimney structure in the Mid-Okinawa Trough (Nakagawa et al. 2005a, b). To this group belongs the recently described species *Nitratiruptor labii*, isolated from the same deep-sea hydrothermal region (Fukushi et al. 2020). Only in the last few years, a new strain EPR55-1 T belonging to the *Nitratiruptoraceae* species was isolated from the East Pacific Rise, being the first thermophilic campylobacterial species which is able to utilize thiosulfate and sulfite as its sole electron acceptor and sulfur source, respectively. This strain possessed lophotrichous flagella, unlike the monotrichous and amphitrichous flagella of *Nitratiruptor labii* and *Nitratiruptor tergarcus*, respectively (Shiotani et al. 2020).

7 Firmicutes

The phylum Firmicutes (today, also known as Bacillota) was originally described by Gibbons and Murray (1978), comprising low G + C Gram-positive bacteria. According to the Taxonomy in NCBI database (Schoch et al. 2020), the phylum consists of nine classes (*Bacilli, Bacillota incertae sedis, Bacillota* sensu stricto *incertae sedis, Clostridia, Culicoidibactera, Desulfuribacilla, Erysipelotrichia, Limnochordia, Negativicutes, Thermolithobacteria,* and *Tissierellia*), however—two of the listed classes have not been validly published according to LPSN (Parte et al. 2020). At least 45 families and 480 genera have been published; consequently, the phylum Firmicutes is highly diverse (Seong et al. 2018). Within the phylum, several thermophilic bacteria are known, such as members of genera *Clostridia, Thermoanaerobacterium, Thermoanaerobacter, Caldanaerobacter,* and *Caldicellulosiruptor*.

7.1 Clostridia

The type genus of the family *Clostridiacea*, *Clostridium*, is a large and diverse group of bacteria, currently containing more than 250 validly published species (Parte et al. 2020). The genus was first described in 1880, and harbors rod-shaped, Gram-positive, spore-forming anaerobic bacteria distributed in diverse habitats, including soil, water, and other ecological niches rich in plant decaying material (Lawson 2016; Wiegel 2015; Figueiredo et al. 2020). The species of genus Clostridium are extremely diverse and their G + C content can vary from 21 to 54%, which is considered to be too extensive for one genus (Lawson 2016). A phylogenetic analysis on *Clostridia* by Collins et al. (1994) leads to the conclusion that more than half of the species assigned to the genus were in fact not closely related to the type species, *C. butyricum*.

Most of the 250 validly described *Clostridium* species are classified as mesophiles, whereas only about 20 are moderately thermophilic, e.g., *C. thermocellum*, *C. thermosuccinogenes*, *C. thermopalmarium*, and *C. stercorarium*. Many of the thermophilic species of *Clostridium* have been extensively studied regarding their ability to degrade complex and lignocellulosic biomass, especially *C. thermocellum* (Akinosho et al. 2014). Some key features of *C. thermocellum* include its complex cellulolytic machinery, high growth rates, and tolerance to high temperatures and ethanol concentrations (Dumitrache et al. 2016; Sudha Rani et al. 1998). In recent years, researchers have been working to engineer *C. thermocellum* to improve its performance and make it a more effective biocatalyst for industrial applications (Sander et al. 2015; Rangel et al. 2020). Several studies have investigated the genetic and biochemical mechanisms underlying the cellulolytic activity of *C. thermocellum*, e.g., Raman et al. (2011) where transcriptomic and proteomic approaches were used to identify key genes and enzymes involved in cellulose degradation.

7.2 *Thermoanaerobacterium*

Bacteria belonging to the genus *Thermoanaerobacterium* are anaerobic, extreme thermophiles with optimal growth temperature ranging from 55 to 70 °C. In general, their cells have Gram-positive cell wall structure; however, several strains of *Thermoanaerobacterium* appear as Gram-negative during Gram staining. The pH range for growth is also broad, ranging from 3.2 to 8.5 (Onyenwoke and Wiegel 2015). The genus was first described in 1993 when two xylan-degrading bacteria were isolated from Frying Pan Springs in Yellowstone National Park (Lee et al. 1993). Today, the genus consists of eight validly published species: *T. aciditolerans, T. aotearoense, T. butyriciformans, T. polysaccharolyticum, T. saccharolyticum, T. thermosaccharolyticum, T. thermostercoris, T. thermosulfurigenes, T. xylanolyticum*, and *T. zeae* isolated from a variety of extreme environment, such as hot springs, hydrothermal vents, and leachate of waste from canning factories (Parte et al. 2020).

Bacteria of *Thermoanaerobacterium* have been studied extensively for their potential use in industrial biotechnology and are known for their abilities to degrade various sugars present in lignocellulosic biomass. Degradation of lignocellulose, and other substrates, results in a variety of end products, such as ethanol, acetate, lactate, and hydrogen. *Thermoanaerobacterium* species are thus promising candidates for bioethanol production from lignocellulose, although the production of mixed end products limits their use—and the level of ethanol is in general low (Ren et al. 2008; Romano et al. 2010; Sveinsdottir et al. 2009; Wu et al. 2021). For the past two decades or so, various studies have been published on genetical modifications of *Thermoanaerobacterium* species to improve the final ethanol titer and more. *T. saccharolyticum* is one of the most studied bacteria of the *Thermoanaerobacterium* genus and was the first thermophilic bacterium to be genetically modified to enhance its ethanol production (Desai et al. 2004). Attempts have been made to knock out both acetate and lactate formation, resulting in increased ethanol yields both from glucose and xylose (Shaw et al. 2008). In a recent study, the *T. saccharolyticum pforA* and *ferredoxin* were transferred to strain of *Clostridium thermocellum*, increasing the maximum ethanol titer by 14% (Hon et al. 2018). Similarly, other species of *Thermoanaerobacterium* are known for effective hydrogen production from lignocellulosic biomass, i.e., *T. aotearoense* (Li et al. 2019).

7.3 Thermoanaerobacter and Caldanaerobacter

The genus *Thermoanaerobacter* is composed of obligately anaerobic thermophiles. All species have a Gram-positive cell wall structure, but in some cases the Gram-stain reaction can vary. The optimal growth temperature of *Thermoanaerobacter* ranges from 35 to 78 °C and pH optimum range of 5.8–8.5. *Thermoanaerobacter* are organoheterothropic and use various fermentation pathways, including the Wood-Ljungdahl pathway. Some species of the genus are also able to grow chemolithoheterotrophically and others are facultative chemolithoautotrophs (Onyenwoke and Wiegel 2015). The genus was first described in 1981 when the type species, *T. ethanolicus* was isolated from a hot spring in Yellowstone National Park (Wiegel and Ljungdahl 1981). Today, the genus contains 15 validly published species and 5 subspecies, isolated from extreme environments such as hot springs and oil fields (Parte et al. 2020). Most species can degrade various substrates and sugars and their main end products are ethanol, hydrogen, lactate, and in some cases alanine. *Thermoanaerobacter* species, such as *T. ethanolicus*, *T. brockii,* and *T. thermohydrosulfuricus*, have been extensively studied for their ethanol production (Lacis and Lawford 1988a, b; Lee et al. 2007; Georgieva and Ahring 2007; Lamed and Zeikus 1980a, b) and high yields of ethanol production from sugar observed. The ethanol yields can however vary greatly depending on both species and culture conditions. For example, *Thermoanaerobacter* strain AK5 was shown to produce up to 1.7 mol ethanol from 1 mol glucose in batch culture but when lowering the pH 2 the yields dropped, and end product formation was switched to acetate rather than ethanol (Brynjarsdottir et al. 2012). Recently, the production of branched-chain alcohols from branched-chain amino acids by *Thermoanaerobacter* has gained increased attention and has potential as a source for biofuel production from protein waste in the future (Scully and Orlygsson 2014).

Members of *Thermoanaerobacter* were recently reclassified and moved to the genus *Caldanaerobacter* based on phylogenetic, phenotypic, and metabolic characteristics (Fardeau et al. 2015). Today, *Caldanaerobacter* consists of only two species and four subspecies (Parte et al. Parte et al. 2020). *Caldanaerobacter* bacteria are strictly anaerobic heterotrophs and ferment glucose to L-alanine in nearly equal molar amounts (Fardeau et al. 2015). In a recent study on the genomes of three species of *C. subterraneus*, a variety of hydrolases were found, concluding bacteria of the genus *Caldanaerobacter* to harbor enzymes with great biotechnological potential (Sant'Anna et al. 2015). Novel thermostable lipases originating from *Calanaerobacter* have furthermore been cloned and expressed in *E. coli*. The two lipases identified showed high thermoactivity and stability (90 °C, pH11) and were also found resistant to various organic solvents, which makes these enzymes greatly interesting for multiple biotechnological processes (Royter et al. 2009).

7.4 *Caldicellulosiruptor*

Members of the genus *Caldicellulosiruptor* are Gram-negative, strictly anaerobic and have optimal growth temperatures ranging from 65 to 75 °C. They have a broad substrate spectrum and degrade mono-, di- and polysaccharides, e.g., cellulose, cellobiose, xylan, and xylose (Rainey 2015). The type species, *C. saccharolyticus*, was described by Rainey et al. (1994), and today, the genus contains ten validly published species: *C. saccharolyticus, C. acetigenus, C. bescii, C. changaiensis, C. diazotrophicus, C. hydrothermalis, C. kristjanssonii, C. kronotskyensis, C. lactoaceticus,* and *C. owensensis* (Parte et al. 2020). All species of *Caldicellulosiruptor* have been isolated from geothermal environments such as hot springs and lake sediments (Rainey et al. 1994; Yang et al. 2010). Unlike the previously mentioned *Thermoanaerobacter* and *Thermoanaerobacterium*, *Caldicellulosiruptor* species do not produce ethanol at high titer. They are, however, excellent candidates for hydrogen production, and some have been extensively studied for their H_2 production from sugar and hydrolysates from lignocellulosic biomass (Kádár et al. 2004; Vrije et al. 2007; Zeidan and van Niel 2010). High ethanol yields have however been observed by one species, *C. bescii*, by genetical modification of the wild type (Chung et al. 2015). The genomes of all validly published *Caldicellulosiruptor* type species have been whole sequenced, revealing enzymes that can be useful for pharmaceutical production, as well as in textile and paper processing (Blumer-Schuette 2020). Furthermore, *Caldicellulosiruptor* species are now being investigated for the formation of metal nanoparticles (Bing et al. 2018) and heavy metal reduction (Bai et al. 2018).

8 Thermotogae

The Thermotogae phylum is composed of generally thermophilic, anaerobic gram-negative eubacteria characterized by the presence of an outer sheath-like envelope, referred to as "toga," the absence of an outer membrane and a rod-like shape (Frock et al. 2010; Bhandari and Gupta 2014a; Counts et al. 2017), able to thrive at different temperature ranges, from mesophilic to thermophilic and extremely thermophilic organisms. This phylum includes four orders *Kosmotogales, Mesoaciditogales, Petrotogales,* and *Thermotogales*. At present, the *Thermotogales* order includes two families, *Fervidobacteriaceae* and *Thermotogaceae* (Bhandari and Gupta 2014a; Schoch et al. 2020).

For the most part, members of the class have been isolated from submarine geothermal features (Jannasch et al. 1988; Boniface et al. 2006; Conners et al. 2006), either shallow or deep (Wery et al. 2001; Alain et al. 2002; Postec et al. 2005, 2010; Nunoura et al. 2007; L'Haridon et al. 2019), oil reservoirs (Ravot et al. 1995; Fardeau et al. 1997; Takahata et al. 2001), continental solfataric (Windberger et al. 1989), and/or acidic springs (Itoh et al. 2016). The majority of the members of this class are able to grow on a myriad of carbon sources, including hexoses, pentoses, disaccharides, glucans, xylans, glucomannan, galactomannan, pectin, chitin, and amorphous cellulose (Conners et al. 2006; Vanfossen et al. 2008; Latif et al. 2015).

All members of the *Thermotogae* class are characterized by the presence of a toga-like outer membrane and are generally rod-shaped cells, although some members can also have cocci-shaped ones (Bhandari and Gupta 2014b; Reysenbach 2015). While most of them are obligate anaerobes, some exhibit oxygen tolerance and can survive under concentrations below 15% (DiPippo et al. 2009). The most distinctive feature of all members of this class is the toga-like outer membrane, which has been reported to have a rather unusual composition regarding other bacteria, leading to their classification as atypical Gram-negative bacteria rather than the prototypical diderm Gram-negative bacteria. Other unusual features of the membranes of the Thermotogae are the presence of long-chain C30, C32, and C34 dicarboxylic acids in lipids from *Kosmotoga*, *Thermotoga*, *Fervidobacterium,* and *Thermosipho*, which are considered an adaptation to grow in high-temperature environments. This consideration is further supported by their absence in the lipids of *Mesotoga prima* (Bhandari and Gupta 2014b).

The members of this phylum are all chemo-organotrophs capable of coupling the degradation of various types of carbon substrates, from simple forms of sugars to more complex organic matter, and to the oxidation of different forms of sulfur. The metabolic pathways involved in sugar degradation are the Entner–Doudoroff and the Embden–Meyerhof–Parnas pathways along with the non-oxidative portion of the pentose-phosphate pathway. The presence of genes that coded for different sugar metabolisms was also observed by (Zhaxybayeva et al. 2009) after functionally annotating large fractions of genomes besides the one studied previously from *T. maritima* (Nelson et al. 1999). One of the main products of glucose fermentation, H_2, inhibits growth for most of the species. However, this effect can be overcome in the presence of sulfur, thiosulfate, or cysteine, depending on the species, and the production of H_2S is thought of as a kind of a detoxification reaction. (Bhandari and Gupta 2014a; Huber and Hannig 2006).

The potentially high yields of H_2 as a metabolic product, as well as the capability to use a myriad of carbon substrates, and its thermophilic and hyperthermophilic nature has sparked great interest for technological applications. Thermotolerant enzymes present in these bacteria can be industrially used to increase reaction velocities and solubility of substrates and they can easily be overproduced in *E. coli* through recombinant vectors (Bhandari and Gupta 2014a). Their capability to use different types of carbon substrates has also been used in large-scale industrial production of sugars, artificial sweeteners, and syrups. Enzymes from *Thermotogae* species have also been applied in the pretreatment of pulp and for the braking of fluids in oil wells. A more detailed description of all the other applications of enzymes isolated from this class can be found in (Bhandari and Gupta 2014a). The high yield of H_2 reported from the degradation of various carbon sources, as well as the fact that both *Thermotoga maritima* and *Thermotoga neapolitana* reach the Thauer limit (i.e., completely efficient fermentation of sugar into the production of H_2) these species represent prime candidates for the production of clean, renewable energy (Auria et al. 2016; Eriksen et al. 2011; Frock et al. 2010; Van Ooteghem et al. 2002, 2004).

8.1 *Thermotogales*

Thermotogales are widespread cosmopolitan bacteria, able to grow at high temperatures (optimal growth temperature around 80 °C and maximum temperature of 90 °C) making them, together with members of the phylum Aquificota (formerly known as Aquificae), the bacteria with the highest growth temperature known now. Its thermophilic nature explains its predominance in volcanically or geothermally heated environments, and its organotrophic nature reflects its capability of degrading complex organic substrates (Huber and Hannig 2006). They also exhibit a broad range of tolerance to salinity conditions, with some species restricted to growth under low salinity ranges while others can thrive at higher salt concentrations. The capability to use a wide variety of carbon sources also explains its biotechnological application to produce clean energy and the degradation of organic pollutants (Blumer-Schuette et al. 2008; Bhandari and Gupta 2014a; Latif et al. 2015). The first isolated organisms from the phylum all belong to the *Thermotogales* order, either to the *Fervidobacteriacea* class *(Fervidobacterium nodosum*, Patel et al. 1985) or the *Thermotogaceae* class (*Termotoga maritima*, Huber et al. 1986). Generally speaking, members of this class are heterotrophic thermophiles capable of using both simple and complex forms of carbon to transform them into H_2, CO_2, and acetate.

The Fervidobacteriaceae family includes two genera, *Fervidobacterium* and *Thermosipho*, both of them composed of thermophilic bacteria, able to grow at temperatures between 65 and 80 °C, neutral pH and low salinity (<1% salt concentration). All members of this family are strictly anaerobic and able to reduce sulfur to sulfide and have been found in hot springs, terrestrial geothermal aquifers, shallow and deep-sea hydrothermal vents, and deep oil reservoirs (Farrell et al. 2021). Their growth is inhibited by the presence of H_2 and they can utilize different types of carbon substrates with various degrees of complexity, from glucose and sucrose to starch, and lactose (Huber et al. 1990; Patel et al. 1985; Kanoksilapatham et al. 2016). The G + C content in the DNA of members of this family ranges between 29.5 and 45.8 mol%. The type genus for this family is *Fervidobacterium*, with its type species being *F. nodosum*, isolated in 1985 by Patel et al. from a hot spring in New Zealand. The other species that compose the *Fervidobacterium* genus are *Fervidobacterium changbaicum, Fervidobacterium gondwanense, Fervidobacterium islandicum, Fervidobacterium pennivorans, Fervidobacterium riparium,* and *Fervidobacterium thailandense*.

On the other hand, the *Thermosipho* genus is made up of eight species *Thermosipho africanus, Thermosipho melanesiensis, Thermosipho japonicus, Thermosipho geolei, Thermosipho atlanticus, Thermosipho affectus, Thermosipho globiformans,* and *Thermosipho activus* (Nesbø et al. 2022). The type species for this genus is *T. africanus*, which was isolated by Huber et al. (1989) from a marine hydrothermal area in Obock, Djibouti (Africa). *T. africanus* differs from other members of the *Thermotogales* order in their GC content (29.5–31.5 mol%), the presence of a divergent RNA polymerase, and lower growth temperatures (from 35 to 77 °C). All species of the *Thermosipho genus* share over 91% of their 16S rRNA with each other and are closely related with genus *Fervidobacterium* (Bhandari and Gupta 2014a). Species have been isolated from shallow and deep-sea hydrothermal vents and deep subsurface oil reservoirs, and all have shown growth inhibition in the presence of H_2.

Thermotogaceae members are thermophilic and hyperthermophilic bacteria, with optimal growth temperatures between 60 and 80 °C and that require anaerobic conditions to grow. They are fermentative and hydrogen-producing rod-shaped bacteria that can use a wide variety of carbon sources (Zhaxybayeva et al. 2019). This family is divided into two, *Thermotoga* and *Pseudothermotoga*, with the type genus being *Thermotoga*, as defined by (Huber et al. 1986). All species from both families are known to occur in geothermally heated sediments, shallow-depth marine hydrothermal vents, hot springs, solfataric springs, oil reservoirs, and bioreactors (Bhandari and Gupta 2014a; Huber and Hannig 2006; Jannasch et al. 1988; Zhaxybayeva et al. 2019). Species from the *Thermotogaceae* family have a G + C content between 38.7 and 51.3 mol, and within the *Thermotoga* genus, the species can be differentiated based on the presence of the lipid 15, 16-dimethyl-30-glyceryloxytriacontanoic acid (Bhandari and Gupta 2014a).

The *Thermotoga* genus is composed of four validly named species, *T. maritima, T. neapolitana, T. naphthophila,* and *T. petrophila,* with *T. maritima* being the type species for the genus, while the *Pseudothermotoga* genus consists of *Pseudothermotoga thermarum, Pseudothermotoga elfii, Pseudothermotoga hypogea, Pseudothermotoga Profunda,* and *Pseudothermotoga caldifontis,* with *P. thermarum* being the type species (Farrell et al. 2021). Members of the *Pseudothermotoga* genus are able to use thiosulfate as an electron acceptor and different carbon substrates, like xylose, glucose, sucrose, and starch, while for the type species, *P. thermarum*, elemental sulfur can inhibit growth, as well as the presence of H_2 (Windberger et al. 1989).

8.2 *Kosmotogales*

The *Kosmotogales* order was proposed by Bhandari and Gupta (2014b) to include all bacteria that were either rod or cocci-shaped and had a thermophilic or mesophilic nature, with the genus *Kosmotoga* as its nomenclature type. Within the *Kosmotogales*, the *Kosmotogaceae* family includes two genera *Kosmotoga* and *Mesotoga* and is characterized by a G + C content between 36.4 and 48.3 mol% (Nesbø et al. 2019). All the members of this order are known to be present in hydrothermal vents, marine sediments, crustal fluids, deep aquifers, oil reservoirs, environments contaminated with hydrocarbons, and other anthropic environments. The genus *Kosmotoga* is composed of three species, *K. olearia, K. arenicorallina,* and *K. shengliensis*, all of them thermophiles, with an optimum growth temperature between 60 and 65 °C (DiPippo et al. 2009; Feng et al. 2010; Nunoura et al. 2010). The name *Kosmotoga* was proposed by DiPippo et al. (2009) to indicate the fact that its members are present in different environments, from oil fields to marine sediments and hot springs. Among the substances that inhibit its growth are sulfite, acetate, lactate, and propionate (DiPippo et al. 2009). The fact that Kosmotoga species can grow at temperatures from 20 °C to 79 °C has been considered as evidence of high phenotypic flexibility in expressing genes associated with energy and carbohydrate metabolic activity at high temperatures, and up-regulation of amino acid production at lower temperatures (Pollo et al. 2017). However, not all species from this family share the same tolerance to anaerobiosis and capability to use alternative electron acceptors. For instance, *K. arenicorallina* is an obligate anaerobe and chemo-organotroph that grows in a very narrow temperature range, between 50 and 65 °C, and unable to reduce thiosulfate (Nunoura et al. 2010). The *Mesotoga* genus includes mesophilic obligate heterotrophs that grow optimally between 37 and 45 °C and is composed of two species *M. prima* (type species, described by Nesbø et al. 2012) and *M. infera*.

8.3 *Mesoaciditogales*

This order was proposed by Itoh et al. (2016) to accommodate the deeply branched, more acidic, low salinity, anaerobic, chemo-organotrophic bacteria that belonged to the Thermotogae. Accordingly, *Mesoaciditogales* are divided into two families, *Athalassotoga* and *Mesoacidotoga*, which include one species each (Itoh et al. 2016; Reysenbach et al. 2013). The only member of the *Athalassotoga* family, *Athalassotoga saccharophila*, has short rod cells that prefer mildly acidic (pH 4.5–7.5) and thermophilic (between 30 and 60 °C) conditions for growth as well as low salinity (< 1.0% NaCl), and anaerobic conditions. They can use different types of complex carbon substrates and can reduce Fe (III), thiosulfate, and L-cystine as electron acceptors. The G + C content is 41 mol% (Itoh et al. 2016). The family *Mesoaciditoga*, composed only of *Mesoaciditoga lauensis* (Reysenbach et al. 2013), is characterized by moderately thermophilic (45 to 65 °C) acidophilic (pH between 5.5 and 5.7) bacteria with cells shaped like short rods or cocci and a G + C content of 45 mol%. They grow under anaerobic conditions and are chemo-organotrophs capable of fermenting a broad range of carbohydrates, proteinaceous substrates, and yeast, reducing elemental sulfur to H_2S. Members of this family, similarly to other Thermotogae, have been isolated from deep-sea hydrothermal sediments (*M. lauensis*) or acidic hot spring waters (*A. saccharophila*).

8.4 Petrotogales

Petrotogales are all thermophilic (growth temperature between 45 and 70 °C) bacteria known to occur in oil reservoirs or deep-sea hydrothermal vents or chimneys. This order has only one family, *Petrotogaceae* that is composed of six genera: *Defluviitoga, Geotoga, Marinitoga, Oceanotoga, Petrotoga,* and *Tepiditoga. Petrotoga* is the type genus of the *Petrotogaceae* family, with *Geotoga petraea* as the type species. All members of the Petrogales order are also anaerobic, able to use elemental sulfur as an electron acceptor, have rod-shaped cells, and can degrade different types of carbon substrates.

Geotoga petraea was isolated and described by Davey et al. (1993) from oil field brines, along with *G. subterranea*, also isolated and characterized first by Davey et al. (1993) from brines in oil fields in Texas and Oklahoma. As all other *Thermotogae*, both are obligate anaerobes, and capable of fermentation and of reducing elemental sulfur to H_2S. They also share some morphological features like rod-shaped cells and a sheath-like outer structure. However, unlike other members of the *Thermotogales* class, they grow optimally at temperatures between 45 and 50 °C (Bhandari and Gupta 2014a; Davey et al. 1993) and are considered as halophilic since they grow optimally at salt concentrations between 3 to 4% and can tolerate salt concentrations of up to 10%.

Other relevant members of this order are the six species of the genus *Marinitoga* (*M. camini, M. piezophila, M. hydrogenotolerans, M. okinawensis, M. litoralis,* and *M. lauensis*). These species are capable of growth at low pH (4.5–5) and high salinity (from 2 to 4%) conditions as well as temperatures between 55 and 65 °C. This is thought to reflect the environmental conditions of their habitats (Alain et al. 2002; Bhandari and Gupta 2014a; L'Haridon et al. 2019; Nunoura et al. 2007; Postec et al. 2005, 2010; Wery et al. 2001). One peculiarity is the fact that *M. hydrogenotolerans* and *M. okinawensis*, unlike most members of the *Marinotoga* genus, and the *Petrotogaceae* family, are able to grow under 100% H_2 in the absence of sulfur (Nunoura et al. 2007).

The *Petrotoga* genus includes seven moderately thermophilic and halophilic species: *P. halophila, P. japonica, P. mexicana, P. miotherma. P. mobilis, P. olearia,* and *P. sibirica*, all isolated from oil brines, oil reservoirs, or oil wells. These bacteria are able to grow at temperatures that range from 35 to 65 °C (optimal temperature between 55 and 60 °C), at a pH range between 5.5 and 9.0, and variable salinity conditions (between 0.5 and 10% NaCl) with an optimum from 2.0 to 4.0% (Bhandari and Gupta 2014b; Davey et al. 1993; Lien et al. 1998; Miranda-Tello et al. 2004, 2007; Purwasena et al. 2014). The other three genera, *Defluviitoga, Oceanotoga,* and *Tepiditoga*, are all composed of just one species each: *D. tunisiensis, O. teriensis,* and *T. spiralis*, all of them exhibiting more of a moderate thermophilic nature, with an optimum growth temperature around 50 °C (Ben Hania et al. 2012; Jayasinghearachchi and Lal 2011; Mori et al. 2020).

9 Thermodesulfobacteria

Thermodesulfobacterias are thermophilic chemolithoautotrophs and heterotrophs that can use oxidized sulfur and iron compounds as terminal electron acceptors (Vick et al. 2010). The phylum contains one class *Thermodesulfobacteria*, one order *Thermodesulfobacteriales*, and one family *Thermodesulfobacteriaceae* with four valid genera: *Thermodesulfobacterium, Thermodesulfatator, Caldimicrobium, Thermosulfurimonas,* and the candidate genus *Geothermobacterium* (Mori 2014). Although there is one family in this phylum, it is one of the deeply branching lineages within the domain Bacteria, with a sequence divergence of the 16S rRNA gene is 11.6 b% (Mori 2014).

Members of this group inhabit different thermal environments. They were isolated from terrestrial hot spring in Yellowstone (U.S.A.) (Kashefi et al. 2002; Zeikus et al. 1983), in Kamchatka (Russia) (Miroshnichenko et al. 2009), in Iceland (Sonne-Hansen and Ahring 1999), from stratal waters of oil deposits (Hatchikian et al. 2001), from deep-sea hydrothermal vents at Guaymas Basin (Jeanthon et al. 2002), Central Indian Ridge (Moussard et al. 2004), mid-Atlantic Ridge (Alain et al. 2010), and Eastern Lau Spreading Center (Slobodkin et al. 2012). In addition, Thermodesulfobacterium species were retrieved from oil field waters from oil production platforms in the Norwegian sector of the North Sea (Christensen et al. 1992).

In general, they have cells morphologically rod-shaped and motile with single or some polar flagella (with some exceptions (Hatchikian et al. 2001; Sonne-Hansen and Ahring 1999)). Their Gram reaction is negative and spore formation is not reported. They are thermophilic, neutrophilic, and strictly anaerobic, with some exceptions. For example, species members of the genus *Thermodesulfobacterium* are chemolithoautotrophs growing using sulfur compounds or Fe(III) as an energy source, and CO_2 as a carbon source. Moreover, organic compounds can be used for growth by some species of the genera *Thermodesulfobacterium* and *Thermodesulfatator* (Mori 2014).

10 Aquificota

The phylum Aquificota (formerly referred to as Aquificae) is widespread in hydrothermal environments, both marine and terrestrial (Gupta 2014). Aquificota is composed of both thermophilic and hyperthermophilic bacteria, with the species *Aquifex aeolicus* holding the record of the highest temperature for the Bacterial domain, being able to grow at above 90 °C (Clarke 2014; Dworkin et al. 2006). The vast majority of members belonging to Aquificota are chemolithoautotrophs, being able to derive energy from inorganic compounds, such as different oxidation states of sulfur, hydrogen, ferrous iron, nitrate, and oxygen, coupled to inorganic carbon fixation using the reductive citrate cycle (rTCA) (Gupta, 2014b). Their versatility in the usage of electron donors and acceptors, as well as their abundance in geothermal environments, makes them key players in the biogeochemical cycles in these ecosystems. Notably, they have been used as a model to infer the evolution of metabolism for different reasons: they inhabit what could be referenced as relic ecosystems, which resemble the conditions present in the early Earth, as well as the fact that they root deeply in the phylogenetic tree of Life (Braakman et al. 2014; Giovannelli et al. 2017). This means that these organisms carry both ancestral and acquired genes. For instance, Braakman et al. (2014), using a novel approach nominated phylometabolics, which merges phylogenetic inference with metabolic reconstructions, was able to show three classes of innovations in metabolic evolution studying the genomic architecture of *Aquifex aeolicus*. The phylum Aquificota is composed of only one class, nominated *Aquificae*, and two orders, the *Aquificales*, and *Desulfurobacterales*. The order *Aquificales* encompasses aerobic and microaerophilic bacteria, while *Desulfurobacteriales*, only strictly anaerobic bacteria. This implies that they are able to colonize different niches in geothermal environments given the disponibility of oxygen in the ecosystem. Within *Aquificales*, two families have been described, *Hydrogenothermaceae* and *Aquificaceae*, while within *Desulfurobacteriales*, only the family *Desulfurobacteriaceae* has been described.

10.1 Aquificae

Aquificae > Aquificales > Hydrogenothermaceae.

The family *Hydrogenothermaceae*, within the phylum Aquificota, is composed by the genera *Hydrogenothermus*, *Persephonella*, *Sulfurihydrogenibium*, and *Venenivibrio*. All of the members belonging to this family are known thermophiles, with optimum growth temperatures ranging from 60 to 73 °C, using a diverse set of electron donors, such as elemental sulfur, thiosulfate, ferrous iron, and hydrogen, coupled to the reduction of oxygen and nitrate. One of the key characteristics of this family is the usage of oxygen as an electron acceptor.

The genus *Hydrogenothermus*, represented by the type strain *Hydrogenothermus marinus* VM1t, was firstly isolated from a marine shallow-water hydrothermal vent in the island of Vulcano, Italy (Stöhr et al. 2001). This organism grows chemolithoautotrophically, through the oxidation of hydrogen, coupled to the reduction of oxygen, and with an optimum temperature of 65 °C. Characterized by being a gram-negative bacterium, *H. marinus* has an optimum growth between pH5 and 7. *Hydrogenothermus marinus* have been reported from diverse geothermal environments, such as the deep-sea hydrothermal vents along the Eastern Lau Spreading Center (Ferrera et al. 2014), the hot springs of Tengchong, China (Briggs et al. 2014), as well as the mariana back-arc venting fluids (Trembath-Reichert et al. 2019). Previous studies have shown that this organism is able to grow on high abundances of perchlorates, naturally found in arid regions such as the Atacama Desert, Chile, with maintained growth patterns up to 200 mM perchlorate. Since high concentrations of perchlorates are also found on Mars, *H. marinus* was suggested to be considered as a model organism for future space experiments (Beblo-Vranesevic et al. 2017b). Given their ecological role in geothermal systems, the capabilities of coupling hydrogen oxidation to oxygen reduction means that this group occupies the aerobic niche in hydrothermal systems, coupling the reduction of hydrothermal fluids with the oxidized sea water, taking the advantage of the thermodynamic equilibrium formed.

The genus *Persephonella*, represented by the type strain *Persephonella marina*, was initially isolated in deep-sea hydrothermal vents in the Pacific Ocean (Gotz 2002). It is characterized by being a strictly chemolithoautotrophic, microaerophilic, hydrogen-oxidizing bacteria, with an optimum growth temperature of 73 °C and pH6. Other species belonging to *Persephonella* genus include *Persephonella guaymasensis*, isolated from the deep-sea hydrothermal vents of the Guaymas Basin (Gotz 2002), *Persephonella hydrogeniphila* (Nakagawa et al. 2003), isolated from the deep-sea hydrothermal vents of the Suyiyo Seamount in the Izu-Bonin Arc, Japan, and *Persephonella atlantica*, isolated from deep-sea hydrothermal chimney collected from the Lucky Strike hydrothermal vent field on the Mid-Atlantic Ridge (François et al. 2021). For this reason, they are thought to be major colonizers of the deep-sea hydrothermal vent ecosystem (Mino et al. 2013). Members of this genus are known to use hydrogen and thiosulfate as electron donors, coupled with the reduction of oxygen and nitrate, and have similar optimum growth temperatures (~70 °C). However, only *Persephonella atlantica* is not able to utilize elemental sulfur as an electron donor. Due to their thermophilic way of life, members belonging to *Persephonella* have been used for carbon sequestration technologies, for instance, by the utilization of their high temperature adapted carbonic anhydrase used in an amine-based absorption process (Kanth et al. 2014).

The genus *Sulfurihydrogenibium*, represented by the type strain *Sulfurihydrogenibium subterraneum*, was initially isolated from a hot subsurface aquifer in a Japanese gold mine. Having an optimum growth temperature of 60–65 ° C, this species grows chemolithoautotrophically, using hydrogen, sulfur, and thiosulfate as electron donors, coupled to oxygen reduction (Takai et al. 2003). Other species belonging to this genus include *Sulfurihydrogenibium azorense*, isolated from terrestrial hot spring in the Azores, Portugal (Aguiar et al. 2004), *Sulfurihydrogenibium kristjanssonii*, isolated from terrestrial Icelandic hot spring (Flores et al. 2008), and *Sulfurihydrogenibium rodmanii* isolated from terrestrial hot springs in the Geyser Valley and the Uzon Caldera, Kamchatka, Russia (O'Neill et al. 2008). These species have been demonstrated to use a diverse set of electron donors, encompassing hydrogen, ferrous iron, sulfide, sulfate, thiosulfate, sulfur, coupled to oxygen as electron acceptor. Their optimum temperature ranges from 55 ° C to 70 °C, and they are morphologically characterized by being gram-negative with motile rods.

The genus *Venenivibrio* is the least characterized genus from the *Hydrogenothermaceae*, with only one described species. The type strain, *Venenivibrio stagnispumantis*, is a thermophilic, hydrogen-oxidizing bacterium, isolated from a hot spring in Waiotapu, New Zealand (Hetzer et al. 2008). This species is able to grow using hydrogen as an electron donor, and oxygen as an electron acceptor, with an optimum temperature for growth at 70 °C, with elemental sulfur and thiosulfate shown to be essential for growth. Additionally, *V. stagnispumantis* is gram-negative, rod-shaped, and motile bacteria, growing with an optimum pH of 5.4.

The family Aquificaceae, together with *Hydrogenothermaceae*, make up the order *Aquificales*. This family encompasses the most thermophilic members belonging to the Aquificota, with optimum growth temperatures ranging from 65 to 90 °C, and spanning five different genera: *Hydrogenivirga*, *Hydrogenobacter*, *Hydrogenobaculum*, *Thermocrinis*, and *Aquifex*. As a whole, members belonging to the *Aquificaceae* are obligate chemolithoautotrophs, using hydrogen as the most preferred electron donor, and oxygen as electron acceptor. However, some species are also able to use sulfur species as electron donors, and nitrate as electron acceptor.

The genus *Hydrogenivirga* is composed of chemolithoautotrophic and thermophilic bacteria. Represented by the type species, *Hydrogenivirga caldilitoris*, a bacterium isolated from a coastal hot spring in Ibusuki, Kagoshima Prefecture, Japan. This species is able to grow at an optimum temperature of 75 °C and optimum pH of 6.5–7, using sulfur and hydrogen as electron donors, and nitrate and oxygen (microaerophilic) as electron acceptors Nakagawa et al. 2004a, b). The other species belonging to this genus is Hydrogenivirga okinawensis, a chemolithoautotrophic, thermophilic bacteria, isolated from a deep-sea hydrothermal field at the Yonaguni Knoll IV, Southern Okinawa Trough (Nunoura et al. 2008a, b). Compared to the other species of this genus, *H. okinawensis* can use thiosulfate as an electron donor, instead of hydrogen, and is able to grow at an optimum temperature of 70–75 °C and at an optimum pH of 6.9–7.5.

The genus *Hydrogenobacter,* represented by the type species *Hydrogenobacter thermophilus*, was initially isolated from hot springs located in Izu and Kyushu, Japan, and was regarded as the first obligate chemolithoautotroph described among the aerobic, hydrogen-oxidizing bacteria (Kawasumi et al. 1984). This thermophilic species grows through the oxidation of hydrogen, coupled to the reduction of oxygen, and has an optimum growth temperature between 70 and 75 °C, at neutral pH. Interestingly, it was on *H. thermophilus* that it was discovered a novel pathway of the reductive TCA cycle, where the cleavage of citrate was carried out by citryl-CoA synthetase (CCS), instead of the known enzyme ATP citrate lyase (ACL) (Aoshima et al. 2004). Other species of this genus include *Hydrogenobacter halophilus,* isolated from a seaside saline hot spring in Izu Peninsula, Japan. This species grows at an optimum temperature of 70 °C (Nishihara et al. 1990), and compared to the other species of the genus, *H. halophilus* can utilize elemental sulfur and thiosulfate as alternative electron donors to hydrogen. Furthermore, the species *Hydrogenobacter hydrogenophilum* (formerly classified as *Calderobacterium hydrogenophilum*) was isolated from the hydrothermal vents of the Kamchatka region (Ludvík and Benada 1994) and is able to grow at an optimum temperature of 74 °C, using hydrogen as an electron donor, and oxygen as an electron acceptor.

The genus *Hydrogenobaculum* was created to accommodate the species *Hydrogenobaculum acidophilum,* which was previously classified as *Hydrogenobacter acidophilus,* belonging to the genus *Hydrogenobacter.* This species was isolated from a solfataric field in Tsumagoi, Japan (Shima and Suzuki 1993), and it is the only thermoacidophile belonging to the Aquificota, with an optimum pH between 3 and 4, and an optimum growth temperature of 65 °C (Stöhr et al. 2001). This species can utilize hydrogen and reduced sulfur species as electron donors, with oxygen as electron acceptor.

The genus *Thermocrinis*, represented by the type species *Thermocrinis ruber*, is a thermophilic bacterium isolated from the Octopus spring in the Yellowstone National Park (H. Huber 2002). This species can grow at 89 °C in neutral to alkaline pH, using hydrogen, thiosulfate, and elemental sulfur as electron donors, and with oxygen as an electron acceptor. *T. ruber* was shown to be able to grow both chemolithoautotrophically, as well as heterotrophically. Additionally, a previous study reported the ability of *Thermocrinis ruber* to utilize arsenite as a sole electron donor, producing arsenate (Härtig et al. 2014). Other species belonging to this genus include *Thermocrinis minervae*, isolated from a Costa Rican terrestrial hot spring, a chemolithoautotrophic bacteria, growing at an optimum temperature of 75 ° C, and 5.9–6.5 pH (Caldwell et al. 2010). Similar to *T. ruber*, this species is able to utilize elemental sulfur, thiosulfate, and hydrogen as electron donors, and oxygen as electron acceptor. Moreover, the species *Thermocrinis jamiesonii* was isolated from the water column of Great Boiling Spring, Nevada, USA (Dodsworth et al. 2015). It can grow chemolithoautotrophically, using hydrogen and thiosulfate as electron donors and oxygen as an electron acceptor, having an optimum growth temperature of 80 °C, and an optimum pH of 7.25. However, the most thermophilic species within the genus *Thermocrinis* is *Thermocrinis albus,* which grows at an optimum temperature of 89 °C, and similar to the other species of this genus, is able use hydrogen, thiosulfate, and sulfur as electron donors, and oxygen as electron acceptor (Eder and Huber 2002).

The genus *Aquifex,* represented by the type species *Aquifex pyrophilus*, was initially isolated from hot marine sediments, retrieved from the Kolbeinsey Ridge, Iceland (R. Huber et al. 1992). This species is characterized by being a strictly chemolithoautotroph, growing at an optimum growth temperature of 85 °C and optimum pH of 6.8. It can use hydrogen, thiosulfate, and elemental sulfur as electron donors, and oxygen (microaerophilic) and nitrate as electron acceptors. The other species belonging to *Aquifex,* and the most studied member within the Aquificota is *Aquifex aeolicus*. As one of the most thermophilic bacteria known (growing at 95 ° C), *A. aeolicus* has sparked the interest of the scientific community. Similar to other members of the genus *Aquifex,* this species can utilize hydrogen as an electron donor, and oxygen (microaerophilic) as electron acceptor (Deckert et al. 1998), but contrary to *A. pyrophilus,* it has not been observed to perform nitrate respiration. Given its observed resistance to higher temperatures, it has been screened for temperature-resistant enzymes which could provide essential biotechnological applications (Guiral et al. 2012).

The family *Desulfurobacteriaceae,* the only described family within the order *Desulfurobacteriales,* is widely distributed in marine geothermal environments (Gupta 2014). It is composed of four genera: *Balnearium, Desulfurobacterium, Phorcysia, and Thermovibrio*. Conversely to the other families of Aquificota, *Desulfurobacteriaceae* is composed entirely of obligate anaerobic, chemolithoautotrophic, thermophilic bacteria (60–75 °C) (Giovannelli et al. 2017). Their inability to use oxygen as a terminal electron acceptor is reflected by the diversity of inorganic electron acceptors used to derive energy for growth, such as sulfur, polysulfide, thiosulfate, sulfite, and nitrate, coupled to hydrogen oxidation. Furthermore, similar to all the Aquificota, members belonging to this family are able to fix inorganic carbon using the reverse citrate cycle. However, they represent the only lineage within Aquificota to retain a complete reductive acetyl-coA pathway, including the oxygen-sensitive enzyme carbon monooxidase (*CodH*) (Giovannelli et al. 2017).

The genus *Desulfurobacterium* is ubiquitous in deep-sea hydrothermal vent environments. Represented by the type species *Desulfurobacterium thermolithotrophum*, it was initially isolated from a deep-sea hydrothermal chimney sample, retrieved from the mid-Atlantic ridge (L'Haridon et al. 1998). Other known species belonging to this genus include *Desulfurobacterium atlanticum,* isolated from East Pacific rise deep-sea hydrothermal vents, *Desulfurobacterium pacificum*, isolated from the mid-ocean ridge, *Desulfurobacterium crinifex*, isolated from the Juan de Fuca Ridge, and *Desulfurobacterium indicum*, isolated from a high temperature, deep-sea hydrothermal vent (Cao et al. 2017; L'Haridon et al. 2006; Alain et al. 2003). Members of this genus can oxidize hydrogen, coupled to a variety of electron acceptors, such as elemental sulfur, thiosulfate, sulfite, and nitrate. Additionally, this genus comprises only thermophilic bacteria, with optimum growth temperatures ranging from 60 °C to 75 °C, and pH 6.

The genus *Thermovibrio* is composed of three known species: *Themovibrio guaymasensis*, isolated from a deep-sea hydrothermal vent chimney at Guaymas Basin, *Thermovibrio ammonificans,* isolated from the East Pacific Rise hydrothermal field, and *Thermovibrio ruber*, isolated from a hydrothermal system off the beach of Lihir Island, Papua New Guinea (H. Huber 2002; Vetriani et al. 2004; L'Haridon et al. 2006). Of all the known species within *Thermovibrio, T. ruber*, the type species, remains the only microorganism that is not obtained from a deep-sea hydrothermal system. It is characterized by being a strictly anaerobic, chemolithoautotrophic, thermophilic bacteria (75 ° C), growing through the oxidation of hydrogen, coupled to the reduction of sulfur or nitrate. Interestingly, compared to the genus *Desulfurobacterium* of the *Desulfurobacteriaceae,* members belonging to *Thermovibrio* possess less flexibility regarding electron acceptors, mainly using sulfur and nitrate to produce H_2S and ammonium, respectively. For this reason, they have been considered key players in the nitrogen cycle on hydrothermal systems (Giovannelli et al. 2017). Additionally, all members of this genus possess similar optimum growth temperatures (75 °C to 80 °C), and optimum pH between 5.5 and 6.0.

The genus *Phorcysia* has only one described species, the type species *Phorcysia thermohydrogeniphila*, isolated from the tube of *Alvinella pompejana* tubeworms, collected from the wall of a sulfide structure on the East Pacific Rise deep-sea hydrothermal vents (Perez et al. 2012). Characterized by being an anaerobic, chemolithoautotrophic bacteria, *P. thermohydrogeniphila* can grow using hydrogen as the sole electron donor, and nitrate and sulfur as electron acceptors, producing ammonium and hydrogen sulfide, respectively.

The genus *Balnearium* is represented by only one identified species, *Balnearium lithotrophicum,* a strictly anaerobic, hydrogen-oxidizing, and chemolithoautotrophic bacteria. This species is a known thermophile, having an optimum growth temperature ranging from 70 to 75 °C, and the lowest optimum pH within the *Desulfurobacteriaceae,* 5.4 (K. Takai 2003). Contrarily to the other genus of this family, *Balnearium lithotrophicum* can grow using hydrogen as a sole electron acceptor, and elemental sulfur as the sole electron acceptor, therefore presenting the lowest metabolic plasticity of the family.

11 Deferribacterota

The phylum Deferribacterota, described in 2001, consists of a family formerly known as *Deferribacteres* (Huber and Stetter 2001) and is widespread in marine environments, deep hydrothermal vents, contaminated soils and oil reservoirs, and the gut mucus of rodents. Ten Gram-negative, anaerobic (rarely microaerophilic), and rod- or vibrio-shaped genera are grouped in the phylum, e.g., *Calditerrivibrio, Deferribacter, Denitrovibrio, Flexistipes, Geovibrio, Deferrivibrio, Limisalsivibrio, Petrothermobacter, Selenivibrio*, and *Mucispirillum*. Of these, only five are thermophilic, able to live and survive above 41 °C, e.g., *Calditerrivibrio, Deferribacters, Flexistipes, Deferrivibrio,* and *Petrothermobacter*. Members belonging to Deferribacterota employ different metabolic strategies, such as chemo-organotrophy and chemolithotrophy. In general, they are capable of deriving energy from the anaerobic respiration of various organic substrates and the use of nitrate, iron (II), manganese (IV), sulfur-reduced compounds, and cobalt (III) as terminal electron acceptors. However, they can also be able to grow heterotrophically by fermentation (Huber and Stetter 2001). The dissimilatory metal reduction is a metabolic pathway that some members of the Deferribacterota phylum can use to thrive on iron and manganese oxides as electron acceptors. The thermophilic genera of this phylum are commonly found in sulfur-rich and deep-sea hydrothermal systems, making them key players in the biogeochemical cycles in these environments (Slobodkin et al. 2019), while the other genera can be found in contaminated soils and oil reservoirs (Hidalgo et al. 2021), with the exception for *Mucispirillum* sp. that showed a unique lifestyle connected to gut mucus of rodents (Robertson et al. 2005). The presence of some genera in deep-sea systems is also supported by their halophilic and halotolerant capabilities.

11.1 Deferribacteres

The genus *Calditerrivibrio* is an important nitrate reducer in nature and was discovered in hot spring waters in Yumata, Nagano, Japan ($T = 60$ °C) (Iino et al. 2008). The only known species is *Calditerrivibrio nitroreducens,* an anaerobic, thermophilic bacterium with an optimum growth temperature between 30 °C and 65 °C, and a pH optimum between 7.0 and 7.5 (Iino et al. 2008). From a metabolic point of view, *C. nitroreducens* can be defined as a chemo-organotrophic, non-fermentative organism. It is involved in the nitrogen biogeochemical cycle due to its ability to use nitrate as the only electron acceptor during growth, having ammonium as the final product. Other types of electron acceptors are not used in their metabolic activities. *C. nitroreducens* use different organic carbon species as electron donors (Iino et al. 2008).

The genus *Deferribacters*, composed of four different species, is of fundamental importance in the biogeochemical cycling of iron and other metal oxides as part of their respiratory metabolism. Only four species belong to this genus: *D. abyssi, D. autotrophicus, D. desulfuricans*, and *D. thermophilus*. All four species are thermophiles able to grow between 40 °C and 70 °C with an optimum temperature of 60 °C. *D. abyssi, D. autotrophicus,* and *D. desulfuricans* have been isolated from deep-sea hydrothermal environments between the Mid-Atlantic Ridge, and the deep-sea hydrothermal vents on the Izo-Bonin Arc in Japan (Miroshnichenko et al. 2003; Takai et al. 2003). Furthermore, *D. autotrophicus* was isolated from the deepest known ocean hydrothermal field at a depth of 4100 m (Slobodkina et al. 2009). *D. thermophilus* was discovered in the North Sea's Beatrice oil field, UK. Metabolically, *D. abyssi* is a chemolithoautotrophic bacteria capable of using molecular hydrogen and inorganic carbon compounds as electron donors, and reduced sulfur, iron (III), and nitrate species as electron acceptors (Miroshnichenko et al. 2003). The same is for *D. autotrophicus* but it adds manganese (IV) to the plethora of electron acceptors. All four species can use hydrogen or organic substances as energy sources and carbon sources, and elemental sulfur and nitrate as electron acceptors (Greene et al. 1997). Additionally, *D. desulfuricans* has been shown to be able to use arsenate as an electron acceptor, making it a major player in the arsenic biogeochemical cycle (Takai et al. 2003).

The genus *Flexistipes*, along with the only known species *F. sinusarabici*, has been isolated from the hot brines of the Atlantis II Deep at the bottom of the Red Sea, one of the most extreme environments on our planet, e.g., temperature of 64 °C and salinity values of 26% NaCl, high concentration of heavy metals, and no free oxygen (Hartmann 1985; Brewer and Spencer 1969). *F. sinusarabici* is a thermophile organism that can grow in a range of temperatures between 30 °Cand 53 °C with an optimum temperature between 45 °C and 50 °C. It is a chemo-organotroph capable of metabolizing complex organic compounds under anaerobic growth using molecular nitrogen, carbon dioxide, molecular hydrogen, and methane. It has been found that *F. sinusarabici* can produce sulfur hydrides in the presence of elemental sulfur (Fiala et al. 1990).

The genus *Deferrivibrio* is a key actor in the biogeochemical cycling of elements in hot and haline mineral water deposits. *D. essentukiensis* is the only known species of this genus, isolated for the first time in the Yessentukskoye mineral water deposit in Russia (Zavarzina et al. 2022). It is a moderate thermophile capable of growing at a temperature between 30 °C and 54 °C and a pH between 6.2 and 7.9. In addition, it is an anaerobic halotolerant organism able to thrive in haline environments with a concentration of NaCl between 0 and 18 g/L. *D. essentukiensis* is a chemo-organotroph that uses carbonic organic compounds as electron donors and ferrihydrite as the sole electron acceptor (Zavarzina et al. 2022).

The single species known as *Petrothermobacter organivorans* makes up the entire genus *Petrothermobacter*. *P. organivorans* has been isolated for the first time in a deep subsurface oil field in Japan. It is an anaerobic, chemoheterotrophic organism capable of oxidizing carbon compounds while reducing iron (III), manganese (IV), nitrate, and sulfate as electron acceptors. It has an optimum of growth at 55 °C and the optimum pH is in the range of 6.0–8.0 (Tamazawa et al. 2017). *P. organivorans* has a relatively broad substrate range and can utilize pyruvate, fumarate, succinate, malate, yeast extract, and peptone for fermentative growth. Due to its metabolic capabilities, it can be considered as an essential organism for the biogeochemical cycling of various metals, such as iron and manganese (Tamazawa et al. 2017).

12 Chlorobi

The phylum Chlorobi has long been regarded as a monophyletic group of strictly anaerobic and anoxygenic, sulfur-oxidizing, photolithoautotroph microorganisms, generally referred to as the Green Sulfur Bacteria (GSB) (Frigaard and Dahl 2009; Fröstl and Overmann 2000; Gregersen et al. 2011; Imhoff 2003; van Niel 1932; Overmann and Tuschak 1997; Trüper Hans and Pfennig 1992). Photosynthesis is carried out due to bacteriochlorophylls and photosynthetic pigments—similar to plant, algal, and cyanobacterial chlorophyll—hosted in the chlorosomes. Sulfide and hydrogen sulfide are used as electron donors to fix CO_2 via reverse TCA cycle (Bello et al. 2022; Trüper Hans and Pfennig 1992).

Chlorobi traditionally included a single monotypic class *Chlorobia*, in turn consisting of one order (*Chlorobiales*) and family (*Chlorobiaceae*) (Gibson et al. 1984; Imhoff 2003), which is further divided into several genera (e.g., *Chlorobium*, *Ancalochloris*, *Chloroherpeton*, *Pelodicyton*, and *Prostheocochloris*) (Fröstl and Overmann 2000; Garrity et al. 2001a, b, c, d, e, f). The only metabolic exception was believed to concern the species *Chlorobium ferroxidans* and *Chlorobium phaeoferrooxidans*, which use ferrous iron rather than sulfur as electron donor in a metabolic pathway termed photoferrotrophy (Beatty et al. 2005; Frigaard and Bryant 2004; Frigaard and Dahl 2009; Garcia et al. 2021; Ghosh and Dam 2009; Hegler et al. 2008; Heising et al. 1999; Hiras et al. 2016; Imhoff 2003). However, later studies demonstrated that the cultured representative GSB are rather part of a wider and more diverse taxonomic group (Hiras et al. 2016). Firstly, a second class termed *Ignavibacteria* was included in the phylum when the anaerobic chemo-organoheterotrophic thermophilic species *Ignavibacterium album* (Iino et al. 2010) and *Melioribacter roseus* (Podosokorskaya et al. 2013) were discovered. Then, metagenomic approaches widened the class *Chlorobi* by revealing uncultured aerobic photoheterotrophic genome species, namely *Thermochlorobacter aerophylum*, and *Chlorobium* sp. GBChlB, and Chlorobi-445 (Liu et al. 2012; Roy et al. 2019); these latter have been proposed as part of the novel family *Chloroherpetonaceae* (Bello et al. 2022). Lastly, the aerobic chemo-organoheterotrophic OPB56 clade, whose clones are reported to be ubiquitous in thermal environments, was proposed to have the same common ancestor as Chlorobi and Ignavibacterium classes based on both concatenated ribosomal protein tree and concatenated single copy genes tree (Hiras et al. 2016; Soo et al. 2014).

12.1 Chlorobia

Besides GSB, the order Chlorobiales encompasses various thermophilic microorganisms whose optimal growth temperature range is 45 °C – 55 °C, and 4.5 and 6 for pH. The species *Chlorobium tepidum* was in fact isolated from an anoxygenic mat in a volcanic area in New Zealand and consists of anaerobic, phototrophic—either autotrophic or heterotrophic—microorganisms (Wahlund et al. 1991). Likewise, *Thermochlorobacter aerophilum* was isolated from microbial mats of alkaline siliceous hot spring at the Yellowstone National Park, but it is an aerobic phototrophic species, unable to oxidize sulfur compounds and to fix both N_2 and CO_2 (Liu et al. 2012). Conversely, the GSB1 species, which has been molecularly correlated to the *Chlorobium* and *Prosthecochloris*, was isolated from water collected at the plume of a 2 km-deep black smoker on the East Pacific rise. GSB1 retrieval is surprising as its growth requires anaerobiosis, sulfide or hydrogen sulfide, CO_2, and light, which in such an environment can only derive from geothermal activity (Beatty et al. 2005).

12.2 Ignavibacteria

The order *Ignavibacteriales* only harbors two thermophilic species, namely *Ignavibacterium album* and *Melioribacter roseus*, isolated from a sulfide-rich hot spring in Japan and water springing from a 2775 m-deep artesian borehole in Western Siberia, respectively. Both isolates have been cultured, thus demonstrating to be able to grow both aero- and anaerobically via chemo-organotrophic metabolism, with their optimal growth temperature spanning between 35 °C and 60 °C (Iino et al. 2010; Podosokorskaya et al. 2013).

Notably, *Melioribacter roseus* has 2 L-asparaginases, whose activity can be widely exploited by various biotechnological fields: these enzymes can both promote apoptosis in cancer cells and catalyze L-asparagine hydrolysis (thus preventing the formation of acrylamide, a human carcinogen). One of the two *M. roseus* enzymes exhibits the highest rate of activity at 70 °C, and thermophilic enzymes have been demonstrated to exhibit higher enzymatic activity, despite increased KM. Adaptation to high temperatures is given by the low abundance of thermolabile residues and the high frequency of thermostable residues (Dumina et al. 2021).

13 Thaumarchaeota

Thaumarchaeota is Gram-negative, short-rod, microbial group, including members of 0.3–0.6 μm in diameter and 0.6–1.0 μm long (Jung et al. 2018). Most of them possess a wider range of cell envelope structures than bacteria, and they differ from bacteria in the absence of peptidoglycan in the cell walls. Thaumarchaeota range among the most abundant archaea on Earth, and in particular in soil systems where they constitute about 5% of all prokaryotic biomass (Schleper and Nicol 2010). Initially classified as "mesophilic Crenarchaeota," comparative genomics has recently revealed that they form a separate and deep-branching phylum within the Archaea (Zhang and He 2012). Everything starts back in 1992, when Jed Fuhrman's team and Ed DeLong reported the discovery of a novel clade of archaeal 16S rRNA sequences from ocean surface waters, which formed a mesophilic sister group to the hyperthermophilic Crenarchaeota (DeLong 1992; Fuhrman et al. 1992). When it became apparent that this novel group contained autotrophic ammonia-oxidizing archaea, these organisms were consequently also referred to as mesophilic Crenarchaeota. When Brochier-Armanet and colleagues analyzed a concatenated data set of 53 ribosomal proteins common to Archaea and Eukarya, they observed that *C. symbiosum* branched off before the separation of Crenarchaeota and Euryarchaeota. Based on this phylogenetic analysis, on gene presence/absence data, and on the diversity and wide distribution of autotrophic ammonia-oxidizing archaea, they proposed that these organisms belong to the phylum Thaumarchaeota (Brochier-Armanet et al. 2008). This novel phylum includes all known archaeal ammonia oxidizers and additionally several clusters of environmental sequences representing microorganisms with unknown energy metabolism (Pester et al. 2011).

Aerobic ammonia oxidation, the first and rate-limiting step in nitrification, is the only biological process converting reduced to oxidized inorganic nitrogen species on Earth (Gruber and Galloway 2008). The Thaumarchaeota are the first example of nitrification in the Archaea kingdom, a metabolism previously thought to be restricted to a few proteobacterial lineages (Könneke et al. 2005). Most of the ammonia-oxidizing archaea that have been identified to date have not been fully characterized because it is extremely difficult to obtain pure cultures of these organisms. Autotrophic CO_2 fixation on low NH_3 concentrations is the primary anabolic process in thaumarchaeal ammonia oxidizers; however, a potential for mixotrophic growth has also been reported (Hatzenpichler 2012; Stahl and de la Torre 2012). They have an extremely high affinity for substrate (Martens-Habbena and Stahl 2011), which means that they are frequently the dominant ammonia-oxidizing organisms in natural environments with low NH_3 concentrations. These environments include oligotrophic open ocean waters and nutrient-poor soils (Verhamme et al. 2011). They also seem to be adapted to growth at low pH, low dissolved oxygen concentrations, and high temperatures (Hatzenpichler 2012) and are the primary ammonia oxidizers in acidic forest soils (Lehtovirta-Morley et al. 2011), coastal waters (Urakawa et al. 2010), in the Atlantic and Pacific Oceans (Aylward and Santoro 2020), marine sediments (Dang et al. 2013), wastewater treatment bioreactors (Badar et al. 2022), in a coffee compost (Papale et al. 2021) and geothermal habitats (Beam et al. 2014; Hedlund et al. 2013; Nishizawa et al. 2016).

The presence of Thaumarchaeota, both in mesophilic and in thermophilic environments (De la Torre et al. 2008), confirms the wide spectrum of phenotypes belonging to this phylum. Different studies show that ammonia-oxidizing archaea are widely distributed in terrestrial geothermal systems (De la Torre et al. 2008) featured by diverse environmental variables. For example, ammonia-oxidizing archaea growth has been demonstrated up to 74 °C (de la Torre et al. 2008) and in situ activity measurements and quantitative studies of these microbial populations have revealed high activity and abundance up to ~81 °C (Cole et al. 2013; Dodsworth et al. 2011). In addition, studies about the distribution and relative abundance of the isoprenoid glycerol dialkyl glycerol tetraether (iGDGT) crenarchaeol, a potential biomarker for ammonia-oxidizing archaea, suggest this microbe's higher abundance compared to other archaea at 45–50 °C (Zhang et al. 2006). We also know that hyperthermophiles constitute the first diverging lineages of the currently described archaeal phyla (Crenarchaeota, Euryarchaeota, Korarchaeota) indicating that the last common ancestor of Archaea might have been a hyperthermophile (Forterre 2002; Woese 1987; Woese et al. 1990). Mesophily in some thaumarchaeal lineages is a derived character that helped their adaptation to colder habitats. The analysis of environmental sequences from Thaumarchaeota indicates that adaptation to mesophilia may have happened through horizontal gene transfer from bacteria or from mesophilic archaea (López-García et al. 2015). Because secondary adaptations to mesophily are also observed in

Euryarchaeota, this would be consistent with the hypothesis of multiple independent adaptations to mesophily in the Archaea from thermophilic or hyperthermophilic ancestors.

There are clear biotechnological applications related to the Thaumarchaeota group, mainly related to its thermostable enzyme properties (Saghatelyan et al. 2021). It is known, for example, that the type IB DNA topoisomerases are very important targets for antitumoral drugs in humans (Pommier 2009). Recent studies suggest that the archaeal type IB topoisomerases are much more similar to the eukaryotic enzymes than the Vaccinia virus type IB topoisomerase, which has been widely used as model system to understand human enzyme (Dahmane et al. 2016). These characteristics indicate that the Cs-TopIB or others thaumarchaeal TopIB might be promising new models for phylogenetic and structural studies of the type IB DNA topoisomerases.

14 Archaeoglobi

Archaeoglobi is a class of archaea that includes two orders, namely the *Archaeoglobales* and the *Desulfurococcales*. Members of the *Archaeoglobales* reduce sulfate, thiosulfate, iron, and nitrate, while they are facultative autotrophs or heterotrophs that reduce sulfate and thiosulfate to hydrogen sulfide. Archaeoglobi are strict anaerobes and hyperthermophiles that thrive in high-temperature environments, typically between 60 °C and 95 °C, and pH 5.5–7.5. They occur singly and in pairs and are gram-negative. Thermophilic archaeoglobi are generally small, spherical, or rod-shaped, and lack motility structures such as flagella. They typically have a cell wall made of pseudomurein, a type of peptidoglycan-like polymer that is unique to the Archaea domain.

Archaeoglobi are found in various locations around the world, such as Iceland, Italy, Japan, New Zealand, Russia, and the United States, and in a wide range of extreme habitats, such as deep-sea hydrothermal vents, hot springs, and oil reservoirs.

There are two orders of marine generalist archaea that are both considered Archaeoglobi: the *Archaeoglobales* in the Euryarchaeota and the *Desulfurococcales* in the Crenarchaeota.

The family *Archaeoglobaceae* is composed of *Archaeoglobi, Geoglobus*, and *Ferroglobus*. Archaeoglobi includes five species with validly published names: *A. fulgidus*, *A. profundus*, *A. veneficus*, *A. infectus*, and *A. sulfaticallidus*. Archaeoglobi can use a wide range of electron donors and acceptors, including sulfate, sulfite, thiosulfate, iron, nitrate, and various organic compounds. Members of the *Archaeoglobales* order, including Archaeoglobi, are able to grow by reduction of sulfite and thiosulfate. *A. sulfaticallidus* is the only species capable of lithoautotrophic growth with sulfate as a terminal electron acceptor.

The cell envelope of Archaeoglobi consists of an S-layer composed of subunits in hexagonal array containing a periodate-Schiff-positive polypeptide. Cells contain phytanyl ether lipids.

A complete oxidative TCA cycle has been shown to function in the cells of Archaeoglobi members such as *A. fulgidus*, *A. profundus*, *F. placidus*, *Geoglobus ahangari*, and *Geoglobus acetivorans*. The recently discovered species of Archaeoglobus, named *A. lithotrophicus*, has been found to be capable of lithoautotrophic growth using hydrogen gas and carbon dioxide as electron donor and acceptor, respectively.

One of the defining characteristics of thermophilic archaeoglobi is their ability to use molecular hydrogen (H_2) as an electron donor and sulfate (SO_4^{2-}) as an electron acceptor for energy production. This process, known as sulfate reduction, produces hydrogen sulfide (H_2S) as a byproduct. The ability of thermophilic archaeoglobi to carry out sulfate reduction is believed to be a key factor in their success in extreme environments, as sulfate is often abundant in such environments. Thermophilic archaeoglobi are also capable of carrying out a number of other metabolic processes, including the oxidation of organic compounds and the reduction of elemental sulfur. Some species are even capable of using carbon monoxide (CO) or methanol as electron donors.

There is growing interest in the biotechnological applications of Archaeoglobi due to their unique metabolic capabilities, which make them potentially useful for a range of industrial processes. Archaeoglobi are known, for example, for their ability to degrade a variety of hydrocarbons, including alkanes and aromatic compounds, under anaerobic conditions. This makes them a potential tool for bioremediation of contaminated environments, such as oil spills or petroleum-contaminated soils. Archaeoglobi can use hydrogen and carbon dioxide to produce methane, making them useful for the production of biogas as a renewable energy source. This archaeal group is also able to produce a range of enzymes that have potential biocatalytic applications. For example, some archaeal enzymes are known to be highly stable and active under extreme conditions of temperature, pH, and salinity, and this is why they are widely used in the production of chiral compounds for pharmaceuticals and agrochemicals, and in various industrial processes that require high stability and activity under extreme conditions. Some species of Archaeoglobi are capable of oxidizing sulfide minerals, making them useful for bioleaching of metal ores. Overall, there is significant potential for biotechnological applications of Archaeoglobi, and research in this area is ongoing.

15 Euryarcheota

The Euryarcheota phyla belongs to the Archaea domain and is composed of organisms that show very diverse physiological traits. These characteristics are known since the greatest number of cultured and diverse archaeal lineages belong to the Euryarchaeota phyla, in particular from the thermophilic classes *Archaeoglobi*,

Methanopyri, and *Thermococci* (Baker et al. 2020). Generally speaking, the Euryarchaeota can occur as rods, cocci, irregular cocci, triangular, square, lancet, spiral, or disk-shaped cells. Its nature with regard to Gram staining can be either positive or negative, based on the presence or absence of pseudomurein in cell walls (Garrity et al. 2001a, b, c, d, e, f). This phylum includes hyperthermophilic, halophilic, and methanogenic organisms. This diversity also poses a sharp contrast among organisms classified as Euryarcheota, since methanogens are the most strict type of anaerobes described, while halophiles are for the most part primarily aerobes (Pesaro and Widmer 2002; Baker et al. 2020). Euryarchaeota thermophiles belong to the *Thermoplasmatales* order, which includes three genera: *Thermoplasma, Picrophilus,* and *Ferroplasma,* with the *Thermoplasma* having the peculiarity of not having cell walls (Garrity et al. 2001a, b, c, d, e, f). There are three key genera of hyperthermophiles, two of them (*Thermococcus* and *Pyrococcus*) make up a distinct taxonomic order, the *Thermococcales* while the other one (*Methanopyrus*) is a methanogen that despite its similarity with the other methanogens has the peculiarity of being able to thrive at high temperatures. The growth temperatures reported for these thermophiles range from 55 °C (*Thermoplasma*) to as high as 110 °C (*Methanopyrus*). Other hyperthermophiles belonging to the *Thermoplasmatales* include *Archaeoglobus* and *Ferroglobus.* Thermophilic and hyperthermophilic Euryarcheaota metabolisms are varied, spanning from methanogenesis like in *Methanopyrus* to chemoheterotrophy, where proteins such as starch or maltose are used as electron donors and Fe, sulfate or S^0 act as a terminal electron acceptor in *Thermococcus, Pyrococcus, Thermoplasma*, and *Picrophilus* (Leigh and Whitman 2013a, b).

15.1 Thermoplasmatales

The key genera of this class are *Thermoplasma, Picrophilus,* and *Ferroplasma.* All members of this class are facultative aerobes capable of heterotrophic growth, under acidic conditions; however, only *Thermoplasma* and *Picrophilus* can thrive under thermophilic conditions. *Thermoplasma* lacks cell walls and grows optimally at temperatures near 55 °C (Garrity et al. 2001a, b, c, d, e, f; Leigh and Whitman 2013a, b). *Thermoplasma* species are facultative aerobes, chemo-organotrophic organisms, able to perform sulfur respiration through the degradation of organic compounds. These archaea are also acidophiles (optimum pH2.0) and are able to cope with the high temperatures and low pH due to the production of lipoglycan inside its cytoplasmic membranes. Members of the *Thermoplasma* have been found in self-heating coal refuse piles, terrestrial solfataras, and acid hot springs, with *T. acidophilum* being the first species described by Darland et al. (1970). Other members of this genus are *T. volcanium* (Segerer et al. 1988) and *T. thiooxidans* (Li et al. 1994).

Picrophilus is the other thermophilic member of the *Thermoplasmatales.* This genus differs from *Thermoplasmata* in the fact that it has a cell membrane.

Picrophilus species have an optimum growth temperature between 47 and 60 °C and is able to grow at pH values below 0, making it an hyperacidophile. It has also been shown that its growth is inhibited by high salt concentrations, which supports the hypothesis of a distribution restricted to terrestrial geothermal settings (Garrity et al. 2001a, b, c, d, e, f). This genus is composed of two species, both of them isolated from solfataric springs in Japan by Schleper et al. (1996) and capable of forming a highly acid-impermeable membrane at pH values below 4.

15.2 *Thermococcales*

The Thermococcales order includes all the hyperthermophilic genera belonging to the Euryarchaeota. They are heterotrophic archaea able to grow optimally at temperatures between 75 and 100 °C using sulfur respiration. They are commonly found in deep and shallow marine hydrothermal vents, although they have also been isolated from terrestrial hot springs (Garrity et al. 2001a, b, c, d, e, f). The genera included in this order are as follows: *Thermococcus* and *Pyrococcus*. *Thermococcus* are obligate chemo-organotrophic anaerobes that can grow metabolizing proteins and other complex C sources due to the reduction of elemental sulfur (Zillig et al. 1983a, b). The optimum growth temperature for these archaea is between 75 and 88 ° C (Zillig et al. 1983a, b; Garrity et al. 2001a, b, c, d, e, f), in the presence of salt (between 2 and 4%). There are at least 15 species formally recognized as members of this genus, most of them isolated from terrestrial hot springs, coastal and marine solfataras, and deep-sea hydrothermal vents (Garrity et al. 2001a, b, c, d, e, f). The other member of the *Thermococcales* order is *Pyrococcus*. Members of this genus have a higher optimum growth temperature, between 70 and 106 °C, with this being the major difference among the two genera. From a metabolic point of view, they are also very similar, since both genera are heterotrophs that couple degradation of complex carbon substrates to the reduction of elemental sulfur to form H_2S. Species belonging to the *Pyrococcus* genus have been isolated from geothermally heated marine sediments such as *P. furiosus* (Fiala and Stetter 1986), deep-sea hydrothermal vents such as *P. abyssi, P. glycovorans, P. chitonophagus, P. horikoshii, P. kukulkanii, P. woesei*, and *P. yayanosii* (Zillig et al. 1987; Erauso et al. 1993; González et al. 1998; Barbier et al. 1999; Lepage et al. 2004; Birrien et al. 2011).

15.3 *Methanopyrus*

Methanopyrus organisms are hyperthermophilic archaea able to grow optimally at temperatures of 98 °C (Garrity et al. 2001a, b, c, d, e, f). Members of this genus have the peculiarity that they share phenotypic properties with both methanogens and hyperthermophiles. They can grow autotrophically through chemolithoautotrophy, coupling the reduction of CO_2 to CH_4, to the oxidation of H_2. The membrane lipids found in the cell membranes of these archaea are an unsaturated form of diphytanoyl tetraethers and are thought to be able to stabilize the cytoplasmic membrane at high temperatures.

16 Crenarchaeota

The phylum Crenarchaeota (Garrity et al., 2001) were identified as a physiologically homogeneous group characterized by thermophilic taxa like *Sulfolobus* spp. (Brock et al. 1972; Woese 1987; Woese et al. 1990; Pesaro and Widmer 2002) but also by hyperthermophiles like *Pyrolobus* spp. (Blöchl et al. 1997). Most Crenarchaeota species were first isolated from submarine and terrestrial environments such as deep-sea hydrothermal vents, terrestrial hot springs, and hot acidic mudpots (Zillig et al. 1983a, b; Nakagawa et al. 2004a, b; Leigh and Whitman 2013a, b). The only class within the Crenarchaeota is named *Thermoprotei* which is divided in further orders among which the *Acidilobales, Desulfurococcales, Fervidicoccales, Sulfolobales,* and *Thermoproteales* (Offre et al. 2013). Energy-generating metabolism known for Crenarchaeota includes autotrophic pathways during which carbon is assimilated from oxidized inorganic compounds, i.e., carbon dioxide or bicarbonate, and reduced to form simple organic molecules (Berg et al. 2010). Some lineages of Crenarchaeota include facultative autotrophic organisms as well as obligate heterotrophs which utilize proteins and sugars as main carbon source (Kletzin 2007). The Crenarchaeota include some of the first cultivated lineages of archaea belonging to the *Sulfolobus* genus that were isolated for the first time from the hot acid springs in Yellowstone National Park by Thomas Brock (Brock et al. 1972) and then become one of the main models of thermophile archaea investigations (Zhang et al. 2018).

16.1 Thermoprotei

Sulfolobus species present an aerobic and microaerophilic way of life, able to thrive at temperatures around 75–80 °C and pH 2–3 (Brock et al. 1972). Most species are chemo-organoheterotrophs, e.g., *S. solfataricus* which uses a wide range of carbon substrates (i.e., sugars, tryptone, peptides, and amino acids) (Hanner et al. 1990; Wolf et al. 2016) while others like *S. acidocaldarius* are considered facultative autotrophic, involved in the respiration of sulfur and the utilization of CO_2 or HCO_3—as source of carbon (Brock et al. 1972; Schönheit and Schäfer 1995; Leigh and Whitman 2013a, b). Among the eight *Sulfolobus* species known in literature, only three (*S. islandicus, S. solfataricus, and S. acidocaldarius*) are well described as model organisms for comparative genomics and genetics, host–virus interactions, investigation of catabolic enzymes, and industrial applications (Held and Whitaker 2009; Reno et al. 2009; Chavan and Deshpande 2013; Bräsen et al. 2014). Indeed, the genus *Sulfolobus* has a key role in several biotechnological applications since it is a source of unique enzymes, biomaterials, and unique catabolic pathways (i.e., the Entner–Doudoroff (ED) pathway and the Weimberg and Dahms pathway) involved in the degradation of pentoses and hexoses for the exploitation of novel products (Siebers and Schönheit 2005; Nunn et al. 2010; Kouril et al. 2013; Besse et al. 2015; Quehenberger et al. 2017).

17 Conclusions

This chapter was intended to give an overview of thermophilic bacteria, found within the prokaryotic Tree of Life. At this present date, more than 1200 species of thermophilic bacteria have been discovered and are found in at least 15 phyla. Thermophilic prokaryotes are found in a wide range of extreme environments, including hot springs, geothermal vents, deep-sea hydrothermal vents, and desert soil. In general, thermophilic bacteria have diverse metabolic activity and are capable of utilizing a wide range of substrates. Their diversity and adaptation to extreme environments make them a promising source of biotechnological applications, including production of enzymes and biofuels as was addressed in the present chapter.

References

Abraham W-R, Rohde M, Bennasar A (2014) The family Caulobacteraceae. In: Rosenberg E, DeLong EF, Lory S et al (eds) The prokaryotes: alphaproteobacteria and betaproteobacteria. Springer, Berlin, pp 179–205

Aguiar P, Beveridge TJ, Reysenbach A-L (2004) *Sulfurihydrogenibium azorense*, sp. nov., a thermophilic hydrogen-oxidizing microaerophile from terrestrial hot springs in the Azores. Int J Syst Evol Microbiol 54:33–39. https://doi.org/10.1099/ijs.0.02790-0

Akinosho H, Yee K, Close D et al (2014) The emergence of *Clostridium thermocellum* as a high utility candidate for consolidated bioprocessing applications. Front Chem 2(Aug):66. https://doi.org/10.3389/fchem.2014.00066

Alain K, Marteinsson VT, Miroshnichenko ML et al (2002) *Marinitoga piezophila* sp. nov., a rod-shaped, thermo-piezophilic bacterium isolated under high hydrostatic pressure from a deep-sea hydrothermal vent. Int J Syst Evol Microbiol 52:1331–1339. https://doi.org/10.1099/00207713-52-4-1331

Alain K, Rolland S, Crassous P et al (2003) *Desulfurobacterium crinifex* sp. nov., a novel thermophilic, pinkish-streamer forming, chemolithoautotrophic bacterium isolated from a Juan de Fuca Ridge hydrothermal vent and amendment of the genus *Desulfurobacterium*. Extremophiles 7:361–370. https://doi.org/10.1007/s00792-003-0329-4

Alain K, Postec A, Grinsard E et al (2010) *Thermodesulfatator atlanticus* sp. nov., a thermophilic, chemolithoautotrophic, sulfate-reducing bacterium isolated from a mid-Atlantic ridge hydrothermal vent. Int J Syst Evol Microbiol 60:33–38. https://doi.org/10.1099/ijs.0.009449-0

Albuquerque L, Simões C, Nobre MF et al (2005) *Truepera radiovictrix* gen. nov., sp. nov., a new radiation resistant species and the proposal of Trueperaceae fam. nov. FEMS Microbiol Lett 247:161–169. https://doi.org/10.1016/j.femsle.2005.05.002

Albuquerque L, Froufe H, Egas C et al (2018) Albidovulum. In: Boone DR, Castenholz RW, Garrity GM (eds) Bergey's manual of systematics of archaea and bacteria. Springer, New York, pp 1–8

Allioux M, Yvenou S, Slobodkina G et al (2020) Genomic characterization and environmental distribution of a thermophilic anaerobe *Dissulfurirhabdus thermomarina* SH388(T) involved in disproportionation of sulfur compounds in shallow sea hydrothermal vents. Microorganisms 8: 1132. https://doi.org/10.3390/microorganisms8081132

Amin A, Ahmed I, Habib N et al (2016) *Microvirga pakistanensis* sp nov., a novel bacterium isolated from desert soil of Cholistan, Pakistan. Arch Microbiol 198:933–939. https://doi.org/10.1007/s00203-016-1251-3

Anderson A, Nordan H, Cain R et al (1956) Studies on a radioresistant *Micrococcus*. I. Isolation, morphology, cultural characteristics and resistance to gamma radiation. Food Technol 10:575–578

Aoshima M, Ishii M, Igarashi Y (2004) A novel enzyme, citryl-CoA synthetase, catalysing the first step of the citrate cleavage reaction in *Hydrogenobacter thermophilus* TK-6: Citryl-CoA synthetase in *H. thermophilus*. Mol Microbiol 52:751–761. https://doi.org/10.1111/j.1365-2958.2004.04009.x

Arif S, Willenberg C, Dreyer A et al (2021) Sasso Pisano geothermal field environment harbours diverse *Ktedonobacteria* representatives and illustrates habitat-specific adaptations. Microorganisms 9(7):1402. https://doi.org/10.3390/microorganisms9071402

Armstrong FA, Belsey NA, Cracknell JA et al (2009) Dynamic electrochemical investigations of hydrogen oxidation and production by enzymes and implications for future technology. Chem Soc Rev 38:36–51. https://doi.org/10.1039/b801144n

Auria R, Boileau C, Davidson S et al (2016) Hydrogen production by the hyperthermophilic bacterium *Thermotoga maritima* part II: modeling and experimental approaches for hydrogen production. Biotechnol Biofuels 19(9):268. https://doi.org/10.1186/s13068-016-0681-0

Aylward FO, Santoro AE (2020) Heterotrophic Thaumarchaea with small genomes are widespread in the dark ocean. MSystems 5:10–128. https://doi.org/10.1128/msystems.00415-20

Badar A, Mukhtar S, Mehnaz S, Malik KA (2022) Microbial diversity analysis of Hudiara Drain wastewater using cultureindependent and culture-dependent approaches. Environ Sustain 5:65–76. https://doi.org/10.1007/s42398-022-00215-y

Bai YN, Lu YZ, Shen N et al (2018) Investigation of Cr (VI) reduction potential and mechanism by *Caldicellulosiruptor saccharolyticus* under glucose fermentation condition. J Hazard Mater 344:585–592. https://doi.org/10.1016/j.jhazmat.2017.10.059

Baker BJ, De Anda V, Seitz KW et al (2020) Diversity, ecology and evolution of archaea. Nat Microbiol 5:887–900. https://doi.org/10.1038/s41564-020-0715-z

Barbier G, Godfroy A, Meunier JR, et al (1999) *Pyrococcus glycovorans* sp. nov., a hyperthermophilic archaeon isolated from the East Pacific rise. Int J Syst Bacteriol 49(4):1829–1837. https://doi.org/10.1099/00207713-49-4-1829

Beam JP, Jay ZJ, Kozubal MA, Inskeep WP (2014) Niche specialization of novel Thaumarchaeota to oxic and hypoxic acidic geothermal springs of Yellowstone National Park. ISME J 8:938–951. https://doi.org/10.1038/ismej.2013.193

Beatty J, Overmann J, Lince MT et al (2005) An obligately photosynthetic bacterial anaerobe from a deep-sea hydrothermal vent. Proc Natl Acad Sci U S A 102:9306–9310. https://doi.org/10.1073/pnas.0503674102

Beblo-Vranesevic K, Huber H, Rettberg P (2017b) High tolerance of *Hydrogenothermus marinus* to sodium perchlorate. Front Microbiol 8:1369. https://doi.org/10.3389/fmicb.2017.01369

Bello S, Howard-Azzeh M, Schellhorn HE et al (2022) Phylogenomic analyses and molecular signatures elucidating the evolutionary relationships amongst the *Chlorobia* and *Ignavibacteria* species: robust demarcation of two family-level clades within the order *Chlorobiales* and proposal for the family *Chloroherpetonaceae* fam. nov. Microorganisms 10(7):1312. https://doi.org/10.3390/microorganisms10071312

Ben Hania W, Godbane R, Postec A, et al (2012) *Defluviitoga tunisiensis* gen. nov., sp. nov., a thermophilic bacterium isolated from a mesothermic and anaerobic whey digester. Int J Syst Evol Microbiol 62:1377–1382. https://doi.org/10.1099/ijs.0.033720-0

Berg IA, Kockelkorn D, Ramos-Vera WH et al (2010) Autotrophic carbon fixation in archaea. Nat Rev Microbiol 8:447–460. https://doi.org/10.1038/nrmicro2365

Bergmann F, Selesi D, Weinmaier T et al (2011) Genomic insights into the metabolic potential of the polycyclic aromatic hydrocarbon degrading sulfate-reducing deltaproteobacterium N47. Environ Microbiol 13:1125–1137. https://doi.org/10.1111/j.1462-2920.2010.02391.x

Berleman JE, Kirby JR (2009) Deciphering the hunting strategy of a bacterial wolfpack. FEMS Microbiol Rev 33:942–957. https://doi.org/10.1111/j.1574-6976.2009.00185.x

Besse A, Peduzzi J, Rebuffat S, Carre-Mlouka A (2015) Antimicrobial peptides and proteins in the face of extremes: lessons from archaeocins. Biochimie 118:344–355. https://doi.org/10.1016/j.biochi.2015.06.004

Bessman MJ, Kornberg A, Lehman IR et al (1956) Enzymic synthesis of deoxyribonucleic acid. Biochim Biophys Acta 21(1):197–198. https://doi.org/10.1016/0006-3002(56)90127-5

Bhandari V, Gupta RS (2014a) Molecular signatures for the phylum (class) Thermotogae and a proposal for its division into three orders (*Thermotogales, Kosmotogales* ord. nov. and *Petrotogales* ord. nov.) containing four families (*Thermotogaceae, Fervidobacteriaceae* fam. nov., *Kosmotogaceae* fam. nov. and *Petrotogaceae* fam. nov.) and a new genus *Pseudothermotoga* gen. nov. with five new combinations. Antonie Van Leeuwenhoek 105: 143–168. https://doi.org/10.1007/s10482-013-0062-7

Bhandari V, Gupta RS (2014b) The phylum Thermotogae. In: Rosenberg E, DeLong EF, Lory S, Stackebrandt E, Thompson F (eds) The prokaryotes. Springer, Heidelberg, pp 989–1015. https://doi.org/10.1007/978-3-642-38954-2_118

Bing W, Sun H, Wang F et al (2018) Hydrogen-producing hyperthermophilic bacteria synthesized size-controllable fine gold nanoparticles with excellence for eradicating biofilm and antibacterial applications. J Mater Chem B 6(28):4602–4609. https://doi.org/10.1039/c8tb00549d

Birrien J-L, Zeng X, Jebbar M, et al (2011) *Pyrococcus yayanosii* sp. nov., an obligate piezophilic hyperthermophilic archaeon isolated from a deep-sea hydrothermal vent. Int J Syst Evol Microbiol 61:2827–2881. https://doi.org/10.1099/ijs.0.024653-0

Bizic M, Brad T, Ionescu D et al (2023) Cave *Thiovulum* (Candidatus *Thiovulum stygium*) differs metabolically and genomically from marine species. ISME J 17:340–353. https://doi.org/10.1038/s41396-022-01350-4

Blöchl E, Rachel R, Burggraf S et al (1997). *Pyrolobus fumarii*, gen. and sp. nov., represents a novel group of archaea, extending the upper temperature limit for life to 113 C Extremophiles 1: 14–21

Blumer-Schuette SE (2020) Insights into thermophilic plant biomass hydrolysis from *Caldicellulosiruptor* systems biology. Microorganisms 8(3):385. https://doi.org/10.3390/microorganisms8030385

Blumer-Schuette SE, Kataeva I, Westpheling J et al (2008) Extremely thermophilic microorganisms for biomass conversion: status and prospects. Curr Opin Biotechnol 19:210–217. https://doi.org/10.1016/j.copbio.2008.04.007

Boden R, Hutt LP (2019a) Thermithiobacillus. In: Boone DR, Castenholz RW, Garrity GM (eds) Bergey's manual of systematics of archaea and bacteria. Springer, New York, pp 1–11

Boden R, Hutt LP (2019b) Bergey's manual of systematics of archaea and bacteria. In: Boone DR, Castenholz RW, Garrity GM (eds) Acidithiobacillus. Springer, New York, pp 1–19

Boesen T, Nielsen LP (2013) Molecular dissection of bacterial nanowires. MBio 4:e00270–e00213. https://doi.org/10.1128/mBio.00270-13

Bond DR, Holmes DE, Tender LM et al (2002) Electrode-reducing microorganisms that harvest energy from marine sediments. Science 295:483–485. https://doi.org/10.1126/science.1066771

Boniface A, Bouhss A, Mengin-Lecreulx D et al (2006) The MurE synthetase from *Thermotoga maritima* is endowed with an unusual D-lysine adding activity. J Biol Chem 281:15680–15686. https://doi.org/10.1074/jbc.M506311200

Braga LPP, Pereira RV, Martins LF et al (2021) Genome-resolved metagenome and metatranscriptome analyses of thermophilic composting reveal key bacterial players and their metabolic interactions. BMC Genomics 22:1–19. https://doi.org/10.1186/s12864-021-07957-9
Bräsen C, Esser D, Rauch B et al (2014) Carbohydrate metabolism in archaea: current insights into unusual enzymes and pathways and their regulation. Microbiol Mol Biol Rev 78:89–175. https://doi.org/10.1128/MMBR.00041-13
Brewer PG, Spencer DW (1969) A note on the chemical composition of the Red Sea brines. In: Hot brines and recent heavy metal deposits in the Red Sea: a geochemical and geophysical account, pp 174–179. https://doi.org/10.1007/978-3-662-28603-6_19
Briggs BR, Brodie EL, Tom LM et al (2014) Seasonal patterns in microbial communities inhabiting the hot springs of Tengchong, Yunnan Province, China: PhyloChip and physicochemical analyses. Environ Microbiol 16:1579–1591. https://doi.org/10.1111/1462-2920.12311
Brochier-Armanet C, Boussau B, Gribaldo S, Forterre P (2008) Mesophilic Crenarchaeota: proposal for a third archaeal phylum, the Thaumarchaeota. Nat Rev Microbiol 6:245–252. https://doi.org/10.1038/nrmicro1852
Brock T, Freeze H (1969) *Thermus aquaticus* gen. n and sp. n., a non-sporulating extreme thermophile. J Bacteriol 98:289–297.
Brock, TD, Brock, KM, Belly, RT et al (1972) *Sulfolobus*: a new genus of sulfur-oxidizing bacteria living at low pH and high temperature. Arch Mikrobiol 84:54–68. doi: https://doi.org/10.1007/BF00408082
Brynjarsdottir H, Wawiernia B, Orlygsson J (2012) Ethanol production from sugars and complex biomass by *Thermoanaerobacter* AK5: the effect of electron-scavenging systems on end-product formation. Energy Fuel 26(7):4568–4574. https://doi.org/10.1021/ef300754q
Caldwell SL, Liu Y, Ferrera I, et al (2010) *Thermocrinis minervae* sp. nov., a hydrogen- and sulfur-oxidizing, thermophilic member of the Aquificales from a Costa Rican terrestrial hot spring. Int J Syst Evol Microbiol 60:338–343. https://doi.org/10.1099/ijs.0.010496-0
Campbell BJ, Jeanthon C, Kostka JE et al (2001) Growth and phylogenetic properties of novel bacteria belonging to the epsilon subdivision of the proteobacteria enriched from *Alvinella pompejana* and deep-sea hydrothermal vents. Appl Environ Microbiol 67(10):4566–4572. https://doi.org/10.1128/AEM.67.10.4566-4572.2001
Campbell BJ, Stein JL, Cary JS (2003) Evidence of chemolithoautotrophy in the bacterial community associated with *Alvinella pompejana*, a hydrothermal vent polychaete. Appl Environ Microbiol 69(9):5070–5078. https://doi.org/10.1128/AEM.69.9.5070-5078.2003
Campbell BJ, Engel AS, Porter ML et al (2006) The versatile ε-proteobacteria: key players in sulphidic habitats. Nat Rev Microbiol 4(6):458–468. https://doi.org/10.1038/nrmicro1414
Cao J, Birien T, Gayet N et al (2017) *Desulfurobacterium indicum* sp. nov., a thermophilic sulfur-reducing bacterium from the Indian Ocean. Int J Syst Evol Microbiol 67:1665–1668. https://doi.org/10.1099/ijsem.0.001837
Cary SC, Shank T, Stein J (1998) Worms bask in extreme temperatures. Nature 391:545–546. https://doi.org/10.1038/35286
Cavaletti L, Monciardini P, Bamonte R et al (2006) New lineage of filamentous, spore-forming, gram-positive bacteria from soil. Appl Environ Microbiol 72:4360–4369. https://doi.org/10.1128/AEM.00132-06
Chavan SB, Deshpande MV (2013) Chitinolytic enzymes: an appraisal as a product of commercial potential. Biotechnol Prog 29:833–846. https://doi.org/10.1002/btpr.1732
Chenevier P, Mugherli L, Darbe S et al (2013) Hydrogenase enzymes: application in biofuel cells and inspiration for the design of noble-metal free catalysts for H-2 oxidation. C R Chim 16:491–505. https://doi.org/10.1016/j.crci.2012.11.006
Chien A, Edgar DB, Trela JM (1976) Deoxyribonucleic acid polymerase from the extreme thermophile *Thermus aquaticus*. J Bacteriol 127:1550–1557. https://doi.org/10.1128/jb.127.3.1550-1557.1976

Chou Y-J, Sheu S-Y, Sheu D-S et al (2006) *Schlegelella aquatica* sp. nov., a novel thermophilic bacterium isolated from a hot spring. Int J Syst Evol Microbiol 56:2793–2797. doi: https://doi.org/10.1099/ijs.0.64446-0

Christensen B, Torsvik T, Lien T (1992) Immunomagnetically captured thermophilic sulfate-reducing bacteria from North Sea oil field waters. Appl Environ Microbiol 58:1244–1248. https://doi.org/10.1128/aem.58.4.1244-1248.1992

Chung AP, Rainey FA, Valente M, et al (2000) *Thermus igniterrae* sp. nov. and *Thermus antranikianii* sp. nov., two new species from Iceland. Int J Syst Evol Microbiol 50:209–217. https://doi.org/10.1099/00207713-50-1-209

Chung D, Cha M, Snyder EN et al (2015) Cellulosic ethanol production via consolidated bioprocessing at 75°C by engineered *Caldicellulosiruptor bescii*. Biotechnol Biofuels 8(1): 163. https://doi.org/10.1186/s13068-015-0346-4

Ciccarelli FD, Doerks T, von Mering C et al (2006) Toward automatic reconstruction of a highly resolved tree of life. Science 311(5765):1283–1287. https://doi.org/10.1126/science.1123061

Clarke A (2014) The thermal limits to life on earth. Int J Astrobiol 13:141–154. https://doi.org/10.1017/S1473550413000438

Cole JK, Gieler BA, Heisler DL et al (2013) *Kallotenue papyrolyticum* gen. nov., sp. nov., a cellulolytic and filamentous thermophile that represents a novel lineage (Kallotenuales ord. nov., Kallotenuaceae fam. nov.) within the class Chloroflexia. Int J Syst Evol Microbiol 63: 4675–4682. https://doi.org/10.1099/ijs.0.053348-0

Collins MD, Lawson PA, Willems A et al (1994) The phylogeny of the genus *Clostridium* - proposal of 5 new genera and 11 new species combinations. Int J Syst Bacteriol 44:812–826. https://doi.org/10.1099/00207713-44-4-812

Conners SB, Mongodin EF, Johnson MR et al (2006) Microbial biochemistry, physiology, and biotechnology of hyperthermophilic *Thermotoga* species. FEMS Microbiol Rev 30:872–905. https://doi.org/10.1111/j.1574-6976.2006.00039.x

Counts JA, Zeldes BM, Lee LL et al (2017) Physiological, metabolic and biotechnological features of extremely thermophilic microorganisms. Wiley Interdiscip Rev Syst Biol Med 9:e1377. https://doi.org/10.1002/wsbm.1377

Cox MM, Battista JR (2005) *Deinococcus radiodurans* — the consummate survivor. Nat Rev Microbiol 3:882–892. https://doi.org/10.1038/nrmicro1264

Dahmane N, Gadelle D, Delmas S et al (2016) topIb, a phylogenetic hallmark gene of *Thaumarchaeota encodes* a functional eukaryote-like topoisomerase IB. Nucl Acids Res 44: 2795–2805. https://doi.org/10.1093/nar/gkw097

Dang H, Zhou H, Yang J et al (2013) Thaumarchaeotal signature gene distribution in sediments of the northern South China Sea: an indicator of the metabolic intersection of the marine carbon, nitrogen, and phosphorus cycles? Appl Environ Microbiol 79:2137–2147. https://doi.org/10.1128/AEM.03204-12

Daims H (2014) The family Nitrospiraceae. In: Rosenberg E, DeLong EF, Lory S, Stackebrandt E, Thompson F (eds) The prokaryotes. Springer, Berlin. https://doi.org/10.1007/978-3-642-38954-2_126

Darland G, Brock TD, Samsonoff W et al (1970) A thermophilic, acidophilic *mycoplasma* isolated from a coal refuse pile. Science 170:1416–1418. https://doi.org/10.1126/science.170.3965.1416

Davey ME, Wood WA, Key R et al (1993) Isolation of three species of *Geotoga* and *Petrotoga*: two new genera, representing a new lineage in the bacterial line of descent distantly related to the "Thermotogales". Syst Appl Microbiol 16:191–200. https://doi.org/10.1016/S0723-2020(11)80467-4

Deckert G, Warren PV, Gaasterland T et al (1998) The complete genome of the hyperthermophilic bacterium *Aquifex aeolicus*. Nature 392:353–358. https://doi.org/10.1038/32831

De la Torre JR, Walker CB, Ingalls AE, Könneke M, Stahl DA (2008) Cultivation of a thermophilic ammonia oxidizing archaeon synthesizing crenarchaeol. Env Microbiol 10:810–818. https://doi.org/10.1111/j.1462-2920.2007.01506.x

DeLong EF (1992) Archaea in coastal marine environments. Proc Nat Acad Sci 89:5685–5689. https://doi.org/10.1073/pnas.89.12.5685

Delgado-Baquerizo M, Oliverio AM, Brewer TE et al (2018) A global atlas of the dominant bacteria found in soil. Science 359:320–325. https://doi.org/10.1126/science.aap9516

Denman S, Hampson K, Patel BKC (1991) Isolation of strains of *Thermus aquaticus* from the Australian Artesian Basin and a simple and rapid procedure for the preparation of their plasmids. FEMS Microbiol Lett 82:73–78. https://doi.org/10.1111/j.1574-6968.1991.tb04842.x

Desai SG, Guerinot ML, Lynd LR (2004) Cloning of L-lactate dehydrogenase and elimination of lactic acid production via gene knockout in *Thermoanaerobacterium saccharolyticum* JW/SL-YS485. Appl Microbiol Biotechnol 65(5):600–605. https://doi.org/10.1007/s00253-004-1575-9

Deweerd KA, Suflita JM (1990) Anaerobic aryl reductive dehalogenation of Halobenzoates by cell extracts of "*Desulfomonile tiedjei*". Appl Environ Microbiol 56:2999–3005. https://doi.org/10.1128/aem.56.10.2999-3005.1990

Di Cesare A, Pjevac P, Eckert E et al (2020) The role of metal contamination in shaping microbial communities in heavily polluted marine sediments. Environ Pollut 265:114823. https://doi.org/10.1016/j.envpol.2020.114823

DiGiacomo J, McKay C, Davila A (2022) ThermoBase: a database of the phylogeny and physiology of thermophilic and hyperthermophilic organisms. PLoS One 17:e0268253. https://doi.org/10.1371/journal.pone.0268253

DiPippo JL, Nesbø CL, Dahle H et al (2009). *Kosmotoga olearia* gen. nov., sp. nov., a thermophilic, anaerobic heterotroph isolated from an oil production fluid. Int J Syst Evol Microbiol, 59: 2991–3000. https://doi.org/10.1099/ijs.0.008045-0

Dodsworth JA, Hungate BA, Hedlund BP (2011) Ammonia oxidation, denitrification and dissimilatory nitrate reduction to ammonium in two US Great Basin hot springs with abundant ammonia-oxidizing archaea. Env Microbiol 13:2371–2386. https://doi.org/10.1111/j.1462-2920.2011.02508.x

Dodsworth JA, Gevorkian J, Despujos F et al (2014) *Thermoflexus hugenholtzii* gen. nov., sp. nov., a thermophilic, microaerophilic, filamentous bacterium representing a novel class in the Chloroflexi, Thermoflexia classis nov., and description of Thermoflexaceae fam. nov. and Thermoflexales ord. nov. Int J Syst Evol Microbiol 64:2119–2127. https://doi.org/10.1099/ijs.0.055855-0

Dodsworth JA, Ong JC, Williams AJ et al (2015) *Thermocrinis jamiesonii* sp. nov., a thiosulfate-oxidizing, autotropic thermophile isolated from a geothermal spring. Int J Syst Evol Microbiol 65:4769–4775. https://doi.org/10.1099/ijsem.0.000647

Dumina M, Zhgun A, Pokrovskaya M et al (2021) Highly active thermophilic L-asparaginase from *Melioribacter roseus* represents a novel large Group of Type II bacterial L-asparaginases from Chlorobi-Ignavibacteriae-Bacteroidetes clade. Int J Mol Sci 22:13632. https://doi.org/10.3390/ijms222413632

Dumitrache A, Akinosho H, Rodriguez M et al (2016) Consolidated bioprocessing of populus using *Clostridium (Ruminiclostridium) thermocellum*: a case study on the impact of lignin composition and structure. Biotechnol Biofuels 9(31). https://doi.org/10.1186/s13068-016-0445-x

Dworkin M, Falkow S, Rosenberg E et al (eds) (2006) The prokaryotes. Springer, New York

Eder W, Huber R (2002) New isolates and physiological properties of the Aquificales and description of *Thermocrinis albus* sp. nov. Extremophiles 6:309–318. https://doi.org/10.1007/s00792-001-0259-y

Edwards TA, Calica NA, Huang DA et al (2013) Cultivation and characterization of thermophilic *Nitrospira* species from geothermal springs in the US Great Basin, China, and Armenia. FEMS Microbiol Ecol 85(2):283–292. https://doi.org/10.1111/1574-6941.12117

Ehrich S, Behrens D, Lebedeva E et al (1995) A new obligately chemolithoautotrophic, nitrite-oxidizing bacterium, *Nitrospira moscoviensis* sp. nov. and its phylogenetic relationship. Arch Microbiol 164:16–23

Elbanna K, Lütke-Eversloh T, Van Trappen S et al (2003) *Schlegelella thermodepolymerans* gen. nov., sp. nov., a novel thermophilic bacterium that degrades poly(3-hydroxybutyrate-co-3-

mercaptopropionate). Int J Syst Evol Microbiol 53:1165–1168. https://doi.org/10.1099/ijs.0.02562-0

Erauso G, Reysenbach A-L, Godfroy A et al (1993) *Pyrococcus abyssi* sp. nov., a new hyperthermophilic archaeon isolated from a deep-sea hydrothermal vent. Arch Microbiol 160:338–349. https://doi.org/10.1007/BF00252219

Eriksen NT, Riis ML, Holm NK et al (2011) H_2 synthesis from pentoses and biomass in *Thermotoga* spp. Biotechnol Lett 33(2):293–300. https://doi.org/10.1007/s10529-010-0439-x

Fardeau ML, Ollivier B, Patel BK et al (1997) *Thermotoga hypogea* sp. nov., a xylanolytic, thermophilic bacterium from an oil-producing well. Int J Syst Bacteriol 47:1013–1019. https://doi.org/10.1099/00207713-47-4-1013

Fardeau ML, Ollivier B, Jl C (2015) *Caldanaerobacter*. In: Trujillo ME, Dedysh S, DeVos P et al (eds) Bergey's manual of systematics of archaea and bacteria. Springer, New York. https://doi.org/10.1002/9781118960608.gbm00745

Farrell AA, Nesbø CL, Zhaxybayeva O et al (2021) Fervidobacteriaceae. In: Trujillo ME, Dedysh S, DeVos P, Hedlund B, Kämpfer P, Rainey FA, Whitman WB (eds) Bergey's manual of systematics of archaea and bacteria. Springer, New York. https://doi.org/10.1002/9781118960608.fbm00360

Feng Y, Cheng L, Zhang X et al (2010) *Thermococcoides shengliensis* gen. nov., sp. nov., a new member of the order Thermotogales isolated from oil-production fluid. Int J Syst Evol Microbiol 60:932–937. https://doi.org/10.1099/ijs.0.013912-0

Ferrera I, Banta AB, Reysenbach A-L (2014) Spatial patterns of Aquificales in deep-sea vents along the eastern Lau spreading Center (SW Pacific). Syst Appl Microbiol 37:442–448. https://doi.org/10.1016/j.syapm.2014.04.002

Fiala G, Stetter KO (1986) *Pyrococcus furiosus* sp. nov. represents a novel genus of marine heterotrophic archaebacteria growing optimally at 100°C. Arch Microbiol 145:56–61. https://doi.org/10.1007/BF00413027

Fiala G, Woese CR, Langworthy TA, Stetter KO (1990) *Flexistipes sinusarabici*, a novel genus and species of eubacteria occurring in the Atlantis II Deep brines of the Red Sea. Arch Microbiol 154:120–126. https://doi.org/10.1007/BF00423320

Figueiredo GGO, Lopes VR, Romano T et al (2020) In beneficial microbes in Agro-ecology. Academic, Cambridge, pp 477–491. isbn:978-0-12-823414-3

Flores GE, Liu Y, Ferrera I et al (2008) *Sulfurihydrogenibium kristjanssonii* sp. nov., a hydrogen- and sulfur-oxidizing thermophile isolated from a terrestrial Icelandic hot spring. Int J Syst Evol Microbiol 58:1153–1158. https://doi.org/10.1099/ijs.0.65570-0

François DX, Godfroy A, Mathien C et al (2021) *Persephonella atlantica* sp. nov.: how to adapt to physico-chemical gradients in high temperature hydrothermal habitats. Syst Appl Microbiol 44: 126176. https://doi.org/10.1016/j.syapm.2020.126176

Forterre P (2002) The origin of DNA genomes and DNA replication proteins. Curr Op Microbiol 5: 525–532. https://doi.org/10.1016/S1369-5274(02)00360-0

Frank YA, Kadnikov VV, Lukina AP et al (2016) Characterization and genome analysis of the first facultatively alkaliphilic *Thermodesulfovibrio* isolated from the deep terrestrial subsurface. Front Microbiol 7:2000. https://doi.org/10.3389/fmicb.2016.02000

Frigaard N-U, Bryant DA (2004) Seeing Green bacteria in a new light: genomics-enabled studies of the photosynthetic apparatus in Green Sulfur bacteria and filamentous anoxygenic phototrophic bacteria. Arch Microbiol 182:265–276. https://doi.org/10.1007/s00203-004-0718-9

Frigaard N-U, Dahl C (2009) Sulfur metabolism in phototrophic Sulfur bacteria. Adv Microb Physiol 54:103–200. https://doi.org/10.1016/S0065-2911(08)00002-7

Frock AD, Notey JS, Kelly RM (2010) The genus *Thermotoga*: recent developments. Environ Technol 31:1169–1181. https://doi.org/10.1080/09593330.2010.484076

Fröstl JM, Overmann J (2000) Phylogenetic affiliation of the bacteria that constitute phototrophic consortia. Arch Microbiol 174:50–58. https://doi.org/10.1007/s002030000172

Fuhrman JA, McCallum K, Davis AA (1992) Novel major archaebacterial group from marine plankton. Nature 356:148–149. https://doi.org/10.1038/356148a0

Fukushi M, Mino S, Tanaka H et al (2020) Biogeochemical Implications of N_2O-reducing thermophilic campylobacteria in deep-sea vent fields, and the description of *Nitratiruptor labii* sp. nov. iScience 23:101462. https://doi.org/10.1016/j.isci.2020.101462

Gaisin VA, Kalashnikov AM, Grouzdev DS et al (2017) *Chloroflexus islandicus* sp. nov., a thermophilic filamentous anoxygenic phototrophic bacterium from a geyser. Int J Syst Evol Microbiol 67:1381–1386. https://doi.org/10.1099/ijsem.0.001820

Garcia SL, Mehshad M, Buck M et al (2021) Freshwater *Chlorobia* exhibit metabolic specialization among cosmopolitan and endemic populations. mSystems 6(3):e01196–e01120. https://doi.org/10.1128/mSystems.01196-20

Garrity GM, Holt JG, Castenholz RW et al (2001a) Phylum BVI. Chloroflexi phy. nov. In: Boone DR, Castenholz RW, Garrity GM (eds) BT - Bergey's manual of systematic bacteriology, Volume one: the archaea and the deeply branching and phototrophic bacteria. Springer, New York, pp 427–446

Garrity GM, Holt JG, Reysenbach A-L et al (2001b) Phylum Al Crenarchaeota phy. nov. In: Boone DR, Castenholz RW, Garrity GM (eds) Bergey's manual of systematic bacteriology, Volume one: the archaea and the deeply branching and phototrophic bacteria. Springer, New York, pp 169–210. https://doi.org/10.1007/978-0-387-21609-6_16

Garrity GM, Holt JG, Battista JR et al (2001c) Phylum BIV. Deinococcus-Thermus. In: Boone DR, Castenholz RW, Garrity GM (eds) Bergey's manual of systematic bacteriology. Springer, New York, pp 395–420

Garrity GM, Holt JG, Overmann J et al (2001d) Phylum BXI. Chlorobi phy. nov. In: Boone DR, Castenholz RW, Garrity GM (eds) Bergey's manual of systematic bacteriology. Springer, New York, pp 601–623. https://doi.org/10.1007/978-0-387-21609-6_28

Garrity GM, Holt JG, Whitman WB et al (2001e) Phylum all. Euryarchaeota phy. nov. In: Boone DR, Castenholz RW, Garrity GM (eds) Bergey's manual of systematic bacteriology, Volume one: the archaea and the deeply branching and phototrophic bacteria. Springer, New York, pp 211–355

Garrity GM, Holt JG, Spieck E et al (2001f) Phylum BVIII Nitrospirae phy. nov. in. In: Boone DR, Castenholz RW, Garrity GM (eds) Bergey's manual of systematic bacteriology. Springer, New York, NY, pp 451–464

Garrity GM, Bell JA, Lilburn T (2015) Burkholderiales ord. nov. In: Boone DR, Castenholz RW, Garrity GM (eds) Bergey's manual of systematics of archaea and bacteria. Springer, New York, p 1

Georgieva T, Ahring BK (2007) Evaluation of continuous ethanol fermentation of dilute-acid corn Stover hydrolysate using thermophilic anaerobic bacterium *Thermoanaerobacter* BG1L1. Appl Microbiol Biotechnol 77:61–68. https://doi.org/10.1007/s00253-007-1149-8

Gerth K, Müller R (2005) Moderately thermophilic myxobacteria: novel potential for the production of natural products isolation and characterization. Environ Microbiol 7:874–880. https://doi.org/10.1111/j.1462-2920.2005.00761.x

Ghimire-Kafle S, Weaver ME Jr, Bollmann A (2023) Ecophysiological and genomic characterization of the freshwater complete ammonia oxidizer *Nitrospira* sp. strain BO4. Appl Environ Microbiol 89(2):e01965–e01922. https://doi.org/10.1128/aem.01965-22

Ghosh W, Dam B (2009) Biochemistry and molecular biology of lithotrophic Sulfur oxidation by taxonomically and ecologically diverse bacteria and archaea. FEMS Microbiol Rev 33(6): 999–1043. https://doi.org/10.1111/j.1574-6976.2009.00187.x

Gibbons NE, Murray RGE (1978) Proposals concerning the higher taxa of bacteria. Int J Syst Bacteriol 28:1–6

Gibson J, Pfennig N, Waterbury JB (1984) *Chloroherpeton thalassium* gen. nov. et spec. nov., a non-filamentous, flexing and gliding Green Sulfur bacterium. Arch Microbiol 138:96–101, https://doi.org/10.1007/BF00413007

Giovannoni SJ, Revsbech NP, Ward DM et al (1987) Obligately phototrophic *Chloroflexus*: primary production in anaerobic hot spring microbial mats. Arch Microbiol 147:80–87. https://doi.org/10.1007/BF00492909

Giovannelli D, Sievert SM, Hügler M et al (2017) Insight into the evolution of microbial metabolism from the deep-branching bacterium, *Thermovibrio ammonificans*. Elife 6:e18990. https://doi.org/10.7554/eLife.18990

González JM, Masuchi Y, Robb FT et al (1998) *Pyrococcus horikoshii* sp. nov., a hyperthermophilic archaeon isolated from a hydrothermal vent at the Okinawa trough. Extremophiles 2:123–130. https://doi.org/10.1007/s007920050051

Gotz D (2002) *Persephonella marina* gen. nov., sp. nov. and *Persephonella guaymasensis* sp. nov., two novel, thermophilic, hydrogen-oxidizing microaerophiles from deep-sea hydrothermal vents. Int J Syst Evol Microbiol 52:1349–1359. https://doi.org/10.1099/ijs.0.02126-0

Greene AC, Patel BK, Sheehy AJ (1997) *Deferribacter thermophilus* gen. nov., sp. nov., a novel thermophilic manganese-and iron-reducing bacterium isolated from a petroleum reservoir. Int J Syst Evol Microbiol 47:505–509. https://doi.org/10.1099/00207713-47-2-505

Gregersen L, Bryant DA, Frigaard NU (2011) Mechanisms and evolution of oxidative sulfur metabolism in green sulfur bacteria. Front Microbiol 2:116. https://www.frontiersin.org/articles/10.3389/fmicb.2011.00116

Gregoire P, Fardeau M-L, Joseph M et al (2011) Isolation and characterization of *Thermanaerothrix daxensis* gen. nov., sp. nov., a thermophilic anaerobic bacterium pertaining to the phylum "Chloroflexi", isolated from a deep hot aquifer in the Aquitaine Basin. Syst Appl Microbiol 34:494–497. https://doi.org/10.1016/j.syapm.2011.02.004

Griffiths E, Gupta RS (2004) Distinctive protein signatures provide molecular markers and evidence for the monophyletic nature of the Deinococcus-Thermus phylum. J Bacteriol 186:3097–3107. https://doi.org/10.1128/JB.186.10.3097-3107.2004

Gruber N, Galloway JN (2008) An Earth-system perspective of the global nitrogen cycle. Nature 451:293–296. https://doi.org/10.1038/nature06592

Guiral M, Prunetti L, Aussignargues C et al (2012) The Hyperthermophilic bacterium *Aquifex aeolicus*. Adv Microb Physiol:125–194

Gupta RS (2014) The phylum aquificae. In: Rosenberg E, DeLong EF, Lory S et al (eds) The prokaryotes. Springer, Berlin, pp 417–445

Hanada S, Hiraishi A, Shimada K et al (1995) *Chloroflexus aggregans* sp. nov., a filamentous phototrophic bacterium which forms dense cell aggregates by active gliding movement. Int J Syst Bacteriol 45:676–681. https://doi.org/10.1099/00207713-45-4-676

Hanada S, Kawase Y, Hirashi A et al (1997) *Porphyrobacter tepidarius* sp. nov., a moderately thermophilic aerobic photosynthetic bacterium isolated from a hot spring. Int J Syst Bacteriol 47:408–413. https://doi.org/10.1099/00207713-47-2-408

Hanada S, Takaichi S, Matsuura K et al (2002) *Roseiflexus castenholzii* gen. nov., sp. nov., a thermophilic, filamentous, photosynthetic bacterium that lacks chlorosomes. Int J Syst Evol Microbiol 52:187–193. https://doi.org/10.1099/00207713-52-1-187

Hanner M, Redl B, Stöffler G (1990) Isolation and characterization of an intracellular aminopeptidase from the extreme thermophilic archaebacterium *Sulfolobus solfataricus*. Biochim Biophys Acta 1033:148–153. https://doi.org/10.1016/0304-4165(90)90005-H

Haouari O, Fardeau ML, Cayol JL et al (2008) *Thermodesulfovibrio hydrogeniphilus* sp. nov., a new thermophilic sulphate-reducing bacterium isolated from a Tunisian hot spring. Syst Appl Microbiol 31(1):38–42. Doi:https://doi.org/10.1016/j.syapm.2007.12.002

Hart BA, Zahler SA (1966) Lytic enzyme produced by *Myxococcus xanthus*. J Bacteriol 92:1632–1637. https://doi.org/10.1128/jb.92.6.1632-1637.1966

Härtig C, Lohmayer R, Kolb S et al (2014) Chemolithotrophic growth of the aerobic hyperthermophilic bacterium *Thermocrinis ruber* OC 14/7/2 on monothioarsenate and arsenite. FEMS Microbiol Ecol 90:747–760. https://doi.org/10.1111/1574-6941.12431

Hatchikian EC, Ollivier B, Garcia JL (2001) Family I. Thermodesulfobacteriaceae fam. nov. In: Boone DR, Castenholz RW, Garrity GM (eds) Bergey's manual of systematic bacteriology, Volume one: The archaea and the deeply branching of phototrophic bacteria, 2nd edn. Springer, New York, pp 390–394

Hatzenpichler R (2012) Diversity, physiology, and niche differentiation of ammonia-oxidizing archaea. Appl Environ Microbiol 78:7501–7510. https://doi.org/10.1128/AEM.01960-12

Hartmann M (1985) Atlantis-II Deep geothermal brine system. Chemical processes between hydrothermal brines and Red Sea deep water. Mar Geol 64:157–177. https://doi.org/10.1016/0025-3227(85)90166-5

Hedlund BP, Dodsworth JA, Cole JK, Panosyan HH (2013) An integrated study reveals diverse methanogens, Thaumarchaeota, and yet-uncultivated archaeal lineages in Armenian hot springs. Antonie Van Leeuwenhoek 104:71–82. https://doi.org/10.1007/s10482-013-9927-z

Hegler F, Posth NR, Jiang J et al (2008) Physiology of phototrophic iron(II)-oxidizing bacteria: implications for modern and ancient environments. FEMS Microbiol Ecol 66:250–260. https://doi.org/10.1111/j.1574-6941.2008.00592.x

Heising S, Richter L, Ludwig W, Schink B (1999) *Chlorobium ferrooxidans* sp. nov., a phototrophic Green Sulfur bacterium that oxidizes ferrous iron in coculture with a '*Geospirillum*' sp. strain. Arch Microbiol 172:116–124, https://doi.org/10.1007/s002030050748

Held NL, Whitaker RJ (2009) Viral biogeography revealed by signatures in *Sulfolobus islandicus* genomes. Environ Microbiol 11:457–466. https://doi.org/10.1111/j.1462-2920.2008.01784.x

Henkel JV, Schulz-Vogt HN, Dellwig O et al (2022) Biological manganese-dependent sulfide oxidation impacts elemental gradients in redox-stratified systems: indications from the Black Sea water column. ISME J 16:1523–1533. https://doi.org/10.1038/s41396-022-01200-3

Henry EA, Devereux R, Maki JS et al (1994) Characterization of a new thermophilic sulfate-reducing bacterium: *Thermodesulfovibrio yellowstonii*, gen. nov. and sp. nov.: its phylogenetic relationship to *Thermodesulfobacterium commune* and their origins deep within the bacterial domain. Arch Microbiol 161:62–69. https://doi.org/10.1007/s002030050022

Hernández M, Vera-Gargallo B, Calabi-Floody M et al (2020) Reconstructing genomes of carbon monoxide oxidisers in volcanic deposits including members of the class Ktedonobacteria. Microorganisms 8(12):1880. https://doi.org/10.3390/microorganisms8121880

Hetzer A, McDonald IR, Morgan HW (2008) *Venenivibrio stagnispumantis* gen. nov., sp. nov., a thermophilic hydrogen-oxidizing bacterium isolated from champagne Pool, Waiotapu, New Zealand. Int J Syst Evol Microbiol 58:398–403. https://doi.org/10.1099/ijs.0.64842-0

Hidalgo KJ, Sierra-Garcia IN, Zafra G, de Oliveira VM (2021) Genome-resolved meta-analysis of the microbiome in oil reservoirs worldwide. Microorganisms 9:1812. https://doi.org/10.3390/microorganisms9091812

Hiraishi A, Imhoff JF (2015) Porphyrobacter. In: Boone DR, Castenholz RW, Garrity GM (eds) Bergey's manual of systematics of archaea and bacteria. Springer, New York, pp 1–6

Hiraishi A, Yonemitsu Y, Matsushita M et al (2002) Characterization of *Porphyrobacter sanguineus* sp. nov., an aerobic bacteriochlorophyll-containing bacterium capable of degrading biphenyl and dibenzofuran. Arch Microbiol 178:45–52. https://doi.org/10.1007/s00203-002-0423-5

Hiras J, Wu YW, Eichorst SA et al (2016) Refining the phylum chlorobi by resolving the phylogeny and metabolic potential of the representative of a deeply branching, uncultivated lineage. ISME J 10:833–845. https://doi.org/10.1038/ismej.2015.158

Hirsch P, Gallikowski CA, Siebert J et al (2004) *Deinococcus frigens* sp. nov., *Deinococcus saxicola* sp. nov., and *Deinococcus marmoris* sp. nov., low temperature and draught-tolerating, UV-resistant bacteria from continental Antarctica. Syst Appl Microbiol 27:636–645. https://doi.org/10.1078/0723202042370008

Holmes DE, Nicoll JS, Bond DR, Lovley DR (2004) Potential role of a novel psychrotolerant member of the family Geobacteraceae, *Geopsychrobacter electrodiphilus* gen. nov., sp. nov., in electricity production by a marine sediment fuel cell. Appl Environ Microbiol 70:6023–6030. https://doi.org/10.1128/AEM.70.10.6023-6030.2004

Hon S, Holwerda EK, Worthen RS et al (2018) Expressing the *Thermoanaerobacterium saccharolyticum* pforA in engineered *clostridium thermocellum* improves ethanol production. Biotechnol Biofuels 11:242. https://doi.org/10.1186/s13068-018-1245-2

Hovanec TA, Taylor LT, Blakis A et al (1998) *Nitrospira*-like bacteria associated with nitrite oxidation in freshwater aquaria. Appl Environ Microbiol 64(1):258–264. https://doi.org/10.1128/AEM.64.1.258-264.1998

Huang L, Bae H-S, Young C et al (2021) Campylobacterota dominate the microbial communities in a tropical karst subterranean estuary, with implications for cycling and export of nitrogen to coastal waters. Environ Microbiol 23:6749–6763. https://doi.org/10.1111/1462-2920.15746

Huber H (2002) *Thermovibrio ruber* gen. nov., sp. nov., an extremely thermophilic, chemolithoautotrophic, nitrate-reducing bacterium that forms a deep branch within the phylum Aquificae. Int J Syst Evol Microbiol 52:1859–1865. https://doi.org/10.1099/ijs.0.02235-0

Huber H, Stetter KO (2001) Class I. Deferribacteres class. nov. Bergey's manual of systematic bacteriology 1:465

Huber R, Hannig M (2006) Thermotogales. In: Dworkin M, Falkow S, Rosenberg E et al (eds) The prokaryotes. Springer, New York. https://doi.org/10.1007/0-387-30747-8_38

Huber R, Langworthy TA, König H et al (1986) *Thermotoga maritima* sp. nov. represents a new genus of unique extremely thermophilic eubacteria growing up to 90°C. Arch Microbiol 144: 324–333. https://doi.org/10.1007/BF00409880

Huber R, Woese CR, Langworthy TA, Fricke H, Stetter KO (1989) *Thermosipho africanus* gen. nov., represents a new genus of thermophilic eubacteria within the "Thermotogales". Syst Appl Microbiol 12:32–37. https://doi.org/10.1016/S0723-2020(89)80037-2

Huber R, Woese CR, Langworthy TA et al (1990) *Fervidobacterium islandicum* sp. nov., a new extremely thermophilic eubacterium belonging to the "Thermotogales". Arch Microbiol 154: 105–111. https://doi.org/10.1007/BF00423318

Huber R, Wilharm T, Huber D et al (1992) *Aquifex pyrophilus* gen. nov. sp. nov., represents a novel group of marine Hyperthermophilic hydrogen-oxidizing bacteria. Syst Appl Microbiol 15:340–351. https://doi.org/10.1016/S0723-2020(11)80206-7

Hug LA, Castelle CJ, Wrighton KC et al (2013) Community genomic analyses constrain the distribution of metabolic traits across the Chloroflexi phylum and indicate roles in sediment carbon cycling. Microbiome 1:22. https://doi.org/10.1186/2049-2618-1-22

Hug LA, Baker BJ, Anantharaman K et al (2016) A new view of the tree of life. Nat Microbiol 1:1–6

Iino T, Nakagawa T, Mori K et al (2008) *Calditerrivibrio nitroreducens* gen. nov., sp. nov., a thermophilic, nitrate-reducing bacterium isolated from a terrestrial hot spring in Japan. Int J Syst Evol Microbiol 58:1675–1679. doi: https://doi.org/10.1099/ijs.0.65714-0.18599715

Iino T, Mori K, Uchino Y et al (2010) *Ignavibacterium album* gen. nov., sp. nov., a moderately thermophilic anaerobic bacterium isolated from microbial Mats at a terrestrial hot spring and proposal of Ignavibacteria classis Nov., for a novel lineage at the periphery of Green Sulfur bacteria. Int J Syst Evol Microbiol 60:1376–1382, https://doi.org/10.1099/ijs.0.012484-0

Iizuka T, Tokura M, Jojima Y et al (2006) Enrichment and phylogenetic analysis of moderately thermophilic myxobacteria from hot springs in Japan. Microbes Environ 21:189–199

Imhoff JF (2003) Phylogenetic taxonomy of the family Chlorobiaceae on the basis of 16S RRNA and Fmo (Fenna-Matthews-Olson protein) gene sequences. Int J Syst Evol Microbiol 53:941–951. https://doi.org/10.1099/ijs.0.02403-0

Imhoff JF, Trüper HG (1977) *Ectothiorhodospira halochloris* sp. nov., a new extremely halophilic phototrophic bacterium containing bacteriochlorophyll b. Arch Microbiol 114:115–121. https://doi.org/10.1007/BF00410772

Imhoff JF, Süling J, Petri R (1998) Phylogenetic relationships among the *Chromatiaceae,* their taxonomic reclassification and description of the new genera *Allochromatium, Halochromatium, Isochromatium, Marichromatium, Thiococcus, Thiohalocapsa* and *Thermochromatium.* Int J Syst Bacteriol 48(Pt 4):1129–1143. https://doi.org/10.1099/00207713-48-4-1129

Itoh T, Onishi M, Kato S et al (2016) *Athalassotoga saccharophila* gen. nov., sp. nov., isolated from an acidic terrestrial hot spring, and proposal of Mesoaciditogales Ord. Nov. and

Mesoaciditogaceae fam. nov. in the phylum Thermotogae. Int J Syst Evol Microbiol 66:1045–1051. https://doi.org/10.1099/ijsem.0.000833

Jannasch HW, Wirsen CO, Molyneaux SJ et al (1988) Extremely thermophilic fermentative archaebacteria of the genus *Desulfurococcus* from deep-sea hydrothermal vents. Appl Environ Microbiol 54:1203–1209. https://doi.org/10.1128/aem.54.5.1203-1209.1988

Jayasinghearachchi HS, Lal B (2011) *Oceanotoga teriensis* gen. nov., sp. nov., a thermophilic bacterium isolated from offshore oil-producing wells. Int J Syst Evol Microbiol 61:554–560. https://doi.org/10.1099/ijs.0.018036-0

Jeanthon C, L'Haridon S, Cueff V et al (2002) *Thermodesulfobacterium hydrogeniphilum* sp. nov., a thermophilic, chemolithoautotrophic, sulfate-reducing bacterium isolated from a deep-sea hydrothermal vent at Guaymas Basin, and emendation of the genus *Thermodesulfobacterium*. Int J Syst Evol Microbiol 52:765–772. https://doi.org/10.1099/00207713-52-3-765

Jørgensen BB, Nelson DC (1988) Bacterial zonation, photosynthesis, and spectral light distribution in hot spring microbial mats of Iceland. Microb Ecol 16(2):133–147. https://doi.org/10.1007/BF02018909

Jung MY, Islam MA, Gwak JH, Kim JG, Rhee SK (2018) *Nitrosarchaeum koreense* gen. nov., sp. nov., an aerobic and mesophilic, ammonia-oxidizing archaeon member of the phylum Thaumarchaeota isolated from agricultural soil. Int J Syst Evol Microbiol 68:3084–3095. https://doi.org/10.1099/ijsem.0.002926

Jurkevitch E, Davidov Y (2007) Phylogenetic diversity and evolution of predatory prokaryotes. In: Jurkevitch E (ed) BT - predatory prokaryotes: biology, ecology and evolution. Springer, Berlin, pp 11–56

Kádár Z, de Vrijek T, van Noorden GE et al (2004) Yields from glucose, xylose, and paper sludge hydrolysate during hydrogen production by the extreme thermophile *Caldicellulosiruptor saccharolyticus*. Appl Biochem Biotechnol 113–116:497–508

Kanoksilapatham W, Pasomsup P, Keawram P, et al (2016) *Fervidobacterium thailandense* sp. nov., an extremely thermophilic bacterium isolated from a hot spring. Int J Syst Evol Microbiol 66:5023–5027. https://doi.org/10.1099/ijsem.0.001463

Kanso S, Patel BKC (2004) *Phenylobacterium lituiforme* sp. nov., a moderately thermophilic bacterium from a subsurface aquifer, and emended description of the genus *Phenylobacterium*. Int J Syst Evol Microbiol 54:2141–2146. https://doi.org/10.1099/ijs.0.63138-0

Kanth BK, Jun S-Y, Kumari S, Pack SP (2014) Highly thermostable carbonic anhydrase from *Persephonella marina* EX-H1: its expression and characterization for CO_2-sequestration applications. Process Biochem 49:2114–2121. https://doi.org/10.1016/j.procbio.2014.10.011

Kashefi K, Holmes DE, Reysenbach A-L et al (2002) Use of Fe(III) as an electron acceptor to recover previously uncultured hyperthermophiles: isolation and characterization of *Geothermobacterium ferrireducens* gen. nov., sp. nov. Appl Environ Microbiol 68:1735–1742. https://doi.org/10.1128/AEM.68.4.1735-1742.2002

Kawaichi S, Ito N, Kamikawa R et al (2013) *Ardenticatena maritima* gen. nov., sp. nov., a ferric iron- and nitrate-reducing bacterium of the phylum "Chloroflexi" isolated from an iron-rich coastal hydrothermal field, and description of Ardenticatenia classis nov. Int J Syst Evol Microbiol 63:2992–3002. https://doi.org/10.1099/ijs.0.046532-0

Kawasumi T, Igarashi Y, Kodama T et al (1984) *Hydrogenobacter thermophilus* gen. nov. , sp. nov. , an extremely thermophilic, aerobic, hydrogen-oxidizing bacterium. Int J Syst Bacteriol 34:5–10. doi: https://doi.org/10.1099/00207713-34-1-5

Khan IU, Habib N, Xiao M et al (2018) *Phenylobacterium terrae* sp. nov., isolated from a soil sample of Khyber-Pakhtun-Khwa, Pakistan. Antonie Van Leeuwenhoek 111:1767–1775. https://doi.org/10.1007/s10482-018-1064-2

Kieft TL, Fredrickson JK, Onstott TC et al (1999) Dissimilatory reduction of Fe(III) and other electron acceptors by a *Thermus* isolate. Appl Environ Microbiol 65:1214–1221. https://doi.org/10.1128/AEM.65.3.1214-1221.1999

Kletzin A (2007) General characteristics and important model organisms. In: Cavicchioli R (ed) Archaea. Wiley, pp 14–92. https://doi.org/10.1128/9781555815516.ch2

Kochetkova TV, Zayulina KS, Zhigarkov VS et al (2020) *Tepidiforma bonchosmolovskayae* gen. nov., sp. nov., a moderately thermophilic Chloroflexi bacterium from a Chukotka hot spring (Arctic, Russia), representing a novel class, Tepidiformia, which includes the previously uncultivated lineage OLB14. Int J Syst Evol Microbiol 70:1192–1202. https://doi.org/10.1099/ijsem.0.003902

Kojima H, Watanabe M, Fukui M (2017) *Sulfurivermis fontis* gen. nov., sp. nov., a sulfur-oxidizing autotroph, and proposal of Thioprofundaceae fam. nov. Int J Syst Evol Microbiol 67:3458–3461. https://doi.org/10.1099/ijsem.0.002137

Könneke M, Bernhard AE, de La Torre JR, Walker CB, Waterbury JB, Stahl DA (2005) Isolation of an autotrophic ammoniaoxidizing marine archaeon. Nature 437:543–546. https://doi.org/10.1038/nature03911

Kouril T, Wieloch P, Reimann J et al (2013) Unraveling the function of the two Entner–Doudoroff branches in the thermoacidophilic crenarchaeon *Sulfolobus solfataricus* P2. FEBS J 280:1126–1138. https://doi.org/10.1111/febs.12106

Kube M, Beck A, Zinder SH et al (2005) Genome sequence of the chlorinated compound–respiring bacterium *Dehalococcoides* species strain CBDB1. Nat Biotechnol 23:1269–1273. https://doi.org/10.1038/nbt1131

L'Haridon S, Cilia V, Messner P et al (1998) *Desulfurobacterium thermolithotrophum* gen. nov., sp. nov., a novel autotrophic, Sulphur-reducing bacterium isolated from a deep-sea hydrothermal vent. Int J Syst Bacteriol 48:701–711. https://doi.org/10.1099/00207713-48-3-701

L'Haridon S, Reysenbach A-L, Tindall BJ et al (2006) *Desulfurobacterium atlanticum* sp. nov., *Desulfurobacterium pacificum* sp. nov. and *Thermovibrio guaymasensis* sp. nov., three thermophilic members of the Desulfurobacteriaceae fam. nov., a deep branching lineage within the bacteria. Int J Syst Evol Microbiol 56:2843–2852. https://doi.org/10.1099/ijs.0.63994-0

L'Haridon S, Gouhier L, John ES et al (2019) *Marinitoga lauensis* sp. nov., a novel deep-sea hydrothermal vent thermophilic anaerobic heterotroph with a prophage. Syst Appl Microbiol 42:343–347. https://doi.org/10.1016/j.syapm.2019.02.006

Lacis LS, Lawford HG (1988a) Effect of growth rate on ethanol production by *Thermoanaerobacter ethanolicus* in glucose-or xylose-limited continuous culture. Biotechnol Lett 10(8). https://doi.org/10.1007/BF01027138

Lacis LS, Lawford HG (1988b) Ethanol production from xylose by *Thermoanaerobacter ethanolicus* in batch and continuous culture. Arch Microbiol 150:48–55. https://doi.org/10.1007/BF00409717

Lamed R, Zeikus JG (1980a) Glucose fermentation pathway of *Thermoanaerobium brockii.* J Bacteriol 141(3):1251–1257. https://doi.org/10.1128/JB.141.3.1251-1257.1980

Lamed R, Zeikus JG (1980b) Ethanol production by thermophilic bacteria: relationship between fermentation product yields of and catabolic enzyme activities in *Clostridium thermocellum* and *Thermoanaerobium brockii.* J Bacteriol 144(2):569–578. https://doi.org/10.1128/JB.144.2.569-578.1980

Meo-Savoie D, Carol A, George W et al (2004) Physicochemical characterization of the microhabitat of the epibionts associated with *Alvinella pompejana*, a hydrothermal vent annelid 1 1Associate editor: P O'Day'. Geochim Cosmochim Acta 68(9):2055–2066. https://doi.org/10.1016/j.gca.2003.10.039

Langwig MV, De Anda V, Dombrowski N et al (2022) Large-scale protein level comparison of deltaproteobacteria reveals cohesive metabolic groups. ISME J 16:307–320. https://doi.org/10.1038/s41396-021-01057-y

Latif H, Sahin M, Tarasova J et al (2015) Adaptive evolution of *Thermotoga maritima* reveals plasticity of the ABC transporter network. Appl Environ Microbiol 81:5477–5485. https://doi.org/10.1128/AEM.01365-15

Lawson PA (2016) The taxonomy of the genus clostridium: current status and future perspectives. Microbiol China 43(05):1070–1074

Lebedeva EV, Alawi M, Maixner F et al (2008) Physiological and phylogenetic characterization of a novel lithoautotrophic nitrite-oxidizing bacterium, 'Candidatus *Nitrospira bockiana*'. Int J Syst Evol Microbiol 58(1):242–250. https://doi.org/10.1099/ijs.0.65379-0

Lebedeva EV, Off S, Zumbrägel S et al (2011) Isolation and characterization of a moderately thermophilic nitrite-oxidizing bacterium from a geothermal spring. FEMS Microbiol Ecol 75(2): 195–204. https://doi.org/10.1111/j.1574-6941.2010.01006.x

Lehtovirta-Morley LE, Stoecker K, Vilcinskas A, Prosser JI, Nicol GW (2011) Cultivation of an obligate acidophilic ammonia oxidizer from a nitrifying acid soil. Proc Nat Acad Sci 108: 15892–15897. https://doi.org/10.1073/pnas.1107196108

Lee YJ, Dashti M, Prange A et al (2007) *Thermoanaerobacter sulfurigignens* sp. nov., an anaerobic thermophilic bacterium that reduces 1 M thiosulfate to elemental sulfur and tolerates 90 mM sulfite. Int J Syst Evol Microbiol 57(7):1429–1434. doi: https://doi.org/10.1099/ijs.0.64748-0

Lee YE, Jain MK, Lee CY et al (1993) Taxonomic distinction of saccharolytic thermophilic anaerobes - description of *Thermoanaerobacterium xylanolyticum* gen-nov, sp-nov, and *Thermoanaerobacterium saccharolyticum* gen-nov, sp-nov - reclassification of *Thermoanaerobium brockii, Clostridium thermosulfurogenes*, and *Clostridium thermohydrosulfuricum* E100–69 as *Thermoanaerobacter brockii* comb-nov, *Thermoanaerobacterium thermosulfurigenes* comb-nov, and *Thermoanaerobacter thermohydrosulfuricus* comb-nov, respectively - and transfer of *Clostridium thermohydrosulfuricum* 39E to *Thermoanaerobacter ethanolicus*. Int J Syst Bacteriol 43:41–51. https://doi.org/10.1099/00207713-43-1-41

Leigh JA, Whitman WB (2013a) Archaeal genetics. In: Brenner's encyclopedia of genetics. Elsevier, pp 188–191. https://doi.org/10.1016/B978-0-12-374984-0.00093-0

Leigh JA, Whitman WB (2013b) Archaeal genetics. In: Hughes K, Maloy S (eds) Brenner's encyclopedia of genetics. Elsevier, pp 188–191

Lepage E, Marguet E, Geslin C et al (2004) Molecular diversity of new *Thermococcales* isolates from a single area of hydrothermal Deep-Sea vents as revealed by randomly amplified polymorphic DNA fingerprinting and 16S rRNA gene sequence analysis. Appl Environ Microbiol 70:1277–1286. https://doi.org/10.1128/AEM.70.3.1277-1286.2004

Letunic I, Bork P (2007) Interactive tree of life (iTOL): an online tool for phylogenetic tree display and annotation. Bioinformatics 23(1):127–128. https://doi.org/10.1093/bioinformatics/btl529

Lien T, Madsen M, Rainey FA et al (1998) *Petrotoga mobilis* sp. nov., from a North Sea oil-production well. Int J Syst Evol Microbiol 48:1007–1013. https://doi.org/10.1099/00207713-48-3-1007

Li S, Zhu Q, Luo J et al (2021) Application progress of *Deinococcus radiodurans* in biological treatment of radioactive uranium-containing wastewater. Indian J Microbiol 61:417–426. https://doi.org/10.1007/s12088-021-00969-9

Li Y, Hu J, Qu C et al (2019) Engineered *Thermoanaerobacterium aotearoense* with nfnAB knockout for improved hydrogen production from lignocellulose hydrolysates. Biotechnol Biofuels 12:214. https://doi.org/10.1186/s13068-019-1559-8

Liu Z, Klatt CG, Ludwig M et al (2012) 'Candidatus *Thermochlorobacter Aerophilum*:' an aerobic Chlorophotoheterotrophic member of the phylum Chlorobi defined by metagenomics and Metatranscriptomics. ISME J 6:1869–1882. https://doi.org/10.1038/ismej.2012.24

Liu J, Häggblom MM (2018) Genome-guided identification of organohalide-respiring deltaproteobacteria from the marine environment. MBio 9:e02471–e02418. https://doi.org/10.1128/mBio.02471-18

Li Y, Liu G, Zhong H (1994) A new species of *Thermoplasma*. Wei Sheng Wu Xue Bao 34:255–260

Liu Y, Zhang J, Zhao L, Zhang X, Xie S (2014) Spatial distribution of bacterial communities in high-altitude freshwater wetland sediment. Limnology 15:249–256. https://doi.org/10.1007/s10201-014-0429-0

Liu R, Wei X, Song W et al (2022) Novel Chloroflexi genomes from the deepest ocean reveal metabolic strategies for the adaptation to deep-sea habitats. Microbiome 10(1):1–17. https://doi.org/10.1186/s40168-022-01263-6

Löffler F, Yan J, Ritalahti K et al (2013) *Dehalococcoides mccartyi* gen. nov., sp. nov., obligately organohalide-respiring anaerobic bacteria relevant to halogen cycling and bioremediation, belong to a novel bacterial class, Dehalococcoidia classis nov., order Dehalococcoidales Ord. Nov. and family. Int J Syst Evol Microbiol 63:625–635. https://doi.org/10.1099/ijs.0.034926-0

Lonergan DJ, Jenter HL, Coates JD et al (1996) Phylogenetic analysis of dissimilatory Fe(III)-reducing bacteria. J Bacteriol 178:2402–2408. https://doi.org/10.1128/jb.178.8.2402-2408.1996

López-García P, Zivanovic Y, Deschamps P, Moreira D (2015) Bacterial gene import and mesophilic adaptation in archaea. Nature Rev Microbiol 13:447–456. https://doi.org/10.1038/nrmicro3485

Lopez-Fernandez M, Broman E, Turner S et al (2018) Investigation of viable taxa in the deep terrestrial biosphere suggests high rates of nutrient recycling. FEMS Microbiol Ecol 94:fiy121. https://doi.org/10.1093/femsec/fiy121

Lovley DR, Malvankar NS (2015) Seeing is believing: novel imaging techniques help clarify microbial nanowire structure and function. Environ Microbiol 17:2209–2215. https://doi.org/10.1111/1462-2920.12708

Lovley DR, Phillips EJ (1988) Novel mode of microbial energy metabolism: organic carbon oxidation coupled to dissimilatory reduction of iron or manganese. Appl Environ Microbiol 54:1472–1480. https://doi.org/10.1128/aem.54.6.1472-1480.1988

Lücker S, Wagner M, Maixner F et al (2010) *Nitrospira* metagenome illuminates the physiology and evolution of globally important nitrite-oxidizing bacteria. Proc Natl Acad Sci U S A 107(30):13479–13484. doi: https://doi.org/10.1073/pnas.1003860107

Ludvík J, Benada O, Mikuh'k K (1994) Ultrastructure of an extreme thermophilic hydrogen-oxidizing bacterium *Calderobacterium hydrogenophilum*. Arch Microbiol 162:267–271

Luther GW, Rozan TF, Taillefert M et al (2001) Chemical speciation drives hydrothermal vent ecology. Nature 410:813–816. https://doi.org/10.1038/35071069

Lütke-Eversloh T, Elbanna K, Cnockaert MC et al (2004) *Caenibacterium thermophilum* is a later synonym of *Schlegelella thermodepolymerans*. Int J Syst Evol Microbiol 54:1933–1935. doi: https://doi.org/10.1099/ijs.0.63204-0

Manaia CM, da Costa MS (1991) Characterization of halotolerant *Thermus* isolates from shallow marine hot springs on S. Miguel, Azores. Microbiology 137:2643–2648. https://doi.org/10.1099/00221287-137-11-2643

Manaia CM, Hoste B, Carmen Gutierrez M et al (1995) Halotolerant *Thermus* strains from marine and terrestrial hot springs belong to *Thermus thermophilus* (ex Oshima and Imahori, 1974) nom. Rev. emend. Syst Appl Microbiol 17:526–532. https://doi.org/10.1016/S0723-2020(11)80072-X

Mardanov AV, Kadnikov VV, Beletsky AV et al (2020) Sulfur and methane-oxidizing microbial community in a terrestrial mud volcano revealed by metagenomics. Microorganisms 8(9):1333. https://doi.org/10.3390/microorganisms8091333

Marteinsson VT (1999) Isolation and characterization of *Thermus thermophilus* Gy1211 from a deep-sea hydrothermal vent. Extremophiles 3:247–251. https://doi.org/10.1007/s007920050123

Martens-Habbena W, Stahl DA (2011) Nitrogen metabolism and kinetics of ammonia-oxidizing archaea. Meth Enzymol 496:465–487. https://doi.org/10.1016/B978-0-12-386489-5.00019-1

Masuda Y, Itoh H, Shiratori Y et al (2017) Predominant but previously-overlooked prokaryotic drivers of reductive nitrogen transformation in Paddy soils, revealed by Metatranscriptomics. Microbes Environ 32:180–183. https://doi.org/10.1264/jsme2.ME16179

McGonigle JM, Lang SQ, Brazelton WJ (2020) Genomic evidence for formate metabolism by Chloroflexi as the key to unlocking deep carbon in lost City microbial ecosystems. Appl Environ Microbiol 86:e02583–e02519. https://doi.org/10.1128/AEM.02583-19

McNichol J, Stryhanyuk H, Sylva SP et al (2018) Primary productivity below the seafloor at deep-sea hot springs. Proc Natl Acad Sci U S A 115:6756–6761. https://doi.org/10.1073/pnas.1804351115
Mehrani MJ, Sobotka D, Kowal P et al (2020) The occurrence and role of *Nitrospira* in nitrogen removal systems. Bioresour Technol 303:122936. https://doi.org/10.1016/j.biortech.2020.122936
Ming Y-Z, Liu L, Lv A-P et al (2021) *Thermaurantiacus tibetensis* gen. nov., sp. nov., a novel moderately thermophilic bacterium isolated from hot spring microbial mat in Tibet. Antonie Van Leeuwenhoek 114:445–455. https://doi.org/10.1007/s10482-021-01530-w
Mino S, Makita H, Toki T et al (2013) Biogeography of *Persephonella* in deep-sea hydrothermal vents of the Western Pacific. Front Microbiol 4:107. https://doi.org/10.3389/fmicb.2013.00107
Mino S, Nakagawa S, Makita H et al (2017) Endemicity of the cosmopolitan mesophilic chemolithoautotroph Sulfurimonas at deep-sea hydrothermal vents. ISME J 11:909–919. https://doi.org/10.1038/ismej.2016.178
Mino S, Yoneyama N, Nakagawa S et al (2018) Enrichment and genomic characterization of a N_2O-reducing Chemolithoautotroph from a Deep-Sea hydrothermal vent. Front Bioeng Biotechnol 6:184. https://doi.org/10.3389/fbioe.2018.00184
Minz D, Flax JL, Green SJ et al (1999) Diversity of sulfate-reducing bacteria in oxic and anoxic regions of a microbial mat characterized by comparative analysis of dissimilatory sulfite reductase genes. Appl Environ Microbiol 65:4666–4671. https://doi.org/10.1128/AEM.65.10.4666-4671.1999
Miranda-Tello E, Fardeau M-L, Joulian C et al (2007) *Petrotoga halophila* sp. nov., a thermophilic, moderately halophilic, fermentative bacterium isolated from an offshore oil well in Congo. Int J Syst Evol Microbiol 57:40–44. https://doi.org/10.1099/ijs.0.64516-0
Miroshnichenko ML, Slobodkin AI, Kostrikina NA et al (2003) *Deferribacter abyssi* sp. nov., an anaerobic thermophile from deep-sea hydrothermal vents of the Mid-Atlantic Ridge. Int J Syst Evol Microbiol 53:1637–1641. https://doi.org/10.1099/ijs.0.02673-0
Miranda-Tello E, Fardeau M-L, Thomas P et al (2004) *Petrotoga mexicana* sp. nov., a novel thermophilic, anaerobic and xylanolytic bacterium isolated from an oil-producing well in the Gulf of Mexico. Int J Syst Evol Microbiol 54:169–174. https://doi.org/10.1099/ijs.0.02702-0
Miroshnichenko ML, Lebedinsky AV, Chernyh NA et al (2009) *Caldimicrobium rimae* gen. nov., sp. nov., an extremely thermophilic, facultatively lithoautotrophic, anaerobic bacterium from the Uzon caldera, Kamchatka. Int J Syst Evol Microbiol 59:1040–1044. https://doi.org/10.1099/ijs.0.006072-0
Miyazaki J, Ikuta T, Watsuji T et al (2020) Dual energy metabolism of the Campylobacterota endosymbiont in the chemosynthetic snail Alviniconcha Marisindica. ISME J 14(5):1273–1289. https://doi.org/10.1038/s41396-020-0605-7
Mori K, Sakurai K, Hosoyama A et al (2020) Vestiges of adaptation to the mesophilic environment in the genome of *Tepiditoga spiralis* gen. nov., sp. nov. Microbes Environ 35:ME20046. https://doi.org/10.1264/jsme2.ME20046
Mori K, Suzuki K (2008) *Thiofaba tepidiphila* gen. nov., sp. nov., a novel obligately chemolithoautotrophic, sulfur-oxidizing bacterium of the Gammaproteobacteria isolated from a hot spring. Int J Syst Evol Microbiol 58:1885–1891. https://doi.org/10.1099/ijs.0.65754-0
Mori K (2014) The family Thermodesulfobacteriaceae. In: Rosenberg E, DeLong EF, Lory S et al (eds) The prokaryotes: firmicutes and Tenericutes. Springer, Berlin, pp 381–388
Moussard H, L'Haridon S, Tindall BJ et al (2004) *Thermodesulfatator indicus* gen. nov., sp. nov., a novel thermophilic chemolithoautotrophic sulfate-reducing bacterium isolated from the central Indian ridge. Int J Syst Evol Microbiol 54:227–233. https://doi.org/10.1099/ijs.0.02669-0
Nakagawa S, Takai K, Horikoshi K et al (2004b) *Aeropyrum camini* sp. nov., a strictly aerobic, hyperthermophilic archaeon from a deep-sea hydrothermal vent chimney. Int J Syst Evol Microbiol 54, 329–335. doi: https://doi.org/10.1099/ijs.0.02826-0
Nakagawa S, Inagaki F, Takai K et al (2005a) '*Thioreductor micantisoli* gen. nov., sp. nov., a novel mesophilic, Sulfur-reducing Chemolithoautotroph within the ε-proteobacteria isolated from

hydrothermal sediments in the mid-Okinawa trough'. Int J Syst Evol Microbiol 55(2):599–605. https://doi.org/10.1099/ijs.0.63351-0
Nakagawa S, Nakamura S, Inagaki F et al (2004a) *Hydrogenivirga caldilitoris* gen. nov., sp. nov., a novel extremely thermophilic, hydrogen- and sulfur-oxidizing bacterium from a coastal hydrothermal field. Int J Syst Evol Microbiol 54:2079–2084. https://doi.org/10.1099/ijs.0.03031-0
Nakagawa S, Takai K, Horikoshi K et al (2003) *Persephonella hydrogeniphila* sp. nov., a novel thermophilic, hydrogen-oxidizing bacterium from a deep-sea hydrothermal vent chimney. Int J Syst Evol Microbiol 53:863–869. https://doi.org/10.1099/ijs.0.02505-0
Nakagawa S, Takai K, Inagaki F et al (2005b) *Nitratiruptor tergarcus* gen. nov., sp. nov. and *Nitratifractor salsuginis* gen. nov., sp. nov., nitrate-reducing chemolithoautotrophs of the epsilon-proteobacteria isolated from a deep-sea hydrothermal system in the mid-Okinawa trough. Int J Syst Evol Microbiol 55:925–933. https://doi.org/10.1099/ijs.0.63480-0
Nakagawa S, Takaki Y, Shimamura S et al (2007) Deep-sea vent epsilon-proteobacterial genomes provide insights into emergence of pathogens. Proc Natl Acad Sci U S A 104:12146–12150. https://doi.org/10.1073/pnas.0700687104
Nakagawa S, Takaki Y (2009) Nonpathogenic epsilonproteobacteria. In: Cullen KE (ed) Encyclopedia of life sciences. Wiley. https://doi.org/10.1002/047001590X
Nakagawa T, Takai K, Suzuki Y et al (2006) Geomicrobiological exploration and characterization of a novel deep-sea hydrothermal system at the TOTO caldera in the Mariana volcanic arc. Environ Microbiol 8:37–49. https://doi.org/10.1111/j.1462-2920.2005.00884.x
Nelson KE, Clayton RA, Gill SR et al (1999) Evidence for lateral gene transfer between archaea and bacteria from genome sequence of *Thermotoga maritima*. Nature 399(6734):323–329. https://doi.org/10.1038/20601
Nesbø C, Farrell AA, Zhaxybayeva O et al (2022) Thermosipho. In: Trujillo ME, Dedysh S, DeVos P, Hedlund B, Kämpfer P, Rainey FA, Whitman WB (eds) Bergey's manual of systematics of archaea and bacteria. Wiley. https://doi.org/10.1002/9781118960608.gbm01272.pub2
Nesbø CL, Bradnan DM, Adebusuyi A et al (2012) *Mesotoga prima* gen. nov., sp. nov., the first described mesophilic species of the Thermotogales. Extremophiles 16:387–393. https://doi.org/10.1007/s00792-012-0437-0
Nesbø CL, Farrell AA, Zhaxybayeva O et al (2019) Thermotoga. In: Trujillo ME, Dedysh S, DeVos P, Hedlund B, Kämpfer P, Rainey FA, Whitman WB (eds) Bergey's manual of systematics of archaea and bacteria. Wiley, pp 1–13. https://doi.org/10.1002/9781118960608.gbm01273.pub2
Nishihara H, Igarashi Y, Kodama T (1990) A new isolate of *Hydrogenobacter*, an obligately chemolithoautotrophic, thermophilic, halophilic and aerobic hydrogen-oxidizing bacterium from seaside saline hot spring. Arch Microbiol 153:294–298. https://doi.org/10.1007/BF00249085
Nishizawa M, Sakai S, Konno U et al (2016) Nitrogen and oxygen isotope effects of ammonia oxidation by thermophilic Thaumarchaeota from a geothermal water stream. Appl Environ Microbiol 82:4492–4504. https://doi.org/10.1128/AEM.00250-16
Nunn CE, Johnsen U, Schönheit P et al (2010) Metabolism of pentose sugars in the hyperthermophilic archaea *Sulfolobus solfataricus* and *Sulfolobus acidocaldarius*. J Biol Chem 285:33701–33709. https://doi.org/10.1074/jbc.M110.146332
Nunoura T, Hirai M, Imachi H et al (2010) *Kosmotoga arenicorallina* sp. nov. a thermophilic and obligately anaerobic heterotroph isolated from a shallow hydrothermal system occurring within a coral reef, southern part of the Yaeyama archipelago, Japan, reclassification of *Thermococcoides shengliensis* as *Kosmotoga shengliensis* comb. nov., and emended description of the genus Kosmotoga. Arch Microbiol 192:811–819. https://doi.org/10.1007/s00203-010-0611-7
Nunoura T, Hirai M, Miyazaki M et al (2013) Isolation and characterization of a thermophilic, obligately anaerobic and heterotrophic marine Chloroflexi bacterium from a Chloroflexi-dominated microbial community associated with a Japanese shallow hydrothermal system,

and proposal for Thermomarinili. Microbes Environ 28:228–235. https://doi.org/10.1264/jsme2.me12193

Nunoura T, Miyazaki M, Suzuki Y et al (2008a) *Hydrogenivirga okinawensis* sp. nov., a thermophilic sulfur-oxidizing chemolithoautotroph isolated from a deep-sea hydrothermal field, southern Okinawa trough. Int J Syst Evol Microbiol 58:676–681. https://doi.org/10.1099/ijs.0.64615-0

Nunoura T, Oida H, Miyazaki M et al (2007) *Marinitoga okinawensis* sp. nov., a novel thermophilic and anaerobic heterotroph isolated from a deep-sea hydrothermal field, southern Okinawa trough. Int J Syst Evol Microbiol 57:467–471. https://doi.org/10.1099/ijs.0.64640-0

Nunoura T, Oida H, Miyazaki M et al (2008b) *Thermosulfidibacter takaii* gen. nov., sp. nov., a thermophilic, hydrogen-oxidizing, sulfur-reducing chemolithoautotroph isolated from a deep-sea hydrothermal field in the southern Okinawa trough. Int J Syst Evol Microbiol 58:659–665. https://doi.org/10.1099/ijs.0.65349-0

Off S, Alawi M, Spieck E (2010) Enrichment and physiological characterization of a novel *Nitrospira*-like bacterium obtained from a marine sponge. Appl Environ Microbiol 76(14): 4640–4646. https://doi.org/10.1128/AEM.00320-10

Offre P, Spang A, Schleper C (2013) Archaea in biogeochemical cycles. Annu Rev Microbiol 67: 437–457. https://doi.org/10.1146/annurev-micro-092412-155614

Okumura T, Takashima C, Yanagawa K et al (2022) Stromatolite formation by Anaerolineae-dominated microbial communities in hot spring travertine in North Sumatra, Indonesia. Sediment Geol 440:106263. https://doi.org/10.1016/j.sedgeo.2022.106263

Onyenwoke RU, Wiegel J (2015) *Thermoanaerobacterium*. In: Trujillo ME, Dedysh S, DeVos P, Hedlund B, Kämpfer P, Rainey FA, Whitman WB (eds) Bergey's manual of systematics of archaea and bacteria. Wiley. https://doi.org/10.1002/9781118960608.gbm00755

Oren A, Garrity GM (2021) Valid publication of the names of forty-two phyla of prokaryotes. Int J Syst Evol Microbiol 71:005056. https://doi.org/10.1099/ijsem.0.005056

Oren A, Xu X-W (2014) The family Hyphomicrobiaceae. In: Rosenberg E, DeLong EF, Lory S et al (eds) The prokaryotes: Alphaproteobacteria and betaproteobacteria. Springer, Berlin, pp 247–281

Oshima T, Imahori K (1971) Isolation of an extreme thermophile and thermostability of its transfer ribonucleic acid and ribosomes. J Gen Appl Microbiol 17:513–517. https://doi.org/10.2323/jgam.17.513

Oshima T, Imahori K (1974) Description of *Thermus thermophilus* (Yoshida and Oshima) comb. nov., a nonsporulating thermophilic bacterium from a Japanese thermal spa. Int J Syst Bacteriol 24:102–112. https://doi.org/10.1099/00207713-24-1-102

Overmann J, Tuschak C (1997) Phylogeny and molecular fingerprinting of Green Sulfur bacteria. Arch Microbiol 167:302–309. https://doi.org/10.1007/s002030050448

O'Neill AH, Liu Y, Ferrera I et al (2008) *Sulfurihydrogenibium rodmanii* sp. nov., a sulfur-oxidizing chemolithoautotroph from the Uzon caldera, Kamchatka peninsula, Russia, and emended description of the genus *Sulfurihydrogenibium*. Int J Syst Evol Microbiol 58:1147–1152. https://doi.org/10.1099/ijs.0.65431-0

Palomo A, Dechesne A, Pedersen AG et al (2022) Genomic profiling of *Nitrospira* species reveals ecological success of comammox *Nitrospira*. Microbiome 10(1):204. https://doi.org/10.1186/s40168-022-01411-y

Papale M, Romano I, Finore I et al (2021) Prokaryotic diversity of the composting thermophilic phase: the case of ground coffee compost. Microorganisms 9:218. https://doi.org/10.3390/microorganisms9020218

Parte AC, Sardà Carbasse J, Meier-Kolthoff JP et al (2020) List of prokaryotic names with standing in nomenclature (LPSN) moves to the DSMZ. Int J Syst Evol Microbiol 70:5607–5612. https://doi.org/10.1099/ijsem.0.004332

Pask-Hughes RA, Williams RAD (1977) Yellow-pigmented strains of *Thermus* spp. from Icelandic hot springs. Microbiology 102:375–383. https://doi.org/10.1099/00221287-102-2-375

Patel BKC, Morgan HW, Daniel RM (1985) *Fervidobacterium nodosum* gen. nov. and spec. nov., a new chemoorganotrophic, caldoactive, anaerobic bacterium. Arch Microbiol 141:63–69. https://doi.org/10.1007/BF00446741

Perez-Rodríguez I, Grosche A, Massenburg L et al (2012) *Phorcysia thermohydrogeniphila* gen. nov., sp. nov., a thermophilic, chemolithoautotrophic, nitrate-ammonifying bacterium from a deep-sea hydrothermal vent. Int J Syst Evol Microbiol 62:2388–2394. https://doi.org/10.1099/ijs.0.035642-0

Pesaro M, Widmer F (2002) Identification of novel Crenarchaeota and Euryarchaeota clusters associated with different depth layers of a forest soil. FEMS Microbiol Ecol 42:89–98. https://doi.org/10.1111/j.1574-6941.2002.tb00998.x

Pester M, Schleper C, Wagner M (2011) The Thaumarchaeota: an emerging view of their phylogeny and ecophysiology. Curr Op Microbiol 14:300–306. https://doi.org/10.1016/j.mib.2011.04.007

Pierson BK, Castenholz RW (1974) A phototrophic gliding filamentous bacterium of hot springs, *Chloroflexus aurantiacus*, gen. and sp. nov. Arch Microbiol 100:5–24. https://doi.org/10.1007/BF00446302

Pierson BK, Giovannoni SJ, Stahl DA et al (1985) *Heliothrix oregonensis*, gen. nov., sp. nov., a phototrophic filamentous gliding bacterium containing bacteriochlorophyll a. Arch Microbiol 142:164–167. https://doi.org/10.1007/BF00447061

Podosokorskaya O, Kadnikov VV, Gavrilov SN et al (2013) Characterization of Melioribacter roseus gen. nov., sp. nov., a novel facultatively anaerobic thermophilic cellulolytic bacterium from the class Ignavibacteria, and a proposal of a novel bacterial phylum Ignavibacteriae. Environ Microbiol 15:1759–1771. https://doi.org/10.1111/1462-2920.12067

Pohl A, Young SAE, Schmitz TC et al (2021) Magnetite-binding proteins from the magnetotactic bacterium *Desulfamplus magnetovallimortis* BW-1. Nanoscale 13:20396–20400. https://doi.org/10.1039/d1nr04870h

Pollo SMJ, Adebusuyi AA, Straub TJ et al (2017) Genomic insights into temperature-dependent transcriptional responses of *Kosmotoga olearia*, a deep-biosphere bacterium that can grow from 20 to 79 °C. Extremophiles 21:963–979. https://doi.org/10.1007/s00792-017-0956-9

Pommier Y (2009) DNA topoisomerase I inhibitors: chemistry, biology, and interfacial inhibition. Chem Rev 109:2894–2902. https://doi.org/10.1021/cr900097c

Postec A, Le Breton C, Fardeau M-L et al (2005) *Marinitoga hydrogenitolerans* sp. nov., a novel member of the order Thermotogales isolated from a black smoker chimney on the mid-Atlantic ridge. Int J Syst Evol Microbiol 55:1217–1221. https://doi.org/10.1099/ijs.0.63550-0

Postec A, Ciobanu M, Birrien J-L et al (2010) *Marinitoga litoralis* sp. nov., a thermophilic, heterotrophic bacterium isolated from a coastal thermal spring on Ile Saint-Paul, southern Indian Ocean. Int J Syst Evol Microbiol 60:1778–1782. https://doi.org/10.1099/ijs.0.017574-0

Prieto MA, Escapa IF, Martínez VV et al (2014) A holistic view of polyhydroxyalkanoate metabolism in pseudomonas putida. Environ Microbiol 18(2):341–357. https://doi.org/10.1111/1462-2920.12760

Purwasena IA, Sugai Y, Sasaki K (2014) *Petrotoga japonica* sp. nov., a thermophilic, fermentative bacterium isolated from Yabase oilfield in Japan. Arch Microbiol 196:313–321. https://doi.org/10.1007/s00203-014-0972-4

Quehenberger J, Shen L, Albers SV et al (2017) Sulfolobus – a potential Key organism in future biotechnology. Front Microbiol 8. https://www.frontiersin.org/articles/10.3389/fmicb.2017.02474. Accessed 7 Apr 2023

Rainey FA, da Costa MS (2001) Order II. Thermales Ord. Nov. In: Garrity GM, Boone DR, Castenholz RW (eds) Bergey's manual of systematic bacteriology, Volume one: the archaea and the deeply branching and phototrophic bacteria, 2nd edn. Springer, New York, p 403

Rainey FA, Ray K, Ferreira M et al (2005) Extensive diversity of ionizing-radiation-resistant bacteria recovered from Sonoran Desert soil and description of nine new species of the genus *Deinococcus* obtained from a single soil sample. Appl Environ Microbiol 71:5225–5235. https://doi.org/10.1128/AEM.71.9.5225-5235.2005

Rainey FA, Silva J, Nobre MF et al (2003) *Porphyrobacter cryptus* sp. nov., a novel slightly thermophilic, aerobic, bacteriochlorophyll a-containing species. Int J Syst Evol Microbiol 53: 35–41. https://doi.org/10.1099/ijs.0.02308-0

Rainey FA (2015) Caldicellulosiruptor. In: Trujillo ME, Dedysh S, Devos P, Hedlund B, Kämpfer P, Rainey FA, Whitman WB (eds) Bergey's manual of systematics of archaea and bacteria. Wiley. https://doi.org/10.1002/9781118960608.gbm00754

Rainey FA, Donnison AM, Janssen PH et al (1994) Description of *Caldicellulosiruptor saccharolyticus* gen. nov., sp. nov: an obligately anaerobic, extremely thermophilic, cellulolytic bacterium. FEMS Microbiol Lett 120:263–266. https://doi.org/10.1016/0378-1097(94)90481-2

Rangel AE, Croft T, González Barrios AF et al (2020) Transcriptomic analysis of a *Clostridium thermocellum* strain engineered to utilize xylose: responses to xylose versus cellobiose feeding. Sci Rep 10(1):14517. https://doi.org/10.1038/s41598-020-71428-6

Raman B, McKeown CK, Rodriguez M et al (2011) Transcriptomic analysis of *Clostridium thermocellum* ATCC 27405 cellulose fermentation. BMC Microbiol 11:134. https://doi.org/10.1186/1471-2180-11-134

Ravot G, Magot M, Fardeau ML et al (1995) *Thermotoga elfii* sp. nov., a novel thermophilic bacterium from an African oil-producing well. Int J Syst Bacteriol 45:308–314. https://doi.org/10.1099/00207713-45-2-308

Reeve W, Ardley J, Tian R et al (2014) Genome sequence of the Listia angolensis microsymbiont *Microvirga lotononidis* strain WSM3557(T). Stand Genomic Sci 9:540–550. 10.4056/sigs.4548266 WE - science citation index expanded (SCI-EXPANDED)

Reichenbach H (1999) The ecology of the myxobacteria. Environ Microbiol 1:15–21. https://doi.org/10.1046/j.1462-2920.1999.00016.x

Ren N, Cao G, Wang A et al (2008) Dark fermentation of xylose and glucose mix using isolated *Thermoanaerobacterium thermosaccharolyticum* W16. Int J Hydro Energy 33(21):6124–6132. https://doi.org/10.1016/j.ijhydene.2008.07.107

Reno ML, Held NL, Fields CJ et al (2009) Biogeography of the *Sulfolobus islandicus* pan-genome. Proc Natl Acad Sci U S A 106:8605–8610. https://doi.org/10.1073/pnas.0808945106

Reysenbach A-L, Liu Y, Lindgren AR et al (2013) *Mesoaciditoga lauensis* gen. nov., sp. nov., a moderately thermoacidophilic member of the order Thermotogales from a deep-sea hydrothermal vent. Int J Syst Evol Microbiol 63:4724–4729. https://doi.org/10.1099/ijs.0.050518-0

Reysenbach AL (2015) Thermotogae class. nov. In: Trujillo ME, Dedysh S, Devos P et al (eds) Bergey's manual of systematics of archaea and bacteria. Wiley. https://doi.org/10.1002/9781118960608.cbm00051

Robertson BR, O'Rourke JL, Neilan BA et al (2005) *Mucispirillum schaedleri* gen. nov., sp. nov., a spiral-shaped bacterium colonizing the mucus layer of the gastrointestinal tract of laboratory rodents. Int J Syst Evol Microbiol 55:1199–1204. https://doi.org/10.1099/ijs.0.63472-0

Rogers TJ, Buongiorno J, Jessen GL et al (2023) Chemolithoautotroph distributions across the subsurface of a convergent margin. ISME J 17:140–150. https://doi.org/10.1038/s41396-022-01331-7

Romano I, Dipasquale L, Orlando P et al (2010) *Thermoanaerobacterium thermostercus* sp., nov, a new anaerobic thermophilic hydrogen-producing bacterium from buffalo-dung. Extremophiles 14(2):233–240. Doi:https://doi.org/10.1007/s00792-010-0303-x

Roy C, BAkshi U, Rameez MJ et al (2019) Phylogenomics of an uncultivated, aerobic and thermophilic, photoheterotrophic member of chlorobia sheds light into the evolution of the phylum chlorobi. Comput Biol Chem 80:206–216. https://doi.org/10.1016/j.compbiolchem.2019.04.001

Royter M, Schmidt M, Elend C et al (2009) Thermostable lipases from the extreme thermophilic anaerobic bacteria *Thermoanaerobacter thermohydrosulfuricus* SOL1 and *Caldanaerobacter subterraneus* subsp. tengcongensis. Extremophiles 13:769–783. https://doi.org/10.1007/s00792-009-0265-z

Russell MJ, Hall AJ (1997) The emergence of life from iron Monosulphide bubbles at a submarine hydrothermal redox and PH front. J Geol Soc 154(3):377–402. https://doi.org/10.1144/gsjgs.154.3.0377

Saghatelyan A, Margaryan A, Panosyan H, Birkeland NK (2021) Microbial diversity of terrestrial geothermal springs in Armenia and Nagorno-Karabakh: a review. Microorganisms 9:1473. https://doi.org/10.3390/microorganisms9071473

Salama NR (2020) Cell morphology as a virulence determinant: lessons from helicobacter pylori. Curr Opin Microbiol 54:11–17. https://doi.org/10.1016/j.mib.2019.12.002

Sander K, Wilson C, Rodriguez M et al (2015) *Clostridium thermocellum* DSM1313 transcriptional responses to redox perturbation. Biotechnol Biofuels 8(1). doi:https://doi.org/10.1186/s13068-015-0394-9

Sant'Anna FH, Lebedinsky AV, Sokolova TG et al (2015) Analysis of three genomes within the thermophilic bacterial species *Caldanaerobacter subterraneus* with a focus on carbon monoxide dehydrogenase evolution and hydrolase diversity. BMC Genomics 16:757. https://doi.org/10.1186/s12864-015-1955-9

Santos MA, Williams RAD, Da Costa MS (1989) Numerical taxonomy of *Thermus* isolates from hot springs in Portugal. Syst Appl Microbiol 12:310–315. https://doi.org/10.1016/S0723-2020(89)80079-7

Schleper C, Nicol GW (2010) Ammonia-oxidising archaea - physiology, ecology and evolution. Adv Microb Physiol 57:1–41. https://doi.org/10.1016/B978-0-12-381045-8.00001-1

Schleper C, Puehler G, Klenk H-P et al (1996) *Picrophilus oshimae* and *Picrophilus torridus* fam. nov., gen. nov., sp. nov., two species of Hyperacidophilic, thermophilic, heterotrophic, aerobic archaea. Int J Syst Evol Microbiol 46:814–816. https://doi.org/10.1099/00207713-46-3-814

Schoch CL, Ciufo S, Domrachev M et al (2020) NCBI Taxonomy: a comprehensive update on curation, resources and tools, Database, Volume 2020, 2020, baaa062, https://doi.org/10.1093/database/baaa062

Schönheit P, Schäfer T (1995) Metabolism of hyperthermophiles. World J Microbiol Biotechnol 11:26–57. https://doi.org/10.1007/BF00339135

Scully SM, Orlygsson J (2014) Branched-chain alcohol formation from branched-chain amino acids by *Thermoanaerobacter brockii* and *Thermoanaerobacter yonseiensis*. Anaerobe 30:82–84. https://doi.org/10.1016/j.anaerobe.2014.09.003

Segerer A, Langworthy TA, Stetter KO (1988) *Thermoplasma acidophilum* and *Thermoplasma volcanium* sp. nov. from solfatara Fields. Syst Appl Microbiol 10:161–171. https://doi.org/10.1016/S0723-2020(88)80031-6

Sekiguchi Y, Yamada T, Hanada S et al (2003) *Anaerolinea thermophila* gen. nov., sp. nov. and *Caldilinea aerophila* gen. nov., sp. nov., novel filamentous thermophiles that represent a previously uncultured lineage of the domain bacteria at the subphylum level. Int J Syst Evol Microbiol 53:1843–1851. https://doi.org/10.1099/ijs.0.02699-0

Sekiguchi Y, Muramatsu M, Imachi H et al (2008) *Thermodesulfovibrio aggregans* sp. nov. and *Thermodesulfovibrio thiophilus* sp. nov., anaerobic, thermophilic, sulfate-reducing bacteria isolated from thermophilic methanogenic sludge, and emended description of the genus *Thermodesulfovibrio*. Int J Syst Evol Microbiol 58(11):2541–2548. doi: https://doi.org/10.1099/ijs.0.2008/000893-0

Selkova P, Vasileva A, Pobegalov G et al (2020) Position of deltaproteobacteria Cas12e nuclease cleavage sites depends on spacer length of guide RNA. RNA Biol 17:1472–1479. https://doi.org/10.1080/15476286.2020.1777378

Seong CN, Kang JW, Lee JH et al (2018) Taxonomic hierarchy of the phylum firmicutes and novel firmicutes species originated from various environments in Korea. J Microbiol 56(1):1–10. https://doi.org/10.1007/s12275-018-7318-x

Seyler LM, Trembath-Reichert E, Tully BJ et al (2021) Time-series transcriptomics from cold, oxic subseafloor crustal fluids reveals a motile, mixotrophic microbial community. ISME J 15:1192–1206. https://doi.org/10.1038/s41396-020-00843-4

Sharma N, Kumar J, Abedin MM et al (2020) Metagenomics revealing molecular profiling of community structure and metabolic pathways in natural hot springs of the Sikkim Himalaya. BMC Microbiol 20:246. https://doi.org/10.1186/s12866-020-01923-3

Shaw AJ, Podkaminer KK, Desai SG et al (2008) Metabolic engineering of a thermophilic bacterium to produce ethanol at high yield. Proc Natl Acad Sci U S A 105:13769–13774

Sheik CS, Anantharaman K, Breier JA et al (2015) Spatially resolved sampling reveals dynamic microbial communities in rising hydrothermal plumes across a back-arc basin. ISME J 9:1434–1445. https://doi.org/10.1038/ismej.2014.228

Shima S, Suzuki K-I (1993) *Hydrogenobacter acidophilus* sp. nov., a Thermoacidophilic, aerobic, hydrogen-oxidizing bacterium requiring elemental Sulfur for growth. Int J Syst Bacteriol 43: 703–708. https://doi.org/10.1099/00207713-43-4-703

Shiotani T, Mino S, Sato W et al (2020) *Nitrosophilus alvini* gen. nov., sp. nov., a hydrogen-oxidizing chemolithoautotroph isolated from a deep-sea hydrothermal vent in the East Pacific rise, inferred by a genome-based taxonomy of the phylum "Campylobacterota". PLoS One 15: e0241366. https://doi.org/10.1371/journal.pone.0241366

Siebers B, Schönheit P (2005) Unusual pathways and enzymes of central carbohydrate metabolism in archaea. Curr Opin Microbiol 8:695–705. https://doi.org/10.1016/j.mib.2005.10.014

Siebert J, Hirsch P (1988) Characterization of 15 selected coccal bacteria isolated from antarctic rock and soil samples from the McMurdo-dry valleys (South-Victoria land). Polar Biol 9:37–44. https://doi.org/10.1007/BF00441762

Singleton CM, Petriglieri F, Kristensen JM et al (2021) Connecting structure to function with the recovery of over 1000 high-quality metagenome-assembled genomes from activated sludge using long-read sequencing. Nat Commun 12:2009. https://doi.org/10.1038/s41467-021-22203-2

Si Y, Zou Y, Liu X et al (2015) Mercury methylation coupled to iron reduction by dissimilatory iron-reducing bacteria. Chemosphere 122:206–212. https://doi.org/10.1016/j.chemosphere.2014.11.054

Skirnisdottir S, Hreggvidsson GO, Holst O et al (2001) Isolation and characterization of a mixotrophic sulfur-oxidizing *Thermus scotoductus*. Extremophiles 5:45–51. https://doi.org/10.1007/s007920000172

Slade D, Radman M (2011) Oxidative stress resistance in *Deinococcus radiodurans*. Microbiol Mol Biol Rev 75:133–191. https://doi.org/10.1128/MMBR.00015-10

Slobodkina GB, Kolganova TV, Chernyh NA, Querellou J, Bonch-Osmolovskaya EA, Slobodkin AI (2009) *Deferribacter autotrophicus* sp. nov., an iron (III)-reducing bacterium from a deep-sea hydrothermal vent. Int J Syst Evol Microbiol 59:1508–1512. https://doi.org/10.1099/ijs.0.006767-0

Slobodkina GB, Kolganova T V, Kopitsyn DS et al (2016) *Dissulfurirhabdus thermomarina* gen. nov., sp. nov., a thermophilic, autotrophic, sulfite-reducing and disproportionating deltaproteobacterium isolated from a shallow-sea hydrothermal vent. Int J Syst Evol Microbiol 66:2515–2519. https://doi.org/10.1099/ijsem.0.001083

Slobodkin AI, Reysenbach A-L, Slobodkina GB et al (2012) *Thermosulfurimonas dismutans* gen. nov., sp. nov., an extremely thermophilic sulfur-disproportionating bacterium from a deep-sea hydrothermal vent. Int J Syst Evol Microbiol 62:2565–2571. https://doi.org/10.1099/ijs.0.034397-0

Slobodkin AI, Reysenbach A-L, Slobodkina GB et al (2013) *Dissulfuribacter thermophilus* gen. nov., sp. nov., a thermophilic, autotrophic, sulfur-disproportionating, deeply branching deltaproteobacterium from a deep-sea hydrothermal vent. Int J Syst Evol Microbiol 63:1967–1971. https://doi.org/10.1099/ijs.0.046938-0

Slobodkin A, Slobodkina G, Allioux M et al (2019) Genomic insights into the carbon and energy metabolism of a thermophilic deep-sea bacterium *Deferribacter autotrophicus* revealed new metabolic traits in the phylum Deferribacteres. Genes 10:849. https://doi.org/10.3390/genes10110849

Sonne-Hansen J, Ahring BK (1999) *Thermodesulfobacterium hveragerdense* sp. nov., and *Thermodesulfovibrio islandicus* sp. nov., two thermophilic sulfate reducing bacteria isolated from a Icelandic hot spring. Syst Appl Microbiol 22:559–564. https://doi.org/10.1016/S0723-2020(99)80009-5

Soo RM, Skennerton CT, Sekiguchi Y et al (2014) An expanded genomic representation of the phylum cyanobacteria. Genome Biol Evol 6(5):1031–1045. https://doi.org/10.1093/gbe/evu073

Sorokin DY, Tourova TP, Mussmann M et al (2008) *Dethiobacter alkaliphilus* gen. nov. sp. nov., and *Desulfurivibrio alkaliphilus* gen. nov. sp. nov.: two novel representatives of reductive sulfur cycle from soda lakes. Extremophiles 12:431–439. https://doi.org/10.1007/s00792-008-0148-8

Spasov E, Tsuji JM, Hug LA et al (2020) High functional diversity among *Nitrospira* populations that dominate rotating biological contactor microbial communities in a municipal wastewater treatment plant. ISME J 14(7):1857–1872. https://doi.org/10.1038/s41396-020-0650-2

Stahl DA, De La Torre JR (2012) Physiology and diversity of ammonia-oxidizing archaea. Annu Rev Microbiol 66:83–101. https://doi.org/10.1146/annurev-micro-092611-150128

Stepanov VG, Vaishampayan P, Venkateswaran K et al (2014) Draft genome sequence of *Deinococcus phoenicis*, a novel strain isolated during the Phoenix Lander spacecraft assembly. Genome Announc 2:e00301–e00314. https://doi.org/10.1128/genomeA.00301-14

Stöhr R, Waberski A, Völker H, Tindall BJ, Thomm M (2001) *Hydrogenothermus marinus* gen. nov., sp. nov., a novel thermophilic hydrogen-oxidizing bacterium, recognition of *Calderobacterium hydrogenophilum* as a member of the genus *Hydrogenobacter* and proposal of the reclassification of *Hydrogenobacter acidophilus* as *Hydrogenobaculum acidophilum* gen. nov., comb. nov., in the phylum'Hydrogenobacter/Aquifex'. Int J Syst Evol Microbiol 51: 1853–1862. https://doi.org/10.1099/00207713-51-5-1853

Sudha Rani K, Swamy MV, Seenayya G (1998) Production of ethanol from various pure and natural cellulosic biomass by *clostridium thermocellum* strains SS21 and SS22. Process Biochem 33(4):435–440. https://doi.org/10.1016/S0032-9592(97)00095-2

Sveinsdottir M, Beck SR, Orlygsson J (2009) Ethanol production from monosugars and lignocellulosic biomass by thermophilic bacteria isolated from Icelandic hot springs. Icel Agric Sci 22: 45–58

Swan BK, Martinez-Garcia M, Preston CM et al (2011) Potential for chemolithoautotrophy among ubiquitous bacteria lineages in the dark ocean. Science 333:1296–1300. https://doi.org/10.1126/science.1203690

Takahata Y, Nishijima M, Hoaki T et al (2001) *Thermotoga petrophila* sp. nov. and *Thermotoga naphthophila* sp. nov., two hyperthermophilic bacteria from the Kubiki oil reservoir in Niigata, Japan. Int J Syst Evol Microbiol 51:1901–1909. https://doi.org/10.1099/00207713-51-5-1901

Takai K, Campbell BJ, Cary SC et al (2005) Enzymatic and genetic characterization of carbon and energy metabolisms by Deep-Sea hydrothermal chemolithoautotrophic isolates of Epsilonproteobacteria. Appl Environ Microbiol 71(11):7310–7320. https://doi.org/10.1128/AEM.71.11.7310-7320.2005

Takai K, Kobayashi H, Nealson KH et al (2003) Sulfurihydrogenibium subterraneum gen. nov., sp. nov., from a subsurface hot aquifer. Int J Syst Evol Microbiol 53:823–827. https://doi.org/10.1099/ijs.0.02506-0

Takai K (2003) *Balnearium lithotrophicum* gen. nov., sp. nov., a novel thermophilic, strictly anaerobic, hydrogen-oxidizing chemolithoautotroph isolated from a black smoker chimney in the Suiyo seamount hydrothermal system. Int J Syst Evol Microbiol 53:1947–1954. https://doi.org/10.1099/ijs.0.02773-0

Tamazawa S, Mayumi D, Mochimaru H et al (2017) *Petrothermobacter organivorans* gen. nov., sp. nov., a thermophilic, strictly anaerobic bacterium of the phylum Deferribacteres isolated from a deep subsurface oil reservoir. Int J Syst Evol Microbiol 67:3982–3986. https://doi.org/10.1099/ijsem.0.002234

Tan S, Liu J, Fang Y et al (2019) Insights into ecological role of a new deltaproteobacterial order Candidatus Acidulodesulfobacterales by metagenomics and metatranscriptomics. ISME J 13: 2044–2057. https://doi.org/10.1038/s41396-019-0415-y

Theodorakopoulos N, Bachar D, Christen R et al (2013) Exploration of Deinococcus-Thermus molecular diversity by novel group-specific PCR primers. Microbiology 2:862–872. https://doi.org/10.1002/mbo3.119

Thompson BG, Murray RG (1981) Isolation and characterization of the plasma membrane and the outer membrane of *Deinococcus radiodurans* strain sark. Can J Microbiol 27:729–734. https://doi.org/10.1139/m81-111

Þorsteinsdóttir GV, Blischke A, Sigurbjörnsdóttir A et al (2020) Gas seepage pockmark microbiomes suggest the presence of sedimentary coal seams in the Oxarfjordur graben of NE-Iceland. Can J Microbiol 66:25–38. https://doi.org/10.1139/cjm-2019-0081

Tian B, Hua Y (2010) Carotenoid biosynthesis in extremophilic Deinococcus–Thermus bacteria. Trends Microbiol 8:512–520. https://doi.org/10.1016/j.tim.2010.07.007

Tonon LAC, Moreira APB, Thompson F (2014) The family Erythrobacteraceae. In: Rosenberg E, DeLong EF, Lory S et al (eds) The prokaryotes: alphaproteobacteria and betaproteobacteria. Springer, Berlin, pp 213–235

Trembath-Reichert E, Butterfield DA, Huber JA (2019) Active subseafloor microbial communities from Mariana back-arc venting fluids share metabolic strategies across different thermal niches and taxa. ISME J 13:2264–2279. https://doi.org/10.1038/s41396-019-0431-y

Trüper Hans G, Pfennig N (1992) The family Chlorobiaceae. In: Balows A et al (eds) The Prokaryotes. Springer, New York, pp 3583–3592. https://doi.org/10.1007/978-1-4757-2191-1_33

Urakawa H, Martens-Habbena W, Stahl DA (2010) High abundance of ammonia-oxidizing Archaea in coastal waters, determined using a modified DNA extraction method. Appl Env Microbiol 76:2129–2135. https://doi.org/10.1128/AEM.02692-09

Vanfossen AL, Lewis DL, Nichols JD et al (2008) Polysaccharide degradation and synthesis by extremely thermophilic anaerobes. Ann NY Acad Sci 1125:322–337. https://doi.org/10.1196/annals.1419.017

van Niel CB (1932) On the morphology and physiology of the purple and Green Sulphur bacteria. Arch Mikrobiol 3(1):1–112. https://doi.org/10.1007/BF00454965

Van Ooteghem SA, Beer SK, Yue PC (2002) Hydrogen production by the thermophilic bacterium *Thermotoga neapolitana*. Appl Biochem Biotechnol 98:177–189. https://doi.org/10.1385/ABAB:98-100:1-9:177

Van Ooteghem SA, Jones A, Van Der Lelie D et al (2004) H2 production and carbon utilization by *Thermotoga neapolitana* under anaerobic and microaerobic growth conditions. Biotechnol Lett 26(15):1223–1232. https://doi.org/10.1023/B:BILE.0000036602.75427.88

Vardanyan A, Khachatryan A, Castro L et al (2023) Bioleaching of sulfide minerals by *Leptospirillum ferriphilum* CC from polymetallic mine (Armenia). Fortschr Mineral 13(2): 243. https://doi.org/10.3390/min13020243

Verhamme DT, Prosser JI, Nicol GW (2011) Ammonia concentration determines differential growth of ammonia-oxidising archaea and bacteria in soil microcosms. ISME J 5:1067–1071. https://doi.org/10.1038/ismej.2010.191

Vetriani C, Speck MD, Ellor SV et al (2004) *Thermovibrio ammonificans* sp. nov., a thermophilic, chemolithotrophic, nitrate-ammonifying bacterium from deep-sea hydrothermal vents. Int J Syst Evol Microbiol 54:175–181. https://doi.org/10.1099/ijs.0.02781-0

Vick TJ, Dodsworth JA, Costa KC et al (2010) Microbiology and geochemistry of little Hot Creek, a hot spring environment in the Long Valley caldera. Geobiology 8:140–154. https://doi.org/10.1111/j.1472-4669.2009.00228.x

Viggi CC, Matturro B, Frascadore E et al (2017) Bridging spatially segregated redox zones with a microbial electrochemical snorkel triggers biogeochemical cycles in oil-contaminated river Tyne (UK) sediments. Water Res 127:11–21. https://doi.org/10.1016/j.watres.2017.10.002

Voordeckers, J, Starovoytov V, Vetriani C (2005) *Caminibacter mediatlanticus* sp. nov., a thermophilic, chemolithoautotrophic, nitrate-ammonifying bacterium isolated from a Deep-Sea hydrothermal vent on the mid-Atlantic ridge'. Int J Syst Evol Microbiol 55 (2): 773–779. https://doi.org/10.1099/ijs.0.63430-0

Vrije GJ, Mars AE, Budde MAW et al (2007) Glycolytic pathway and hydrogen yield studies of the extreme thermophile *Caldicellulosiruptor saccharolyticus*. Appl Microbiol Biotechnol 74: 1358–1367. https://doi.org/10.1007/s00253-006-0783-x

Wächtershäuser G (1990) The case for the chemoautotrophic origin of life in an iron-Sulfur world. Orig Life Evol Biosph 20(2):173–176. https://doi.org/10.1007/BF01808279

Wahlund T, Woese CR, Castenholz RW et al (1991) A thermophilic green sulfur bacterium from New Zealand hot springs, *Chlorobium tepidum* sp. nov. Arch Microbiol 156:81–90. https://doi.org/10.1007/BF00290978

Waite DW, Vanwonterghem I, Rinke C et al (2017) Comparative genomic analysis of the class Epsilonproteobacteria and proposed reclassification to Epsilonbacteraeota (phyl. nov.). Frontiers Microbiol 8:682. https://doi.org/10.3389/fmicb.2017.00682

Wang Z, Cao Y, Zhu-Barker X et al (2019) Comammox *Nitrospira* clade B contributes to nitrification in soil. Soil Biol Biochem 135:392–395. https://doi.org/10.1016/j.soilbio.2019.06.004

Wang M, Sha C, Wu J et al (2021) Bacterial community response to petroleum contamination in brackish tidal marsh sediments in the Yangtze River estuary, China. J Environ Sci (China) 99: 160–167. https://doi.org/10.1016/j.jes.2020.06.015

Ward LM, McGlynn SE, Fischer WW (2018) Draft genome sequences of two basal members of the Anaerolineae class of Chloroflexi from a sulfidic hot spring. Genome Announc 6(25):e00570–e00518. https://doi.org/10.1128/genomeA.00570-18

Watson SW, Waterbury JB (1971) Characteristics of two marine nitrite oxidizing bacteria, *Nitrospina gracilis* nov. gen. nov. sp. and *Nitrococcus mobilis* nov. gen. nov. sp. Arch Mikrobiol 77:203–230. https://doi.org/10.1007/BF00454947

Watson SW, Bock E, Valois FW et al (1986) *Nitrospira marina* gen. nov. sp. nov.: a chemolithotrophic nitrite-oxidizing bacterium. Arch Microbiol 144:1–7. doi: https://doi.org/10.1007/BF00454947

Weisburg WG, Giovannoni SJ, Woese CR (1989) The Deinococcus-Thermus phylum and the effect of rRNA composition on phylogenetic tree construction. Syst Appl Microbiol 11:128–134. https://doi.org/10.1016/s0723-2020(89)80051-7

Weon H-Y, Kim B-Y, Schumann P et al (2007) *Deinococcus cellulosilyticus* sp. nov., isolated from air. Int J Syst Evol Microbiol 57(8):1685–1688. https://doi.org/10.1099/ijs.0.64951-0

Wery N, Lesongeur F, Pignet P et al (2001) *Marinitoga camini* gen. nov., sp. nov., a rod-shaped bacterium belonging to the order Thermotogales, isolated from a deep-sea hydrothermal vent. Int J Syst Evol Microbiol 51:495–504. https://doi.org/10.1099/00207713-51-2-495

Wiegel J, Ljungdahl LG (1981) *Thermoanaerobacter ethanolicus* gen. nov., spec. nov., a new, extreme thermophilic, anaerobic bacterium. Arch Microbiol 128:343–348. https://doi.org/10.1007/BF00405910

Wiegel J (2015) Clostridiaceae. In: Trujillo ME, Dedysh S, DeVos P, Hedlund B, Kämpfer P, Rainey FA, Whitman WB (eds) Bergey's manual of systematics of archaea and bacteria. Wiley. https://doi.org/10.1002/9781118960608.fbm00129

Windberger E, Huber R, Trincone A et al (1989) *Thermotoga thermarum* sp. nov. and *Thermotoga neapolitana* occurring in African continental solfataric springs. Arch Microbiol 151:506–512. https://doi.org/10.1007/BF00454866

Wirsen CO, Sievert SM, Cavanaugh CM et al (2002) Characterization of an autotrophic Sulfide-oxidizing marine arcobacter Sp. that produces filamentous Sulfur. Appl Environ Microbiol 68(1):316–325. https://doi.org/10.1128/AEM.68.1.316-325.2002

Woese CR, Kandler O, Wheelis ML (1990) Towards a natural system of organisms: proposal for the domains archaea, bacteria, and Eucarya. Proc Natl Acad Sci U S A 87:4576–4579. https://doi.org/10.1073/pnas.87.12.4576

Woese CR (1987) Bacterial evolution. Microbiol Rev 51:221–271. https://doi.org/10.1128/mr.51.2.221-271.1987

Wolf J, Stark H, Fafenrot K, Albersmeier A, Pham TK, Müller KB et al (2016) A systems biology approach reveals major metabolic changes in the thermoacidophilic archaeon *Sulfolobus solfataricus* in response to the carbon source L-fucose versus D-glucose. Mol Microbiol 102: 882–908. https://doi.org/10.1111/mmi.13498

Wu M, Jiang Y, Liu Y et al (2021) Microbial application of thermophilic *Thermoanaerobacterium* species in lignocellulosic biorefinery. Appl Microbiol Biotechnol 105:5739–5749. https://doi.org/10.1007/s00253-021-11450-4

Xian W-D, Liu Z-T, Li M-M et al (2020) Rhodobacter flagellatus sp. nov., a thermophilic bacterium isolated from a hot spring. Int J Syst Evol Microbiol 70:1541–1546. https://doi.org/10.1099/ijsem.0.003929

Xu L, Wu Y-H, Zhou P et al (2018) Investigation of the thermophilic mechanism in the genus Porphyrobacter by comparative genomic analysis. BMC Genomics 19:385. https://doi.org/10.1186/s12864-018-4789-4

Yabe S, Aiba Y, Sakai Y et al (2010) *Thermosporothrix hazakensis* gen. nov., sp. nov., isolated from compost, description of *Thermosporotrichaceae* fam. nov. within the class Ktedonobacteria Cavaletti et al. 2007 and emended description of the class Ktedonobacteria. Int J Syst Evol Microbiol 60:1794–1801. https://doi.org/10.1099/ijs.0.018069-0

Yabe S, Sakai Y, Yokota A (2016) *Thermosporothrix narukonensis* sp. nov., belonging to the class Ktedonobacteria, isolated from fallen leaves on geothermal soil, and emended description of the genus Thermosporothrix. Int J Syst Evol Microbiol 66:2152–2157. https://doi.org/10.1099/ijsem.0.001004

Yamada T, Imachi H, Ohashi A et al (2007) *Bellilinea caldifistulae* gen. nov., sp. nov. and *Longilinea arvoryzae* gen. nov., sp. nov., strictly anaerobic, filamentous bacteria of the phylum Chloroflexi isolated from methanogenic propionate-degrading consortia. Int J Syst Evol Microbiol 57:2299–2306. https://doi.org/10.1099/ijs.0.65098-0

Yamada T, Sekiguchi Y, Hanada S et al (2006) *Anaerolinea thermolimosa* sp. nov., *Levilinea saccharolytica* gen. nov., sp. nov. and *Leptolinea tardivitalis* gen. nov., sp. nov., novel filamentous anaerobes, and description of the new classes Anaerolineae classis nov. and Caldilineae classis nov in the phylum Chloroflexi. Int J Syst Evol Microbiol 56:1331–1340. https://doi.org/10.1099/ijs.0.64169-0

Yamada T, Sekiguchi Y (2018) Anaerolineaceae. In: Trujillo ME, Dedysh S, DeVos P, Hedlund B, Kämpfer P, Rainey FA, Whitman WB (eds) Bergey's manual of systematics of archaea and bacteria. Wiley, pp 1–5

Yang S, Kataeva I, Wiegel J et al (2010) Classification of *Anaerocellum thermophilum* strain DSM 6725 as *Caldicellulosiruptor bescii* sp. nov. Int J Syst Evol Microbiol 60:2011–2015. https://doi.org/10.1099/ijs.0.017731-0

Yasir M, Qureshi AK, Srinivasan S et al (2020) Domination of filamentous anoxygenic phototrophic bacteria and prediction of metabolic pathways in microbial mats from the Hot Springs of Al Aridhah. Folia Biol (Praha) 66:24–35

Yeh H-Y, Awad A, Rothrock MJ (2021) Detection of *Campylobacter jejuni* diversity by clustered regularly interspaced short palindromic repeats (CRISPR) from an animal farm. Vet Med Sci 7: 2381–2388. https://doi.org/10.1002/vms3.622

Yoo S-H, Weon H-Y, Kim S-J et al (2009) *Deinococcus aerolatus* sp. nov. and *Deinococcus aerophilus* sp. nov., isolated from air samples. Int J Syst Evol Microbiol 60:1191–1195. https://doi.org/10.1099/ijs.0.016030-0

Yuan C-G, Chen X, Jiang Z et al (2017) *Altererythrobacter lauratis* sp. nov. and *Altererythrobacter palmitatis* sp. nov., isolated from a Tibetan hot spring. Antonie Van Leeuwenhoek 110:1077–1086. https://doi.org/10.1007/s10482-017-0882-y

Yvenou S, Allioux M, Slobodkin A et al (2021) Genetic potential of *Dissulfurimicrobium hydrothermale*, an obligate sulfur-disproportionating thermophilic microorganism. Microorganisms 10. https://doi.org/10.3390/microorganisms10010060

Zavarzina DG, Prokofeva MI, Pikhtereva VA et al (2022) *Deferrivibrio essentukiensis* sp. nov., gen. nov., a Representative of Deferrivibrionaceae fam. nov., Isolated from the subsurface aquifer of caucasian mineral drinking waters. Microbiology 91:143–159. https://doi.org/10.1134/S0026261722020114

Zeidan AA, van Niel EWJ (2010) A quantitative analysis of hydrogen production efficiency of the extreme thermophile *Caldicellulosiruptor owensensis* OLT. Int J Hydro Energy 35:1128–1137. https://doi.org/10.1016/j.ijhydene.2009.11.082

Zeikus JG, Dawson MA, Thompson TE et al (1983) Microbial ecology of volcanic Sulphidogenesis: isolation and characterization of *Thermodesulfobacterium commune* gen. nov. and sp. nov. Microbiology 129:1159–1169. https://doi.org/10.1099/00221287-129-4-1159

Zhang L, He J (2012) A novel archaeal phylum: Thaumarchaeota-a review. Wei Sheng wu xue Bao Acta Microbiol Sin 52:411–421. PMID: 22799205

Zhang CL, Pearson A, Li YL, Mills G, Wiegel J (2006) Thermophilic temperature optimum for crenarchaeol synthesis and its implication for archaeal evolution. Appl Environ Microbiol 72: 4419–4422. https://doi.org/10.1128/AEM.00191-06

Zhang C, Phillips APR, Wipfler RL et al (2018) The essential genome of the crenarchaeal model *Sulfolobus islandicus*. Nat Commun 9:4908. https://doi.org/10.1038/s41467-018-07379-4

Zhaxybayeva O, Nesbø CL, Farrell AA et al (2019) Thermotogaceae. In: Trujillo ME, Dedysh S, Devos P, Hedlund B, Kämpfer P, Rainey FA et al (eds) Bergey's manual of systematics of archaea and bacteria. Wiley. https://doi.org/10.1002/9781118960608.fbm00253.pub2

Zhaxybayeva O, Swithers KS, Lapierre P et al (2009) On the chimeric nature, thermophilic origin, and phylogenetic placement of the Thermotogales. Proc Natl Acad Sci U S A 106(14): 5865–5870. https://doi.org/10.1073/pnas.0901260106

Zheng Y, Saitou A, Wang C-M et al (2019) Genome features and secondary metabolites biosynthetic potential of the class Ktedonobacteria. Front Microbiol 10:893. https://doi.org/10.3389/fmicb.2019.00893

Zhen-Li Z, Xin-Qi Z, Nan W et al (2014) *Amphiplicatus metriothermophilus* gen. nov., sp. nov., a thermotolerant alphaproteobacterium isolated from a hot spring. Int J Syst Evol Microbiol 64: 2805–2811. https://doi.org/10.1099/ijs.0.062471-0

Zillig W, Gierl A, Schreiber G et al (1983a) The archaebacterium *Thermofilum pendens* represents, a novel genus of the thermophilic, anaerobic Sulfur respiring Thermoproteales. Syst Appl Microbiol 4:79–87. https://doi.org/10.1016/S0723-2020(83)80035-6

Zillig W, Holz I, Janekovic D et al (1983b) The archaebacterium *Thermococcus celer* represents, a novel genus within the thermophilic branch of the archaebacteria. Syst Appl Microbiol 4:88–94. https://doi.org/10.1016/S0723-2020(83)80036-8

Zillig W, Holz I, Klenk H-P et al (1987) *Pyrococcus woesei*, sp. nov., an ultra-thermophilic marine archaebacterium, representing a novel order, *Thermococcales*. Syst Appl Microbiol 9:62–70. https://doi.org/10.1016/S0723-2020(87)80057-7

Molecular Basis for Thermostability

Sean Michael Scully

Abstract Thermophilic bacteria have always fascinated scientists because of their tolerance and ability to thrive at extreme temperatures where no other living organisms survive. This has led to immense knowledge of the molecular mechanisms these bacteria possess mostly by comparing psychrophilic and mesophilic molecules with molecules of high-temperature origin. This chapter deals with how thermophilic and extremophilic bacteria adapt their cellular membranes, nucleic acids, and proteins to thrive and survive extreme heat.

1 Introduction

Life has adapted to tolerate and thrive at extremes of temperature, pressure, radiation, water activity, osmolality, and pH across all three domains of life. Of particular interest are the adaptations of organisms to thermal extremes, particularly above 50 °C. While high-temperature environments may be less common on earth than in earlier eons, ecosystems that exhibit high temperatures are still very common. It has been suggested, based on the analysis of the highly conserved proteins using sequence alignment and the thermophily index, that the universal common ancestor was likely a thermophile or hyperthermophile (Di Giulio 2001), making the question of the molecular means of thermal adaptation an evolutionary question in addition to important to understanding the molecular basis for thriving at life at high temperatures.

The extremes of temperature, summarized in Fig. 1, are described as psychrophilic, for those organisms living near freezing, whereas those thriving at elevated temperatures are thermophiles. While no strict definition for a thermophile exists, it is generally accepted that thermophilic boundary is in the range of 50–60 °C given the relative rarity of environments on earth with temperatures greater than 60 °C (Brock 1986). It has been suggested that a thermophile should be defined as an

S. M. Scully (✉)
Faculty of Natural Resource Sciences, University of Akureyri, Akureyri, Iceland
e-mail: scully@unak.is

S. M. Scully, J. Orlygsson (eds.), *Thermophilic Anaerobes*, Emerging Issues and Trends, https://doi.org/10.1007/978-3-031-41720-7_3

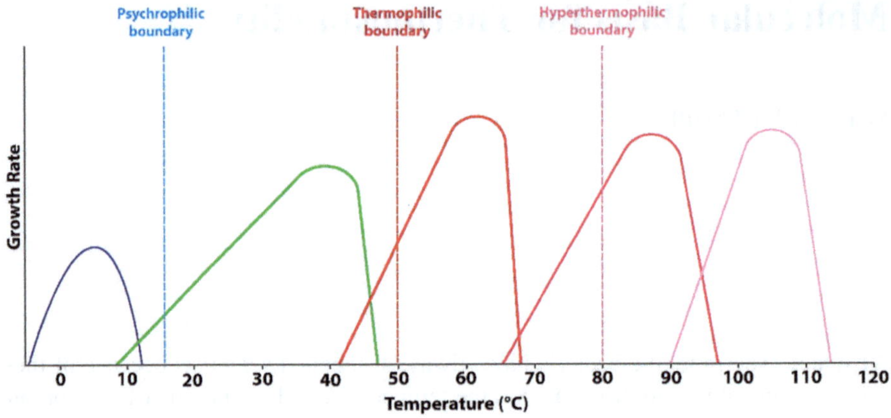

Fig. 1 Distribution of organisms across the terrestrial temperature gradient (Madigan and Martinko 2010)

Table 1 Extremophiles, their environments, and representative genera; taxon in parentheses have some representatives of the phenotype (Hough and Danson 1999; Abe et al. 2004)

Phenotype	Environment	Typical genera
Thermophilic	55–80 °C	*Methanobacterium, Thermoplasma, Thermus, (bacillus)*
Hyperthermophilic	80–113 °C	*Aquifex, Hydrogenobacter, Methanothermus, Pyrococcus, Pyrodictium, Pyrolobus, Sulfolobus, Thermococcus, Thermotoga*
Psychrophilic	−2–20 °C	*Alteromonas, Psychrobacter*
Halophilic	2–5 M NaCl	Haloarcula, *halobacterium, Haloferax, Halorubrum*
Acidophilic	pH < 4	*Acidianus, Desulfurolobus, Sulfolobus, thiobacillus*
Alkaliphilic	pH > 9	*Natronobacterium, Natronococcus, (bacillus)*
Piezophiles (psychrophilic)	30–94 MPa, 2–15 °C	*Colwellia hadiensis, Moritella japonica, photobacterium profundum (Shewanella)*
Piezophiles (thermophilic)	20–45 MPa 80–103 °C	*Methanocaldococcus jannaschii, Palaeococcus ferrophilus, Pyrococcus abyssi, Thermococcus*

organism capable of living at or near that upper-temperature boundary within its taxonomic group (Brock 1986). Hyperthermophiles are organisms that thrive at temperatures above 80 °C. At present, nearly all described hyperthermophiles are archaea (Vieille and Zeikus 1996). Organisms isolated from permafrost have been reported to live at temperatures as low as −20 °C, whereas others isolated from hydrothermal vents and hot springs can survive at 121 °C (Barton and Northup 2011).

To date, many of the Archaea that have been described in the scientific literature are extremophiles (Whitaker et al. 2003) although there are many examples of bacteria that have adapted to extreme conditions (Table 1). While there are examples of eukaryotes and cyanobacteria to be tolerant of mild thermal extremes, they are less common than examples of thermophily among archaea and bacteria. That said, a few interesting examples of eukaryotes that are classified as thermophilic include

Myceliophthora thermophila, which can grow at temperatures up to 62 °C (Apnis 1963), the fungi *Curvularia protuberata* (Nelson and Hodges 1965), and some human-infecting amoebas, such as the so-called brain-eating *Naegleria fowleri* (Huizinga and McLaughlin 1990).

Thermotolerant microbes are a group of microorganisms that can survive high temperatures. These microbes can be found at various places and grow at temperatures of or above 45 °C. These microbes are of great interest for microbiologists because of their potential biotechnological applications. These microbes have thermotolerant proteins to withstand high temperatures and can be used in a wide range of research and biotechnological applications ranging from the classical example of DNA application to organic synthesis (Zeikus 1979; Yamini et al. 2022). Thermotolerant bacteria have been observed in a wide variety of environments, e.g., saltwater, soil, grains, irrigation water, and fecal matter of animals and humans (Kumar et al. 2007; Coorevits et al. 2008; Kurapova et al. 2012; Paruch and Mæhlum 2012; Pachepsky et al. 2016; Sandona et al. 2019). Thermophiles and hyperthermophiles have been isolated from hot springs, deep-sea hydrothermal vents, oil reservoirs, and more (Chaban et al. 2006).

Typically, a 10 °C increase in temperature (within the enzyme's tolerance range) results in the doubling of enzyme activity as defined by the 10° temperature quotient (Q_{10}) as shown in the equation below. Conversely, lowering the temperature by 10° results in a two- to fourfold decrease in enzyme activity (Feller and Gerday 2003).

$$Q_{10} = \frac{U_{\mathrm{T}} + U_{\mathrm{T}+10\,^{\circ}\mathrm{C}}}{U_{\mathrm{T}}}$$

While enzyme-catalyzed reactions may run at increased rates at higher temperatures, the generation times of thermophiles are lower than their mesophilic counterparts due to the need to repair proteins damaged by elevated temperatures (i.e., thermal denaturation and deamination).

The upper thermal limit for life has drawn some attention. To date, the two most heat-resistant organisms are *Pyrolobus fumarii* and "Strain 121," both of which are archeons. "Strain 121" ("*Geogemma barossii*") garners its name as it can grow at 121 °C, which is a noteworthy achievement as this is the most common sterilization temperature used by many autoclaves although this strain ceases growing at 130 °C still it retains (Kashefi and Lovley 2003). *Pyrolobus fumarii* has a maximum growth temperature of 113 °C. Ultimately, the upper temperature boundary may be dependent upon the thermostability of the ribosome (Atlas and Bartha 1998, p. 295).

As the molecular basis of life is the same for all organisms as yet described, understanding the subtle changes to the structure of these molecules is important to understand how organisms function at elevated temperatures. As elevated temperatures pose challenges for major biomolecules, namely denaturation, hydrolysis, and oxidative reactions, it is important to understand the molar basis of thermophily. The following overview will focus on the main biological macromolecules (cell membranes and related lipids, proteins, and nucleic acids) in thermophilic microorganisms and describe the major trends that have been observed in adapting these molecules to function at high temperatures.

2 Adaptations to the Cellular Membrane

The cell membrane provides a number of functions essential for metabolism, and the challenges posed by high temperature are not trivial as the membrane must be kept in a liquid-crystalline state to maintain its core functions of separating the internal and external environments and regulating the transport of materials across the membrane. A delicate balance is needed to maintain enough structural integrity to prevent the membrane from becoming "leaky" while remaining fluid enough for the transport of molecules across the bilayer. Foremost among its functions, the membrane separates the internal volume of the cell from the external environment and in the process serves as a selective barrier requiring molecules that pass through to either be non-polar in nature or transported across the membrane via specific transport systems as is the case for many polar or ionic solutes (i.e., ion channels and ABC transporters). The cell membrane is also critical for energy transduction via proton motive force and the maintenance of other ion gradients associated with bioenergetics. Additionally, the membrane has several roles in cell signaling. Thus, understanding the structural basis for the thermostability of the lipid bilayer is critical.

The properties of the cellular membrane are directly linked to the nature of the chemical and electrostatic properties of its constituent lipids as exhibited in Fig. 2. The lipid bilayers of bacteria are typically composed of fatty acid esters of glycerol (typical *sn*-glycerol-3-phosphate moieties) (Boucher 2007) although other "backbone" molecules have been reported (such as sphingosine). Ester linkages can be hydrolyzed under extremely acidic and basic conditions, a problem exacerbated at elevated temperatures. Archaea, however, compose their cellular membranes of a mono-layer of di- and tetra-isoprenoid ethers linked to a *sn*-glycerol-1-phosphate backbone (Boucher 2007). By comparison, these isoprenoid monolayers are less prone to hydrolysis under the influence of increased temperatures (Wharton 2002). Additionally, ethers are more stable to extremes of pH and temperature (Wuts and Greene 2007) and likely explain why hyperthermophiles are almost exclusively

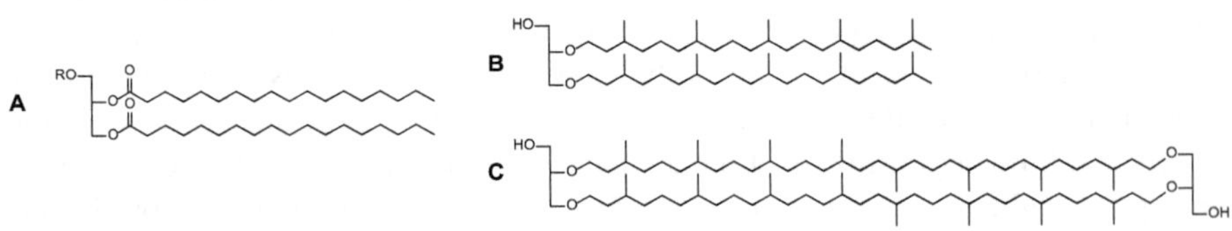

Fig. 2 Structures of common lipids within the cellular membrane in bacteria and archaea where the R group is polar group. (**a**) a diacylglycerol (DAG) containing acyl groups consisting of 18 carbon atoms; (**b**) a diether lipid (archaeol) composed of C20 side chains; (**c**) a tetraether lipid (caldarchaeol) containing isoprenoid units linked between two glycerol molecules

archaea. Indeed, archaeal membranes maintain a liquid-crystalline state over a much wider temperature range than their fatty acyl ester counterparts (De Rosa et al. 1986).

Lipids of bacterial (and eukaryotic) cell membranes are commonly mono-, di-, or tri-acyl glycerides containing C10–C30 fatty acids although C12–C18 is typical. In contrast, the lipids of the archaeal cell membrane are composed of either diether lipids (archaeol) or tetraether lipids (caldarchaeol). Some archaea have also been found to contain cyclic lipids (including macrocyclic ethers and pentacyclic rings), which may be another layer of adaptation for survival at higher temperatures (Boucher 2007).

The structure of the tetraether lipids spans the entire cell membrane, which likely confers additional mechanical and thermal stability to an archaea's bilayer while also being more resistant to oxidation (Benvegnu et al. 2004). The structure and biosynthesis of archaeal membrane lipids are a topic of intense curiosity, and interested readers are directed to reviews on the topic herein (Benvegnu et al. 2004; Jain et al. 2014).

Generally, there are four strategies for regulating membrane fluidity in microorganisms: modulating chain length, branching, and degree of saturation, or the addition of plasticizers (e.g., sterols such as cholesterol) to the membrane environment. The length of the acyl group and its degree of branching and unsaturation directly impact its melting point (T_m) and glass transition temperature (T_g) and thus the fluidity of the membrane at a given temperature. Similarly, the addition of branching alters the fluidity of membrane lipids with the addition of a methyl group generally decreasing the melting point of a lipid as evidenced by the shifts of the iso-fatty acids. These general trends can be observed in Fig. 3. The position of the branch point is also critical; the addition of alkyl groups toward the end of a lipid is typically observed.

The introduction (or removal) of double bonds is another means for organisms to regulate membrane fluidity. Higher degrees of unsaturation can maintain cellular membrane fluidity at lower temperatures as compared to highly saturated lipids (Boucher 2007), while saturated lipids will maintain more membrane integrity at higher temperatures. As an example, numerous 20-carbon fatty acids with various degrees of unsaturation can be found in nature (Fig. 4), which highlights the importance of both the degree of unsaturation and position of the double bond on its fluidity; arachidic acid has a melting point of 75 °C but the addition of a single double bond decreases the melting point to 23 °C (in the case of eicosenoic acid/11-eicosenoic acid) or 13 °C in the case of paullinic acid/13-eicosenoic acid).

It should be noted that unsaturated fatty acids are prone to oxidation, particularly at higher temperatures. Not surprisingly, thermophilic bacteria have lipid bilayers that are dominated by saturated fatty acids, which are less prone to oxidation than more fluid unsaturated fatty acids (Wharton 2002). However, as saturated fatty acids have higher melting points than unsaturated fatty acids, thermophiles may experience a loss in membrane fluidity, limiting the exchange of materials across the bilayer, and even solidification if the environmental temperature drops low enough (Wharton 2002).

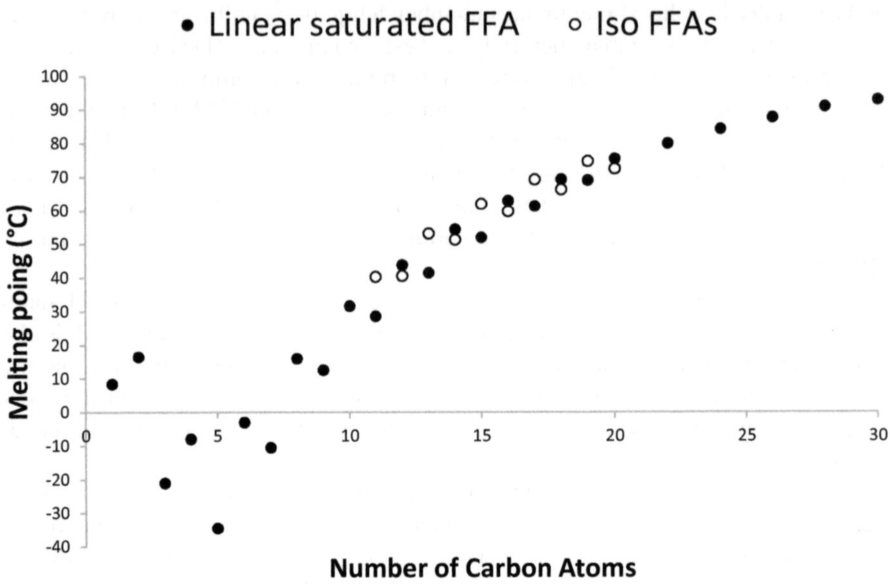

Fig. 3 Melting point of linear and iso-free fatty acids data from Knothe and Dunn (2009)

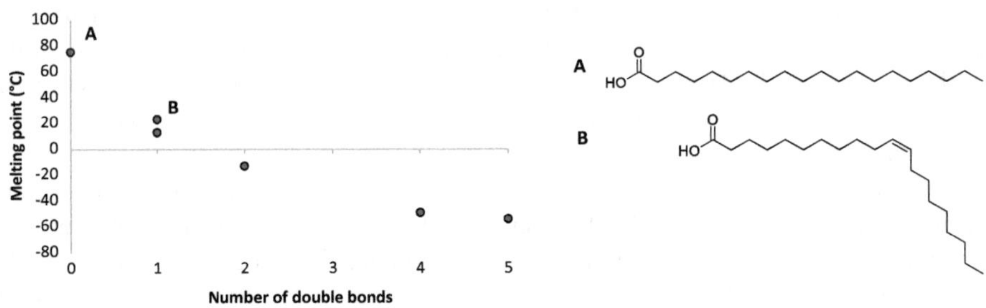

Fig. 4 Influence of unsaturation degree and position on the melting point of C20 fatty acids; note the structure of the arachidic acid (**a**) and 11-eicosenoic acid (**b**)

The final method of membrane stabilization is remodeling via the inclusion of cyclic lipids such as sterols in the case of eukaryotes or hopanoids in bacteria (Fig. 5). Sterols, such as cholesterol and ergosterol, found in vertebrates and plants, respectively, consist of a skeleton of four fused rings. Membrane sterols serve to regulate the elasticity or stiffness of the lipid bilayer. Additionally, sterols contribute to the formation (or suppression) of localized three-dimensional protrusions from the membrane bilayer as a means of regulating membrane stress (Kawakami et al. 2017). The addition of sterols to membranes is a phenomenon largely limited to eukaryotes and is rarely found in bacteria. One of the most notable groups of bacteria to use cholesterol are Mollicutes such as various mycoplasma (Dahl 1993) and bacteria within the genera of *Borrelia* and *Helicobacter* (Huang and London 2016). While

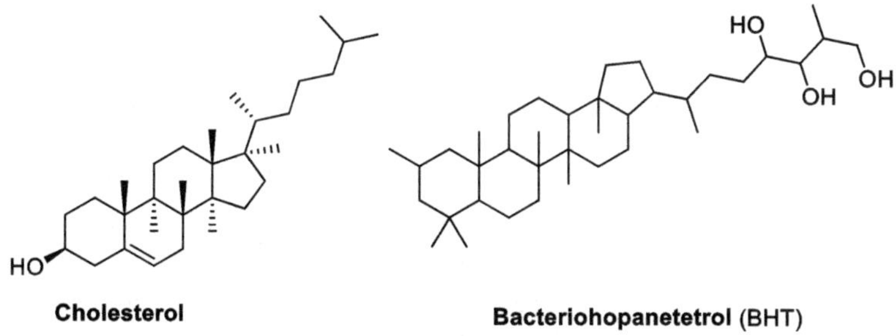

Fig. 5 Sterol and sterol-like structures used for the regulation of membrane fluidity

sterol biosynthesis and utilization are thought to be a feature of eukaryotes, genes associated with sterol synthesis have been found among a large number of bacterial genomes although the mechanism for modification is generally absent (Wei et al. 2016).

Similar to sterols, hopanoids are a class of planar pentacyclic triterpenoids, which contain a diverse range of side chains and are widely produced by Gram-negative and Gram-positive bacteria and cyanobacteria. Hopanoids function as direct analogs to sterols like cholesterol as they interact with the lipids of the outer membrane (Sáenz et al. 2015). The sheer abundance of hopanoids has even led them to be found as a part of the geological record (Ourisson and Albrecht 1992; Ourisson and Rohmer 1992) suggesting that these "molecular fossils" have long been a component of bacteria's strategy for regulating membrane fluidity. Unlike sterols, hopanoids lack a hydroxyl group on ring A but often contain multiple hydroxyl groups associated with the branched aliphatic tail attached to ring E. As a result, this gives them a polarity that is the inverse of their sterol counterparts (Dufourc 2008).

Alicyclobacillus (formerly *Bacillus*) *acidocaldarius*, an acidophilic thermophile isolated from Yellowstone National Park in the 1970s (Darland and Brock 1971), includes hopanoids with extended side chains to a high degree ($\leq$16% of the membrane lipid content) (Langworthy and Mayberry 1976). The production of hopanoids is thought to stabilize the membrane of *Burkholderia cenocepacia*, a bacteria capable of thriving in a wide range of environmental conditions including within the hostile environment of macrophages (Schmerk et al. 2011). Interestingly, proteins associated with the lipid bilayer have also been implicated in regulating membrane integrity in bacteria that otherwise lack regulatory lipids such as hopanoids (Kaiser 2011). The biosynthetic pathways leading to hopanoid formation and more detailed structural characteristics have been carefully detailed elsewhere (Rezanka et al. 2010; Dufourc 2008; Kannenberg and Poralla 1999; Belin et al. 2018).

3 Adaptations to Nucleic Acids

Nucleic acids serve a variety of roles within the cell but are probably best known for their role in the flow of genetic information. Deoxyribonucleic acid (DNA) is remarkably stable with a purported half-life on the order of 500 years (Willerslev and Cooper 2005; Kaplan 2012). As is the case with other major biomolecules, increased temperatures pose a number of challenges to the components of nucleic acids, namely denaturation, hydrolysis, deamination, the excision of purine, and pyrimidine bases via depurination and depyrimidination, respectively, as well as oxidation. These challenges are non-trivial under ideal circumstances and become even more deleterious in the conditions in which many extremophiles thrive. While some degree of denaturation is required for basic functions such as DNA replication, the genetic material must maintain sufficient integrity to regulate gene transcription.

It is well established that DNA molecules enjoy greater stability in aqueous solution than RNAs (Allentoft et al. 2012; Dabney et al. 2013). While DNA is more inherently resistant to hydrolysis than RNA by virtue of the fact that it lacks a 2′ hydroxyl group, both of these polynucleic acids suffer from increased rates of hydrolysis at higher temperatures. Furthermore, the hydrolysis of polynucleic acids is facilitated by divalent metal ions (Lindahl 1993). Additionally, the base pairs themselves can undergo oxidation reactions ultimately resulting in mutations (Burrows and Muller 1998). Unsurprisingly, elevated temperatures cause more oxidative damage in part due to the generation of more reactive oxygen species, which are ultimately responsible for the oxidation of bases. As an example, guanine can be oxidized to 8-oxoguanine (Bruskov et al. 2002).

Grosjean and Oshima (2007) laid out seven basic adaptations that allow thermophiles and hyperthermophiles to adapt their nucleic acids, which can be broadly clustered into three broad categories: intrinsic properties, extrinsic properties, and detect and repair mechanisms. Intrinsic changes to nucleic acids that permit higher thermostability include having a relatively high G + C content in the RNA (as opposed to the genomic DNA) and stabilizing nucleic acids via covalent modification (such as methylation). Some examples of extrinsic changes to nucleic acids include the stabilization of genetic materials via small ligand binding and forming compact tertiary structures, which help exclude water from attacking the backbone of polynucleic acids. Complex DNA repair mechanisms also play a key role in fostering DNA stability at high temperatures and controlling the damage to RNAs at high temperature can be mitigated by increasing the rate of RNA turnover, thus quickly removing damaged RNAs.

The deamination of cytosine results in uracil, while 5-methylcytosine results in thymine (Fig. 6). Similar nitrogenous bases undergo similar reactions, such as guanine, which yields xanthine, while adenine produces hypoxanthine. These alterations can alter the hydrogen bonding patterns of DNA and cause errors in replication and transcription (Wang and Hu 2016). Depyrimidination involves the loss of a pyrimidine base from the ribose or deoxyribose. Interestingly, depurination-induced

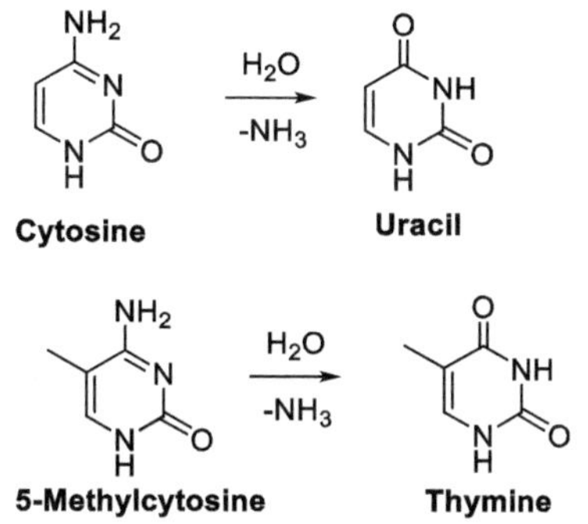

Fig. 6 Examples of deamination reactions occurring among nitrogenous bases found in nucleic acids

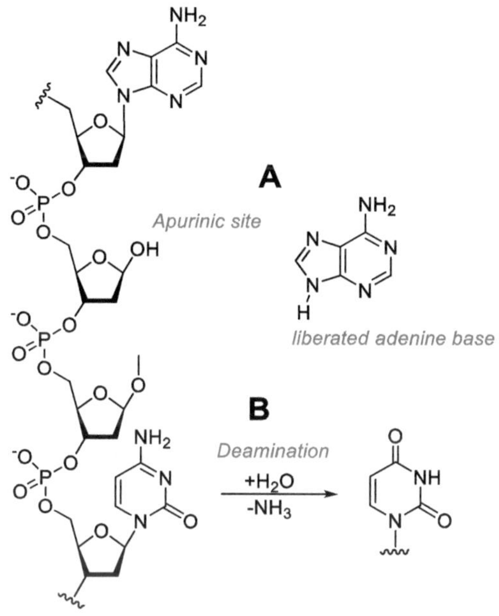

Fig. 7 Two common means of DNA damage. (**a**) the products of depurination (an apurinic site and a liberated adenine base); (**b**) deamination of a cytosine resulting in the loss of ammonia and conversion onto a uracil residue

deamination is speculated to have played an important evolutionary role (Lewis et al. 2016; Fryxell and Zuckerkandl 2000; Ehrlich et al. 1986).

The conversion of 5-methylcytosine into thiamine by the loss of an amine is among the most common mutations to occur and is often corrected by thymine–DNA–glycosylase. Similarly, guanine can be converted into xanthine and adenine can be converted into hypoxanthine. Given the common occurrence of the conversion of adenine and thymine, it might be expected that thermophiles favor G–C-rich sequences.

One of the most common forms of DNA damage is the spontaneous hydrolysis of purine attached to the C1' position of deoxyribose via a *N*-glycosidic linkage, while RNA is much less susceptible to the bases (Fig. 7). Purines are excellent leaving groups and the rate of this reaction is not trivial with depurinations occurring on the

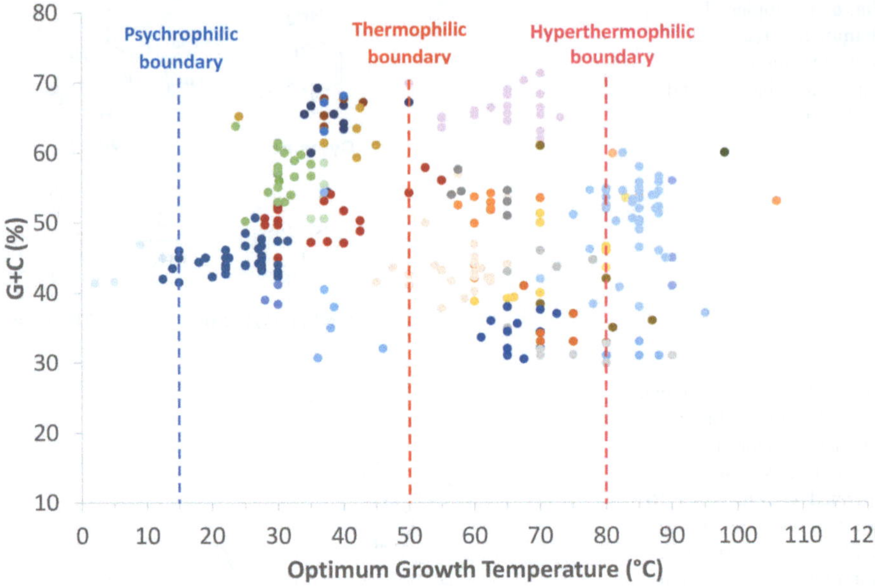

Fig. 8 The G + C content of selected bacteria and archaea as a function of optimum growth temperature

order of 10^3 per cell per hour. The rates of depurination are increased by increasing temperature and low pH conditions making depurination a particular challenge for thermophilic organisms. Similarly, depyrimidination involves the liberation of a pyrimidine base (C or T), resulting in a free C1' hydroxyl group although this occurs at lower rates than depurination. As a result, the creation of a free hydroxyl group can lead to hydrolysis of the phosphodiester backbone (Lindahl 1993; Dabney et al. 2013). The loss of purine groups can be avoided by complexation with polycations, such as chitosan and spermine, by altering the pK_a values of the amino groups and thus inhibiting protonation (An et al. 2017).

The primary composition of polynucleic acids directly influences their ability to form secondary structures and their melting points; the base pairing of an A and T contributes 2 °C to the melting point, while the three hydrogen bonds of a G–C pairing contribute 3 °C. It can be surmised that thermophiles typically have higher proportions of guanine and cytosine in their genomes as a response to their higher environmental temperatures thus to contributing the thermostability of the genetic material (Atlas and Bartha 1998). Indeed, a survey of the G + C content of microorganisms reveals a general trend that supports this (Fig. 8). Paradoxically, hyperthermophiles have lower G:C ratios typically less than 40 mol % (Atlas and Bartha 1998, p. 295).

Under lower temperatures, DNA–DNA interactions are strengthened and unfavorable secondary structures in RNA (i.e., loops); this interferes with transcription and translation processes, respectively. Some cold-adapted nucleic acid-binding molecules, such as RNA helicase, have been observed in psychrophiles growing near zero, and other psychrophiles incorporate post-transcriptional modifications to improve RNA flexibility (Feller and Gerday 2003).

While it seems that the G + C content trends upward as the thermophilic boundary is approached, this trend seems to disperse at greater temperatures. Recalling one of the "rules" of nucleic acid adaptations (Grosjean and Oshima 2007), namely that high G + C content is favored in RNAs, this stabilization or reversal of the G + C content trend begins to make sense as a high genomic G + C content corresponds to a high A + T content in mRNA transcripts.

4 Adaptations to Proteins

The other major biomolecule that should be considered in the context of adaptation to high temperatures are proteins, which serve a diverse range of functions and need to be correctly folded to maintain their structure and functionality. As with other major biomolecules, proteins are susceptible to denaturation at low and high temperatures and a number of chemical reactions, such as deamination, which are exacerbated at higher temperatures. An enzyme is considered thermostable if it meets one of two criteria: a highly defined transition temperature (T_m), typically above the thermophilic boundary (55 °C), or a long half-life at a selected temperature (Turner et al. 2007).

The three-dimensional conformation of proteins, and thus their functionality, can be altered by changes to the physical and chemical environment; variations in temperature, pH, ionic strength, or pressure can give rise to changes in the folded conformation. Such changes are often reversible although covalent modifications of the protein primary structure may lead to irreversible changes that can alter protein function (Creighton 1993). Proteins functioning in environments where such conditions are extreme have adapted to maintain structural resilience.

The ability of the proteins of thermophilic and hyperthermophilic bacteria and archaea to withstand the high temperatures in which they thrive has been of interest to physiologists for many decades. The free energy or stabilization (ΔG_{stab}) of most mesozyme is between 5 and 15 kcal/mol (Vieille and Zeikus 1996). Based upon several comparisons of enzymes from thermophiles to their mesophile counterparts, the ΔG_{stab} of thermophilic proteins is typically only 5–20 kcal/mole higher at 25 °C (Li et al. 2005). This means that relatively few changes are needed to an amino acid's primary sequence to increase an enzymes' thermostability. Thermozymes are highly analogous to their mesophilic counterparts in that the primary sequence of amino acids is 40–85% similar, their three-dimensional structures are superposable, and the catalytic mechanism by which they operate is conserved (Vieille and Zeikus 1996). Analysis of proteins from thermophiles and their mesophilic homologs has revealed that there is no general strategy for increasing thermal stability (Sadeghi et al. 2006). However, a number of features are generally attributed to increases in thermostability; these features can be explained in terms of changes to primary, secondary, and

Table 2 Features common in thermostable enzymes [modified from Turner et al. (2007)]

Feature for Internal Stabilization	Contributing Factor
Helix stabilization	– Low frequency of branched amino acid residues
	– Proline residues at ends
Stabilizing interactions	– Disulfide bridges
	– Hydrogen bonds
	– Hydrophobic interactions
	– Aromatic interactions (π–π stacking)
	– Ion-pair networks
	– Loose end docking
Interactions between domains	– Oligomer formation (ion-pair networks)
Dense packing	– Hydrophobic cores
	– Filled cavities
Stable surface-exposed residues	– Decreased instances of residues prone to deamination (Gln, Aln)
	– Decreased instances of residues prone to oxidative degradation (Cys, met)

tertiary structure, thermodynamic properties, and interactions among functional groups within a protein. Turner et al. (2007) summarized some common features of thermostable enzymes as presented in Table 2.

Maintaining protein functionality at high temperatures seems to involve several factors like higher core hydrophobicity (Schumann et al. 1993), increased packing density (Vetriani et al. 1998; Russell et al. 1997), additional network of hydrogen bonds (Jaenicke and Böhm 1998), decreased length of surface loops (Thompson and Eisenberg 1999), stabilization by heat stable chaperones (Haslbeck et al. 2005), an increase in disulfide bond formation (Beeby et al. 2005), and general shortening of length (Tekaia et al. 2002).

One of the most important trends observed among thermophilic enzymes is that they are often more rigid than their mesophilic counterparts owing to the more efficient packing of their hydrophobic cores (Vieille and Zeikus 1996; Li et al. 2005). Proteins that have adapted to function at high temperature derive their stability from alterations of these interactions relative to their mesophile counterparts (Li et al. 2005). From a thermodynamic standpoint, a buried methylene group can contribute 1.3 kcal/mol to the stability of an enzyme (Vieille and Zeikus 1996), which is an important strategy to counter the increased mobility of proteins at higher temperature. Similarly, closely associated aromatic rings can contribute approximately 1 kcal/mol of stability through pi-pi stacking (Vieille and Zeikus 1996). Alterations to protein structure can also be accomplished through the accumulation of small changes to stability. Hydrogen bonding stabilizes internal peptide chains and coordinates water from the aqueous medium (Li et al. 2005). The stabilization energy of a single buried hydrogen bond can contribute a modest 0.6 kcal/mol with multiple hydrogen bonds having an additive effect on stability.

Like other major biomolecules, proteins exposed to high temperature can undergo several chemical reactions with consequences. For example, at extremes of temperature and pH, asparagine and glutamine undergo deamination to aspartate and glutamate residues, respectively (Creighton 1993). Thiols, such as cysteine, are easily oxidized in air, particularly in the presence of divalent metal cations (Creighton 1993). Disulfide bonds are readily heat-liable and undergo reduction. As such, the enzymes of thermophiles often minimize these residues, particularly those that are exposed to the outer surfaces of the protein (Turner et al. 2007).

5 Future Directions

The molecular basis of thermal adaptation gives us critical insights into the molecular evolution of major biomolecules. Thermal adaptations of organisms have many applications in bioprocessing as explored in subsequent chapters. Additionally, understanding the molecular basis of the thermostability of proteins has obvious applications in engineering proteins to be more (or less) thermal stable through genetic modification. Furthermore, modifying membrane fluidity can potentially be used to fine-tune membrane properties. One potential application of using lipid adaptations is through exploiting the role of hopanoids in conferring drug resistance to bacterial strains, which may be an interesting route to track drug resistance (Sáenz et al. 2015).

It should be noted that many of the trends used to draw conclusions about the nature of thermostability predominately come from our understanding of mesophiles and "true" thermophiles (i.e., those with temperature optima above 65 °C). An understanding of moderately thermophilic bacteria (those growing optimally under 65 °C) is underrepresented, and more careful scrutiny of these taxa may offer additional insights into thermophily. Another notable knowledge gap is of species that thrive at the intersection of multiple extremes such as high pressure and salinity.

6 Conclusions

Microorganisms living at thermal extremes have adopted a number of changes to their major biomolecules in order to adapt to their environments. The lipids of cell membranes of thermophiles are typically more saturated as compared to their lower-temperature counterparts. Organisms at high temperatures also adapt their nucleic acids by altering their composition, methylating residues, and RNA turnover. The proteins of thermophiles often have reduced unpaired ionic residues and more condensed hydrophobic cores, which confer greater rigidity.

References

Abe F, Kato C, Horikoshi K (2004) Extremophiles: pressure. In: Bull AT (ed) Microbial diversity and bioprospecting. ASM Press, Washington, pp 154–159

Allentoft ME, Collins M, Harker D et al (2012) The half-life of DNA in bone: measuring decay kinetics in 158 dated fossils. Proc R Soc B 279(1748):1–12. https://doi.org/10.1098/rspb.2012.1745

An R, Dong P, Komiyama M et al (2017) Inhibition of nonenzymatic depurination of nucleic acids by polycations. FEBS Open Bio 7(11):1707–1714. https://doi.org/10.1002/2211-5463.12308

Apnis AE (1963) Occurrence of thermophilic microfungi in certain alluvid soils near Nottingham. Nova Hedwigia 5:57–78

Atlas RM, Bartha R (1998) Microbial ecology, 4th edn. Benjamin Cummings, New York

Barton LL, Northup DE (2011) Microbial ecology. Wiley-Blackwell, Hoboken

Beeby M, O'Connor BD, Ryttersgaard C et al (2005) The genomics of disulfide bonding and protein stabilization in thermophiles. PLoS Biol 3(9):1549–1558. https://doi.org/10.1371/journal.pbio.0030309

Belin BJ, Busset N, Giraud E et al (2018) Hopanoid lipids: from membranes to plant-bacteria interactions. Nat Rev Microbiol 16(5):304–315. https://doi.org/10.1038/nrmicro.2017.173

Benvegnu T, Brard M, Plusquellec D (2004) Archaeabacteria bipolar lipid analogues: structure, synthesis and lyotropic properties. Curr Opin Colloid Interface Sci 8(6):469–479. https://doi.org/10.1016/j.cocis.2004.01.005

Boucher Y (2007) Lipids: biosynthesis, function, and evolution. In: Cavicchioli R (ed) Archaea: molecular and cellular Biology. ASM Press, Washington, D.C., pp 341–354

Brock TD (1986) Introduction: an overview of the thermophiles. In: Brock TD (ed) Thermophiles: general, molecular, and applied microbiology. Wiley, New York, pp 1–16

Bruskov VI, Malakhova LV, Masalimov ZK et al (2002) Heat-induced formation of reactive oxygen species and 8-oxoguanine, a biomarker of damage to DNA. Nucleic Acids Res 30(6):1354–1363. https://doi.org/10.1093/nar/30.6.1354

Burrows CJ, Muller JG (1998) Oxidative nucleobase modifications leading to strand scission. Chem Rev 98(3):1109–1152. https://doi.org/10.1021/cr960421s

Chaban B, Ng SYM, Jarrell KF (2006) Archaeal habitats – from the extreme to the ordinary. Can J Microbiol 52(2):73–116. https://doi.org/10.1139/w05-147

Coorevits A, De Jonghe V, Vandroemme J et al (2008) Comparative analysis of the diversity of aerobic spore-forming bacteria in raw milk from organic and conventional dairy farms. Syst Appl Microbiol 31(2):126–140. https://doi.org/10.1016/j.syapm.2008.03.002

Creighton TE (1993) Proteins, 2nd edn. W.H. Freeman and Company, New York

Dabney J, Meyer M, Pääbo S (2013) Ancient DNA damage. Cold Spring Harb Perspect Biol 5(7):1–7. https://doi.org/10.1101/cshperspect.a012567

Dahl J (1993) The role of cholesterol in mycoplasma membranes. In: Rottem S, Kahane I (eds) Mycoplasma cell membranes. Springer, Berlin, pp 167–188. https://doi.org/10.1007/978-1-4615-2924-8_5

Darland G, Brock TD (1971) *Bacillus acidocaldarius* sp. nov., an acidophilic thermophilic spore-forming bacterium. Microbiology 67(1):9–15. doi:https://doi.org/10.1099/00221287-67-1-9

De Rosa M, Gambacorta A, Gliozzi A (1986) Structure, biosynthesis, and physicochemical properties of archaebacterial lipids. Microbiol Rev 50(1):70–80. https://doi.org/10.1128/mr.50.1.70-80.1986

Di Giulio M (2001) The universal ancestor was a thermophile or a hyperthermophile. Gene 281(1–2):11–17

Dufourc EJ (2008) Sterols and membrane dynamics. J Chem Biol 1(1–4):63–77. https://doi.org/10.1007/s12154-008-0010-6

Ehrlich M, Norris KF, Wang RY et al (1986) DNA cytosine methylation and heat-induced deamination. Biosci Rep 6(4):387–393. https://doi.org/10.1007/BF01116426

Feller G, Gerday C (2003) Psychrophilic enzymes: hot topics in cold adaptation. Nat Rev Microbiol 1(3):200–208. https://doi.org/10.1038/nrmicro773

Fryxell KJ, Zuckerkandl E (2000) Cytosine deamination plays a primary role in the evolution of mammalian isochores. Mol Biol Evol 17(9):1371–1383. https://doi.org/10.1093/oxfordjournals.molbev.a026420

Grosjean H, Oshima T (2007) How nucleic acids cope with high temperature. In: Physiology and biochemistry of extremophiles. ASM Press, Washington, DC, pp 39–56. https://doi.org/10.1128/9781555815813.ch4

Haslbeck M, Franzmann T, Weinfurtner D et al (2005) Some like it hot: the structure and function of small heat-shock proteins. Nat Struct Mol Biol 12(10):842–846. https://doi.org/10.1038/nsmb993

Hough DW, Danson MJ (1999) Extremozymes. Curr Opin Chem Biol 3:39–46

Huang Z, London E (2016) Cholesterol lipids and cholesterol-containing lipid rafts in bacteria. Chem Phys Lipids 199(1):11–16. https://doi.org/10.1016/j.chemphyslip.2016.03.002

Huizinga HW, McLaughlin G (1990) Thermal ecology of *Naegleria fowleri* from a power plant cooling reservoir. Appl Environ Microbiol 56(7):2200–2205. https://doi.org/10.1128/aem.56.7.2200-2205.1990

Jaenicke R, Böhm G (1998) The stability of proteins in extreme environments. Curr Opin Struct Biol 8(6):738–748

Jain S, Caforio A, AJM D (2014) Biosynthesis of archaeal membrane ether lipids. Front Microbiol 5:1–16. https://doi.org/10.3389/fmicb.2014.00641

Kaiser HJ (2011) Molecular convergence of bacterial and eukaryotic surface order. J Biol Chem 286:40631–40637. https://doi.org/10.1074/jbc.M111.276444

Kannenberg EL, Poralla K (1999) Hopanoid biosynthesis and function in bacteria. Naturwissenschaften 86(4):168–176. https://doi.org/10.1007/s001140050592

Kaplan M (2012) DNA has a 521-year half-life. Nature 1–2. https://doi.org/10.1038/nature.2012.11555

Kashefi K, Lovley DR (2003) Extending the upper temperature limit for life. Science 301(5635): 934–934

Kawakami LM, Yoon BK, Jackman JA et al (2017) Understanding how sterols regulate membrane remodeling in supported lipid bilayers. Langmuir 33(51):14756–14765. https://doi.org/10.1021/acs.langmuir.7b03236

Knothe G, Dunn RO (2009) A comprehensive evaluation of the melting points of fatty acids and esters determined by differential scanning calorimetry. J Am Oil Chem Soc 86(9):843–856. https://doi.org/10.1007/s11746-009-1423-2

Kumar M, Leon V, Materano AD et al (2007) A halotolerant and thermotolerant *Bacillus* sp. degrades hydrocarbons and produces tensio-active emulsifying agent. World J Microbiol Biotechnol 23(2):211–220. https://doi.org/10.1007/s11274-006-9215-4

Kurapova AI, Zenova GM, Sudnitsyn II et al (2012) Thermotolerant and thermophilic actinomycetes from soils of Mongolia desert steppe zone. Microbiology 81(1):98–108. https://doi.org/10.1134/S0026261712010092

Langworthy TA, Mayberry WR (1976) A 1,2,3,4-tetrahydroxypentane-substituted pentacyclic triterpene from *Bacillus acidocaldarius*. Biochem Biophys Acta 431(3):570–577. https://doi.org/10.1016/0005-2760(76)90221-6

Lewis CA, Crayle J, Zhou S et al (2016) Cytosine deamination and the precipitous decline of spontaneous mutation during Earth's history. Proc Natl Acad Sci U S A 113(29):8194–8199. https://doi.org/10.1073/pnas.1607580113

Li WF, Zhou XX, Lu P (2005) Structural features of thermozymes. Biotechnol Adv 23(4):271–281. https://doi.org/10.1016/j.biotechadv.2005.01.002

Lindahl T (1993) Instability and decay of the primary structure of DNA. Nature 362(6422): 709–715. https://doi.org/10.1038/362709a0

Madigan MT, Martinko JM (2010) Brock biology of microorganisms, 13th edn. Benjamin Cummings, New York

Nelson RR, Hodges CS (1965) A new species of *Curvularia* with a protuberant conidial hilum. Mycologia 57(5):822–825. https://doi.org/10.2307/3757020
Ourisson G, Albrecht P (1992) Hopanoids. 1. Geohopanoids: the Most abundant natural products on earth? Acc Chem Res 25(9):398–402. https://doi.org/10.1021/ar00021a003
Ourisson G, Rohmer M (1992) Hopanoids. 2. Biohopanoids: a novel class of bacterial lipids. Acc Chem Res 25(9):403–408. https://doi.org/10.1021/ar00021a004
Pachepsky Y, Shelton D, Dorner S et al (2016) Can *E. coli* or thermotolerant coliform concentrations predict pathogen presence or prevalence in irrigation waters? Crit Rev Microbiol 42(3): 384–393. https://doi.org/10.3109/1040841X.2014.954524
Paruch AM, Mæhlum T (2012) Specific features of *Escherichia coli* that distinguish it from coliform and thermotolerant coliform bacteria and define it as the most accurate indicator of faecal contamination in the environment. Ecol Indic 23:140–142. https://doi.org/10.1016/j.ecolind.2012.03.026
Rezanka T, Siristova L, Melzoch K et al (2010) Hopanoids in bacteria and cyanobacteria – their role in cellular biochemistry and physiology, analysis and occurrence. Mini Rev Org Chem 7(4): 300–313. https://doi.org/10.2174/157019310792246436
Russell RJM, Ferguson JMC, Hough DW et al (1997) The crystal structure of citrate synthase from the hyperthermophilic archaeon *Pyrococcus furiosus* at 1.9 Å resolution. Biochemistry 36(33): 9983–9994. https://doi.org/10.1021/bi9705321
Sadeghi M, Naderi-Manesh H, Zarrabi M et al (2006) Effective factors in thermostability of thermophilic proteins. Biophys Chem 119(3):256–270. https://doi.org/10.1016/j.bpc.2005.09.018
Sáenz JP, Grosser D, Bradley AS et al (2015) Hopanoids as functional analogues of cholesterol in bacterial membranes. Proc Natl Acad Sci U S A 112(38):11971–11976. https://doi.org/10.1073/pnas.1515607112
Sandona K, Tobias TLB, Hutchinson MI et al (2019) Diversity of thermophilic and thermotolerant fungi in corn grain. Mycologia 111(5):719–729. https://doi.org/10.1080/00275514.2019.1631137
Schmerk CL, Bernards MA, Valvano MA (2011) Hopanoid production is required for low-pH tolerance, antimicrobial resistance, and motility in *Burkholderia cenocepacia*. J Bacteriol 193(23):6712–6723. https://doi.org/10.1128/JB.05979-11
Schumann J, Bohm G, Schumacher G et al (1993) Stabilization of creatinase from *Pseudomonas putida* by random mutagenesis. Protein Sci 2:1612–1620
Tekaia F, Yeramian E, Dujon B (2002) Amino acid composition of genomes, lifestyles of organisms, and evolutionary trends: a global picture with correspondence analysis. Gene 297(1–2):51–60. https://doi.org/10.1016/S0378-1119(02)00871-5
Thompson MJ, Eisenberg D (1999) Transproteomic evidence of a loop-deletion mechanism for enhancing protein thermostability. J Mol Biol 290(2):595–604. https://doi.org/10.1006/jmbi.1999.2889
Turner P, Mamo G, Karlsson EN (2007) Potential and utilization of thermophiles and thermostable enzymes in biorefining. Microb Cell Fact 6:9. https://doi.org/10.1186/1475-2859-6-9
Vetriani C, Maeder DL, Tolliday N (1998) Protein thermostability above 100°C: a key role for ionic interactions. Proc Natl Acad Sci U S A 95:12300–12305
Vieille C, Zeikus JG (1996) Thermozymes: identifying molecular determinants of protein structural and functional stability. Trends Biotechnol 14:183–190
Wang S, Hu A (2016) Comparative study of spontaneous deamination of adenine and cytosine in unbuffered aqueous solution at room temperature. Chem Phys Lett 653:207–211. https://doi.org/10.1016/j.cplett.2016.05.001
Wei JH, Yin X, Welander PV (2016) Sterol synthesis in diverse bacteria. Front Microbiol 7:1–19. https://doi.org/10.3389/fmicb.2016.00990
Wharton DA (2002) Life at the limits: organisms in extreme environments. Cambridge University Press, New York
Whitaker RJ, Grogan DW, Taylor JW (2003) Hyperthermophilic Archaea 301:2002–2004

Willerslev E, Cooper A (2005) Ancient DNA. Proc R Soc B Biol Sci 272(1558):3–16. https://doi.org/10.1098/rspb.2004.2813

Wuts PGM, Greene TW (2007) Protective groups in organic synthesis, 4th edn. Wiley, Hoboken

Yamini C, Sharmila G, Muthukumaran C et al (2022) Proteomic perspectives on thermotolerant microbes: an updated review. Mol Biol Rep 49(1):629–646. https://doi.org/10.1007/s11033-021-06805-z

Zeikus JG (1979) Thermophilic bacteria: ecology, physiology and technology. Enz Microbial Technol 1(4):243–252. https://doi.org/10.1016/0141-0229(79)90043-7

Cultivation Techniques and Molecular Methods of Identification of Thermophilic, Anaerobic Bacteria

Sean Michael Scully and Johann Orlygsson

Abstract The cultivation and identification of strictly anaerobic thermophilic microorganisms present a number of challenges owing to the oxygen-sensitive nature of many of these species. This chapter reviews the nature of the anaerobic environment and the techniques currently employed to cultivate both aerotolerant and strictly anaerobic bacteria. Additionally, molecular methods of identifying thermophilic bacteria without cultivation will be addressed.

1 Introduction

Despite our oxygen-centric perspective on life, organisms that inhabit anaerobic environments such as sediments, hot springs, and waste streams, represent a significant contribution to the carbon, nitrogen, and sulfur cycles in the biosphere (Shu and Huang 2021; Bolhuis et al. 2014). Notwithstanding their noteworthy contributions to the biosphere, the cultivation of obligatory and strictly anaerobic microorganisms is seldom covered in any detail in standard undergraduate microbiology textbooks despite such organisms being of clinical and ecological relevance and often of tremendous biological potential. As such, students often encounter mentions of the techniques used to cultivate strict anaerobes under anoxic conditions but seldom are the auxiliary concepts necessary to master the technique in practice alluded to. While many of the techniques for the study of anaerobic microorganisms have been for the past 40 years, there have been several noteworthy advances that have impacted the study of anaerobes in very profound ways. As an example, earlier works describing the cultivation of strictly anaerobe organisms, such as the *Anaerobe Laboratory Manual* (Holdeman et al. 1977), include instructions on the use of heated copper catalyst setups for removing impurities from gases and the use of cannulas for directing gas flow.

S. M. Scully (✉) · J. Orlygsson
Faculty of Natural Resource Sciences, University of Akureyri, Akureyri, Iceland
e-mail: scully@unak.is

S. M. Scully, J. Orlygsson (eds.), *Thermophilic Anaerobes*, Emerging Issues and Trends, https://doi.org/10.1007/978-3-031-41720-7_4

Anaerobic microorganisms are ubiquitous in the environment and may have thrived on early earth prior to the great oxygenation event (Hsia et al. 2013). A distinction must be made between *obligate* anaerobes, which cannot utilize molecular oxygen but may tolerate various degrees of oxygen in the environment, and *strict* anaerobes, which are killed in the presence of oxygen. Strictly anaerobic bacteria cannot grow in the presence of greater than 5 μM of dissolved oxygen (Baughn and Malamy 2004). The clinical relevance and biotechnological potential of many anaerobic bacteria necessitate that their study is carried out under strictly oxygen-free conditions due to their dislike of oxygen due to the toxicity of molecular oxygen and its related species. The cultivation of strictly anaerobic bacteria is even more stringent and requires special precautions to remove the presence of oxygen as does the examination of many of their cofactor-requiring enzyme systems and to lower the redox potential of the medium to create a sufficiently reducing environment (Selmer 2005).

Louis Pasteur's pioneering work on the fermentative activities of life in the absence of oxygen developed the term "anaerobic" in the mid-nineteenth century (Durre 2005 and references therein). In the 150 years since, techniques to highly manipulate oxygen-sensitive organisms have been developed allowing the examination of strictly anaerobic organisms leading to the large-scale production of solvents, a better understanding of their physiology and microbial ecology.

The historical applications of strict anaerobes have long eluded us due to perhaps a lack of methodologies to study them. An excellent example of this is the microbial production of indigo pigment from woad (*Isatis tinctoria*), which has been used for centuries (Durre 2001). The process involves the conversion of pigments (Istan B and Indican) to indigo aerobically by *Enterobacter agglomerans,* which is subsequently converted to water-soluble leuco-indigo at temperatures of up to 52 °C under anaerobic conditions after the addition of urea and potash (Durre 2001). It was discovered that the organism responsible for the last step is *Clostridium isatidis*, a moderately thermophilic, strictly anaerobic organism that has been found associated with dying vats (Padden et al. 1999).

The following theme in this chapter will be divided into two main subchapters, one focusing on general techniques to cultivate thermophilic anaerobes, with an emphasis on strictly anaerobic bacteria, and the other on the identification of anaerobic, thermophilic microbes with various molecular methods.

2 The Trouble with Oxygen

While the use of oxygen as a terminal electron acceptor is ubiquitous in nature, many aerobic organisms have developed specialized mechanisms to negate the inherent toxicity of oxygen. Oxygen is a potent oxidizing agent and in the presence of light or a transition metal such as iron, manganese, or cobalt can undergo radical reactions generating superoxides, peroxides, and hydroxyl radicals (Auten and Davis 2009). These reactive oxygen species in turn can oxidize sensitive biomolecules, including

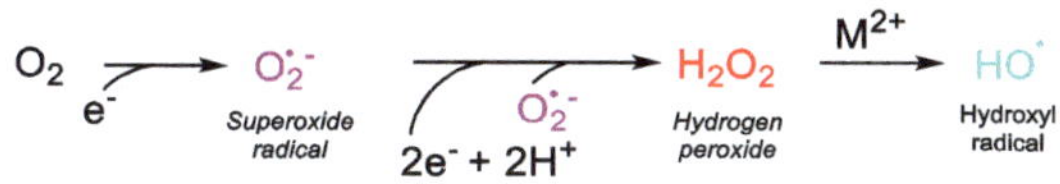

Fig. 1 Formation of reactive oxygen species from molecular oxygen

unsaturated fatty acids. To negate oxygen radicals and radical-forming species, many oxygen-tolerant microorganisms rely on various enzymes like oxidases, peroxidases, and catalases to neutralize these species before cellular damage occurs (Ezraty et al. 2017). Reactions involving molecular oxygen can give rise to singlet oxygen species, which are highly reactive and often react with biomolecules via the transfer of radicals (Mitchell 2001). The reactions below show (Fig. 1) the main toxic compounds formed in the presence of oxygen although many secondary reactive oxygen species (ROS) can form.

ROS are of particular concern to biological systems due to their ability to cause oxidative damage to major biomolecules such as lipids, proteins, and nucleic acids (Choe and Min 2006; Auten and Davis 2009). As such, many aerobic and aerotolerant organisms have developed specific mechanisms to mitigate the formation of reactive oxygen species. Below are someone of the most commonly employed mechanisms for dealing with deleterious ROS:

Catalase: $H_2O_2 + H_2O_2 \rightarrow 2H_2O + O_2$

Peroxidase: $H_2O_2 + NADH + H^+ \rightarrow 2H_2O + NAD^+$

Superoxide dismutase: $O_2^{\cdot-} + O_2^{\cdot-}\ 2H^+ \rightarrow H_2O_2 + O_2$

Superoxide dismutase/catalase in combination: $4\ O_2^{\cdot-} + 4H^+ \rightarrow 2H_2O + 3O_2$

Superoxide reductase: $O_2^{\cdot-} + 2H^+$ Cytochrome $C_{reduced} \rightarrow H_2O_2$ Cytochrome $C_{oxidized}$

Strictly anaerobic bacteria often lack these enzymes, with the notable exception of lactic acid bacteria (LAB) which while they are strictly anaerobic are highly aerotolerant due to their production of highly active catalase and peroxidases, thus creating their own anaerobic environment. Additionally, there have been descriptions of LABs that can perform respiratory metabolism (Pedersen et al. 2012).

Many anaerobes have varying degrees of oxygen tolerance; some simply will not grow in the presence of oxygen but may tolerate periods of oxygenation, while others are the so-called microaerophilic bacteria that thrive when the concentration of oxygen is low (< ppm). Organisms within Class *Clostridia* are all strictly anaerobic bacteria although a number of aerotolerant species have been reported. Strictly anaerobic *Firmicutes*, such as organisms within Class *Clostridia* and *Bacilli*, may rely on endospore formation during periods of oxygen exposure (Mitchell 2001).

Strict anaerobes have been isolated from a diverse number of environments, which typically share the trait of having very low oxygen concentrations such as geothermal features, muds and sediments, the rumen and digestive tract of organisms, and highly stratified bodies of water. Sediment environments are often highly reduced with traces of oxygen that are often quickly removed through various reactions (such as the oxidation of sulfides to sulfur oxides such as sulfate), while biological scavenging of oxygen by facultative anaerobe digestive systems keeps the oxygen concentrations low, thus conferring additional protection to strict anaerobes in the system. Within thermal environments, the among of dissolved oxygen is negligible above 50 °C and only relevant for organisms living at the air–water interface although the presence of highly reduced molecules (such as sulfide) can also serve as oxygen scavengers.

Despite their sensitivity to oxygen, a number of strictly anaerobic bacteria have significant roles in biotechnology (Table 1). Examples of strictly anaerobic bacteria that cause diseases are *C. botulinum* (botulism), *C. difficile* (enterocolitis), *C. perfringens* (gas gangrene), *C. tetani* (tetanus), and *P. gingivalis* (the causative agent of gingivitis) (Mitchel 2001). A number of thermophilic anaerobes have also demonstrated biotechnological potential in a diverse range of areas as highlighted in Table 1.

3 Cultivation of Anaerobic Bacteria

3.1 The Importance of Redox Potential

A major consideration when cultivating anaerobic bacteria is the redox potential (ORP, redox potential, or E_h) of the medium, which relates to not only the availability of oxygen, but also the reducing potential of the culture environment (Liu et al. 2013). In some cases, strictly anaerobic bacteria will not grow if the redox potential of the medium is not sufficiently low enough, often necessitating the addition of reducing agents to the medium.

The measurement of dissolved oxygen and culture redox potential (CRP) is often critical as some anaerobes will not grow unless the reducing potential is sufficiently low (Liu et al. 2013). The CRP is the result of complex interactions contributed by multiple redox couples present in the culture medium. While the measurement of dissolved oxygen (dO_2) via several sensor types is widely available, the measurement of CRP is even more facile and can be conveniently measured using a platinum electrode.

In small batch culture, however, the measurement of redox potential via electrode is cumbersome and it is thus more common to rely on one or more redox-sensitive chromophores. A wide number of redox-sensitive dyes are commercially available (Table 2). The useful range of each of these redox dyes is typically ±50 mV of their

Table 1 Selected biotechnologically relevant anaerobic organisms. Thermophilic microorganisms are bolded

Anaerobe	Features	O_2 relationship	Notes	Refs
Cl. isatidis	Indigo production	Strict anaerobe	Moderate thermophile	Padden et al. (1999)
Cl. acetobutylicum	Butanol	Strict anaerobe	Mesophilic	Delarouzee et al. (2023)
Clostridium beijerinckii	Butanol	Strict anaerobe	Mesophilic	Krishna et al. (2022)
T. ethanolicus	Ethanol from a wide range of hexoses and pentoses	Strict anaerobe	Thermophile	Wiegel and Ljungdahl (1981)
Cl. thermocellum	Cellulose-degrading	Strict anaerobe	Produces cellulosome for biomass desconstruction	Akinosho et al. (2014)
Caldicellulosiruptor saccarolyticus	Cellulose-degrading, hydrogen production	Strict anaerobe		Willquist et al. (2011)
P. gingivalis	Pathogen	Strict anaerobe	Causative agent of gingivitis	Lamont and Jenkinson (1998), Takada and Hirasawa (1998)
Thermoanoaerbacterium thermosaccharolyticum	Ethanol from a wide range of hexoses and pentoses	Strict anaerobe	Isolated from spoiled canned goods	McClung (1935), Collins et al. (1994)
Methanobacterium formicicum	Methane production	Strict anaerobe	Important for interspecies hydrogen transfer	Chellapandi et al. (2018)
Streptococcus pyogenes	Pathogen	Aerotolerant	Upper respiratory tract	Avire et al. (2021)
Spirillum volutans	Fresh water	Microaerophile	Gram negative	Padgett et al. (1982)

midpoint potential (E_m). Chromophores with high molar extinction coefficients are preferable such that minimal quantities can be used in the medium and detected easily by visual examination.

Methylene blue is routinely used in combination with glucose to detect oxygen (or the lack thereof) in anaerobic jars and pouches. In its oxidized form, methylene blue is an intensely blue chromophore although when it is reduced to its leuco form it is colorless (Fig. 2). While this system is common and often sufficient for anaerobic

Table 2 Selected midpoint potentials and other features of commonly used redox indicators. Modified from Jacob (1970), Srinivas et al. (1988)

Dye	Midpoint potential (E_m, mV)	Notes
Methylene blue	11	Commonly available, inexpensive
Resorufin	−51	Low toxicity, used in viability assays
Indigo trisulfonate	−81	
Nile blue	−142	
Cresyl violet	−167	
Brilliant alizarin blue	−173	
Neutral blue	−192	
Phenosafranine	−252	
Safranine-T	−289	
Neutral red	−325	
Benzyl viologen	−359	Inexpensive, respiratory irritant
Methyl viologen	−440	Inexpensive, acutely toxic
Standard hydrogen electrode	−421	

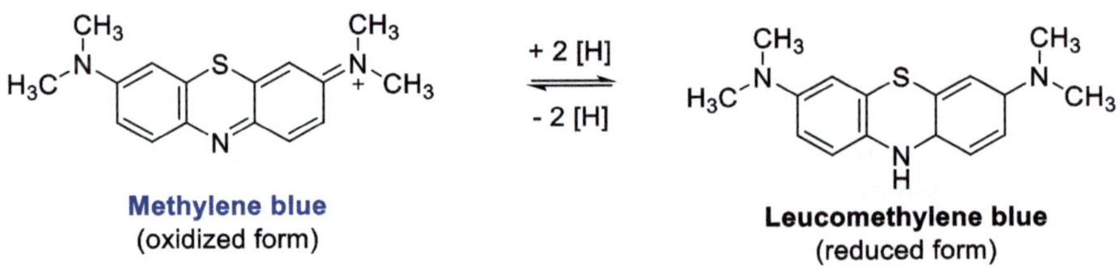

Fig. 2 Methylene blue and its reduction to leucomethylene blue

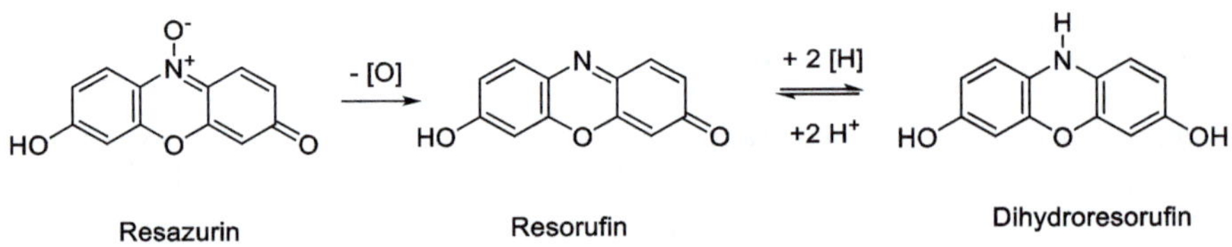

Fig. 3 Resazurin is a commonly used redox indicator for anaerobic media that undergoes a irreversible reduction from *blue* to *pink* (resorufin), which can be reversibly reduced to a colorless form (dihydroresorufin)

systems intended for fairly oxygen-sensitive organisms, its midpoint potential of 11 mV is not sufficient for the cultivation of strictly anaerobic bacteria.

Another commonly used chromophore systems is resazurin (Fig. 3), a fluorescent phenazine dye, which has the notable advantage of having a lower midpoint potential upon its irreversible reduction to resorufin (−51 mV). Resazurin and its reduced forms are commonly used in cell viability assays where resorufin is highly fluorescent, while its reduced form lacks fluorescence. When the redox potential of the medium reaches −110 mV, the reversible reduction of resorufin to dihydroresorufin takes place yielding a colorless molecule.

3.2 *Preparation of Media for Strictly Anaerobic Bacteria*

While several media for the cultivation of anaerobes are available, such as reinforced clostridial medium (RCM), the use of mineral medium is often preferred due to the greater control afforded over its composition and minimizing or eliminating non-essential components to reduce the formation of metabolic end products that are not under study. Unlike the media used for aerobic bacteria, special care must be taken to remove traces of oxygen from the media and lower the redox potential sufficiently to protect cells whether the media is for cultivation or for long-term cell storage (such as 30% v/v glycerol or DMSO stocks).

Typically, the first step in preparing anaerobic medium on the benchtop is to reduce the oxygen load of the liquid medium itself. This can be accomplished using several methods (boiling, membrane filtering, sonication, vacuum, and sparging with an inert gas such as nitrogen, argon, or helium). Heating is facile as it does not require equipment rarely found in laboratories. After boiling, the medium is rapidly cooled under nitrogen flushing (or another inert gas) to prevent oxygen from dissolving in the medium.

Cool, oxygen-free medium is then transferred to serum bottles or Hungate tubes under active nitrogen flushing, and the gas is allowed to sparge through the medium briefly before being sealed with a butyl rubber septa and then sealed with an aluminum crimp cap. It should be noted that long needles that reach deep into the vessel being sparged will give a better result. General advice from DSMZ on the amount of liquid medium to be dispensed into bottles is no more than 25% of the nominal capacity as many anaerobes produce substantial quantities of gas generating substantial overpressure due to the production of hydrogen, carbon dioxide, and, in some cases, methane if a methanogen is present. Gas accumulation in glass vessels can lead to catastrophic failure of the vessel and harm to unprotected researchers. For this reason, it is often wise to limit the amount of carbon source in the medium to avoid producing more than a few atmospheres of overpressure (DSMZ 2012).

It is worth noting that the liquid–gas (L-G) ratio is of great importance with some organisms in terms of the distribution of end products as a function of the partial pressures of gaseous end products such as H_2 and CO_2 (Jessen and Orlygsson 2012; Scully and Orlygsson 2020) and variations in L-G ratio can shift end product formation if the organism is sensitive to changes in the partial pressure of hydrogen (pH_2). It is therefore recommended that a L-G ratio of 1:1 be used to ensure consistent fermentation results.

Table 3 Commonly used reducing agents in anaerobic media. Modified from Breznak and Costilow (2007 and refs therein)

Reducing agent	Typical concentrations	Reduction potential (mV)
Ti^{3+}		−660
Dithionate ($S_2O_4^{2-}$)		−480
Ascorbic acid	0.05% w/v	+58
$Na_2S \cdot 9H_2O$	0.05% w/v	−243
FeS (amorphous)	4 ug/mL	≤270
Cysteine-HCl	0.05% w/v	−325
Sodium thioglycolate ($HSCH_2COONa$)	0.05% w/v	−140
Titanium (III) citrate	1–4 mM	−480
$H_2 + PdCl_2$	Variable	−413
Dithiothreitol	1 mM	−330
Cysteine-HCl		

Autoclaving procedures are carried out in seal vessels at 121 °C for *at least* 15 min. Refer to the subsequent section for considerations relating to sterilization of media and disposal of spent cultures.

Prior to inoculation, other media components that do not survive autoclaving (carbon sources, vitamins, trace elements, oxygen scavengers) must be added. Such solutions are often sterilized using syringe filtering into sterile nitrogen-flushed serum bottles. It is often helpful to reduce the oxygen load in these solutions by preparing them with rigorously degassed distilled water or degassing them after preparation by sonication (with or without vacuum), helium, or nitrogen sparging, prior to syringe filtering them. While carbon sources such as glucose can be autoclaved for short periods in the absence of amines (such as amino acids), heating of sugar solutions often leads to the formation of inhibitory compounds such as 5-hydroxy-2-methylfurfuraldehdye (5-HMF) making syringe filtering a preferable option (Einarsson et al. 1988). The last step prior to media inoculation is the addition of reducing agents, such as sodium sulfide or dithionite, to scavenge oxygen and lower the reduction potential (E_h) of the medium (Table 3).

3.3 *Sterilization of Media and Cultivation Vessels*

One of the often-touted advantages of working with thermophilic bacteria is the decreased risk of mesophilic contamination. However, the corollary of this is that there is an increased risk of thermophilic contamination. It is common practice for the routine cultivation of microorganisms to autoclave media at 121 °C for 15 min with the general expectation that no vegetative organisms or endospores will survive. *Geobacillus stearothermophilus*, a spore-forming thermophilic facultative anaerobe, is routinely used to test the efficacy of autoclave systems and is commercially available with Merck's Sterikon® all-in-one bioindicator ampules being one

Table 4 *D*-values of selected mesophilic and thermophilic spore-forming bacteria (Data from Hyun et al. (1983) and references therein)

Microorganism	T_{opt} (°C)	*D*-value at 121 °C (min)
Clostridium thermocellum	55	0.5
Thermoanaerobacterium thermosulfurogenes	60–70	2.5
Thermoanaerobacter pseudethanolicus	65	11
Desulfotomaculum nigrificans	55	5.6
Bacillus subtilis	35	0.9
Geobacillus stearothermophilus (strain FS1518)	55	3.0
Clostridium sporogenes	37	1.3

example. However, among thermophilic bacteria, there are several examples of organism with decimal reduction times (*D*-values) greater than that of *G. stearothermophilus,* which may warrant a greater degree of caution when sterilizing media and spent cultures (Table 4).

It should be pointed out that comprehensive studies on the *D*-values of many thermoanaerobes have not been carried out so the results in Table 4. may not be representative. Even so, as an example, if a culture of *Thermoanaerobacter pseudethanolicus* with a reported $D_{121°C}$ of 11 min reached a cell density of 1.0×10^9 CFU per mL, it would take 99 min at this temperature for the culture to become completely sterilized, which is nearly 7 times longer than a tradition 15-min cycle used for routine sterilization. For this reason, strict protocols for the sterilization of media and spent cultures should be in place to prevent contamination.

For these reasons, the author's group routinely sterilizes media for 60 min at 121 ° C while autoclaving spent cultures and glassware for 120 min on the backend followed by dry heat sterilization (250 °C, at least 4 h). Great care is also taken to ensure that the risk of cross-contamination is minimized by employing rigorous surface sterilization techniques. We are aware that several research groups use tyndallization heat treatment in their laboratories with the work of thermophilic bacteria.

3.4 Cultivation of Aerotolerant Anaerobic Bacteria

The batch cultivation of aerotolerant anaerobes is reasonably easy. For liquid culture, a common technique involves overlaying liquid medium with an oxygen-impermeable material such as mineral oil or Valspar (one part petroleum jelly to one part paraffin). Cultivations vessels may also become anaerobic if they are tightly sealed and deep, thus preventing atmospheric oxygen from reaching the cultivation broth, or having limited headspace. Another technique involves the use of an airlock (commonly associated with brewing with facultative anaerobes such as *S. cerevisiae* and other yeasts).

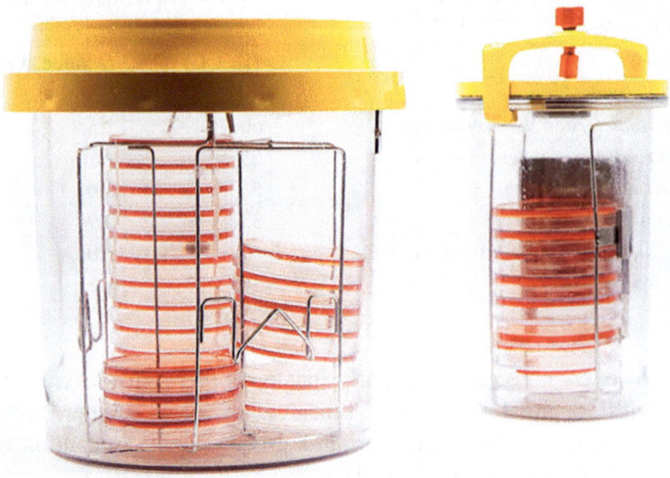

Fig. 4 Commercially available anaerobic jars used for the cultivation of anaerobic microorganisms

For solid cultures such as agar plates, it is common to utilize cultivation cabinets with controlled headspace gas (typically CO_2 or N_2) or to utilize anaerobic jars or pouches to create an anaerobic environment. Anaerobic jars (Fig. 4) are sealable containers which can maintain an oxygen-free atmosphere for cultivating organisms, which require low levels of oxygen (Shahin et al. 2003). Typically, anaerobic jars available from commercial suppliers can accommodate 10–20 petri dishes. The jar's atmosphere is made anaerobic by means of a palladium catalyst, which converts oxygen to water and hydrogen gas. One major drawback of working with anaerobic jars as that the use of this technique is insufficient for working with strict anaerobes. An alternative is an anaerobic pouch. 519F

Anaerobic pouches are often ideal for field work where samples must be transferred to an anaerobic environment quickly.

Anaerobic chambers and glove boxes are typically used in laboratories that specialize in anaerobic culture work (Selmer 2005). They are large plastic tents with an incubator and equipment for culturing the anaerobes. The atmosphere inside the chamber is usually filled with a mixture of carbon dioxide and nitrogen gas. The chamber has an airlock that can be emptied or refilled with nitrogen or oxygen-free gas. It is through this airlock that it is possible to place the culture media inside or remove it from the chamber. The oxygen remaining in the chamber is removed by its reaction with hydrogen in the presence of a palladium catalyst. It is possible to work inside the chamber by extending your arms into specialized gloves attached to the chamber walls; this is why it is also called a glove box. Anaerobic chambers can be very economical if properly constructed because the cost of gases for operating the system is minimal and it allows the use of conventional plating media (Engelkirk et al. 1992).

3.5 Cultivation of Strictly Anaerobic Bacteria

Unlike cultivating aerotolerant anaerobes, the cultivation of strictly anaerobic bacteria requires great measures to protect organisms from oxygen. The seminal work in this field, the so-called "Hungate technique," traces back to Robert Hungate's description of techniques for studying anaerobic cellulose-degrading organism from the rumen of cattle (Hungate 1947). His highly cited 1969 paper, "A roll tube method for cultivation of strict anaerobes," clearly laid out detailed protocols (Hungate 1969). In hindsight, many of the aspects described are old-fashioned due to the modern availability of high-purity (i.e. oxygen-free) gases although many of the central features of Hungate's techniques (such as the use of butyl rubber septa and reducing agents) have carried through to this day.

The current recommended procedures for the anaerobic technique, an evolution of Hungate's earlier work, were largely developed in the laboratory of Ralph S. Wolfe during the mid-1970s and are generically referred to as "the Balch technique." For example, roll tubes are still one of the easier ways to isolate anaerobes on agar surfaces or in agar using the agar-shake roll tube method as compared to the endpoint dilution technique.

These techniques use specialized user-friendly glassware and equipment to easily manipulate anaerobes or any microorganism requiring a defined gas phase. Various laboratories working with anaerobes have developed their own modifications depending on preferences or needs. Simpler approaches to anaerobic microbiology are covered in the literature (Willis 1969), with much of it targeted toward medically important clostridia that are in general less sensitive to air or oxygen than most clostridia. Indeed, the oxygen tolerance of clostridial species ranges from relative insensitivity, such as for *Lacrimispora* (formerly *Clostridium*) *aerotolerans* (Van Gylswyk and Van der Toorn 1987) to extreme sensitivity, such as for some of the hydrogen-oxidizing acetogens. The removal of oxygen and lowering of the redox potential of culture media by the addition of a reducing agent are the two crucial parts of the technique. The removal of oxygen is usually achieved by boiling the medium. This will often be done with the medium under a stream of anoxic gas for a small batch of medium, or by steam sterilization in an autoclave for a large batch (above 1–2 l). Steam sterilization is done with the vessel covered but still open to the atmosphere. After autoclaving, the medium is cooled under a stream of sterile, anoxic gas (Hungate 1969) and then sealed with aluminum capsules (Balch et al. 1979) (Fig. 5). Finally, a sterile reducing agent is added to reduce the medium's redox potential. The reducing agents used are, for example, sulfide, titanium (III) ion, or dithionate. For preparing or dispensing anaerobic media in small batches such as in various tubes or serum bottles, the media are made anoxic by boiling, cooled to below 40 °C, reduced by adding the appropriate amount of reducing agent, and then, under the stream of oxygen-free gas, distributed into the tubes or bottles, which then are sealed with butyl stoppers and aluminum crimps and autoclaved. The butyl rubber is most often used because of its limited permeability to oxygen. In laboratories dealing with a lot of soil samples or various spore-forming strains, autoclaving

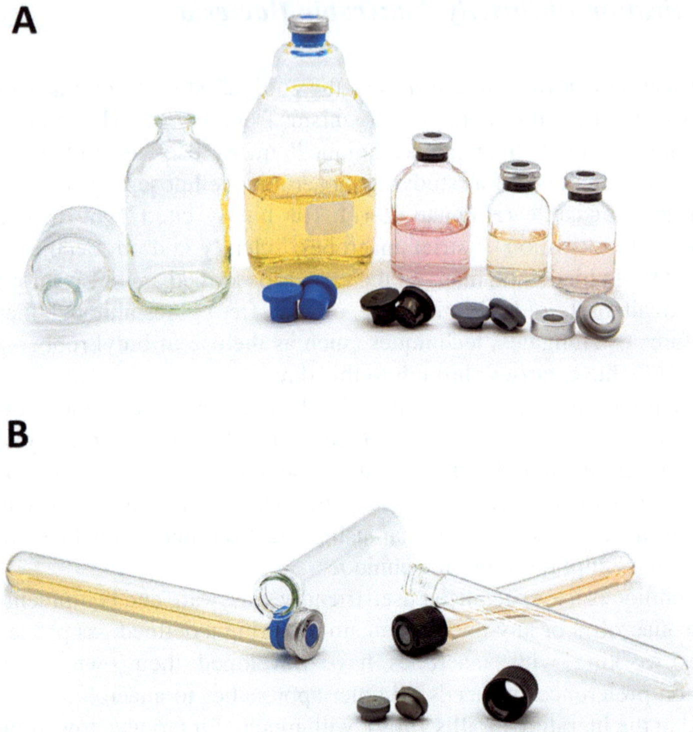

Fig. 5 Serum bottles (**a**) and Hungate tubes (**b**) filled with basal mineral medium

time at 121 °C should be extended to 30–40 min to ensure that heat-resistant spores are killed. In some instances, autoclaving the media a second time after incubating for 36–48 h at 30 °C or 60 °C is recommended for laboratories primarily working with thermophiles and thermobiotic soils and sediments.

4 Culture-Independent Techniques for Identifying Anaerobic Bacteria

In the last few decades, impressive achievements have been made in modern microbiology to isolate and cultivate microbes under laboratory conditions, leading to our understanding of their physiological properties. These achievements have been the basis for the successful development of both fundamental microbiology and applied microbiology. Still, there is a clear paradox in various ecological niches since the total number of bacteria that can be cultivated using traditional methods is a small fraction of those that are revealed using sequence-independent methods; this phenomenon is termed the great plate count anomaly (Kaeberlein et al. 2002; Zengler et al. 2002; Buerger et al. 2012). As such, it is generally accepted that only 1% of prokaryotes may be cultivated in the laboratory although this "cultivation gap" may largely be a matter of finding appropriate conditions for fastidious members of a community. Regardless, this leads to the inescapable conclusion that the remaining "uncultivated fraction" is likely to include many interesting bacteria that play an important part in their microbial communities and may also be of utility in biotechnology.

The remainder of this chapter deals with the identification of thermophilic anaerobes without direct isolation of the organism. The application of several molecular approaches has elucidated the diversity of microbial communities in different environments. The methods used to identify the diversity and activity of microorganism in environmental samples are divided into two main groups: partial and whole community analysis. Firstly, the partial analysis of microbial communities is performed using several molecular methods, such as denaturing gradient gel electrophoresis (DGGE), temperature gradient gel electrophoresis (TGGE), single-strand conformation polymorphism (SSC), DNA microarrays, real-time polymerase chain reaction (qPCR), and fluorescence in situ hybridization (FISH). Secondly, whole community analysis is performed using techniques such as amplicon-based metabarcoding, sequence-based metagenomics, G + C fractionation, functional metagenomics, metatranscriptomics, and metaproteomics (Mohammadali and Davies 2018).

The term metagenome was first proposed by Handelsman and coworkers to describe "the genomes of total microbiota found in nature" (Handelsman et al. 1998). In short, metagenomics and culture-independent genomic analysis of microbial genomes in environmental samples are based on DNA that is isolated. The first microorganisms that were used as a base for preparing metagenomic libraries were from picoplankton (Schmidt et al. 1991). In this study, the total DNA was isolated from the environment, and it was randomly fragmented and sequenced. By this approach, it was possible to identify genes and metabolic processes performed not only from pure cultures isolated, but from other uncultivated microorganisms as well. Later, with better sequencing techniques, it was possible to obtain total DNA from environmental samples, a technique now called environmental DNA-based community analysis. Today, metagenomics studies can be classified as functional metagenomics and shotgun-sequence metagenomics. Instead of a full review of the different methods used for both partial and full analysis of microbial community, some examples of both concerning thermophilic anaerobes are given below.

4.1 Partial Analysis of Thermophilic, Anaerobic Bacteria by Molecular Methods

Denaturing gradient gel electrophoresis (DGGE) is a technique used to separate short- to medium-length DNA fragments based on their melting characteristics. Saghatelyan and coworkers did an analysis of microbial diversity of nine terrestrial geothermal springs in Armenia using DGGE methods (Saghatelyan et al. 2021). Their main finding was that the main bacterial phyla of Proteobacteria, Bacteroidetes, Cyanobacteria, and Firmicutes were the predominant microbes in the springs studied. Some interesting difference was in the presence of archaea in the springs tested. Temperature seemed to be the most important factor in shaping the microbial communities in these springs, but overall the diversity and richness of the microbiota were negatively affected by an increase in temperature. More than 130 bacterial and archaeal strains were reported, among them several thermophilic heterotrophic anaerobic bacteria with potential use as producers of thermostable enzymes and biomolecules of biotechnological interest.

Temperature gradient gel electrophoresis (TGGE) is a form of electrophoresis in which a temperature gradient is used to denature molecules as they move through either acrylamide or agarose gel. TGGE can be applied to analyze DNA, RNA, protein–DNA complexes, and, less commonly, proteins. Young and coworkers investigated anaerobic digestion sludge cultivated in an electrochemical bioreactor and "conventional bioreactor"and studied the methanogenic diversity in both reactors by using TGGE (Jeon et al. 2009). Their main observation was that the methanogenic diversity was higher in the electrochemical bioreactor.

Single-stranded conformation polymorphism (SSCP) analysis is a widely used screening method that allows you to identify different genomic variants in many samples and in a broad range of organisms, from microorganisms to humans. The techniques have been used in a broad field of organisms, ranging from animals, birds, fishes, plants, and with microbes. Examples are analysis of 300-bp fragments of DNA from thermophilic methanogens (Daffonchio et al. 1998) and analysis of structural diversity in a biogas reactor of both mesophilic and thermophilic methanogens (Dohrmann et al. 2011).

DNA microarrays are a tool used to determine whether the DNA from a particular individual contains a mutation in genes (Williams et al. 2007) and are thus widely used for the analysis of microbial genes. As an example, the thermophilic anaerobe *Clostridium thermocellum* was investigated and its genes were identified with the help of DNA microarray tests (Wilson et al. 2013). Other applications of microarrays include applications in transcriptomics to obtain either a targeted functional overview or even a full transcriptome. This approach is particularly useful in microbial ecology for distinguishing between active microbial community (i.e., mRNA-producing) and inactive transient or dead organisms, all of which are typically included in eDNA-based analyses resulting in less than representative picture of the community.

Real-time polymerase chain reaction (RT-PCR) is molecular biology method based on the polymerase chain reaction and is widely used in a broad spectrum of sciences. An example of the use of RT-PCR is in a thermophilic, chemolithotrophic hydrogen-oxidizing bacterium, *Hydrogenobacter thermophilus* strain TK-6 (Ueda et al. 2007). The authors used RT-PCR to show four clusters of hydrogenase genes.

Fluorescence in situ hybridization (FISH) is a cytogenetic technique that uses fluorescent DNA probes to target specific chromosomal locations, resulting in colored signals that can be detected using a fluorescent microscope. The technique is suitable for locating specific DNA sequences, diagnosis of genetic diseases, gene mapping, and identification of novel oncogenes or genetic aberrations contributing to various types of cancers. Mostly FISH has been used on pathogen diagnosis although some studies have been performed on thermophiles. The microbial community of a volcanic mud spring in the Philippines was assessed using 16S rRNA-based approaches (Lantican et al. 2011). The DNA was extracted from solfataric soils and sediments taken from mud springs, and the 16S rDNA was PCR amplified using universal (519F–1392R) and archaeal-specific (23FPL–1391R) primer pairs, cloned, and sequenced. Fluorescence in situ hybridization (FISH) analysis revealed that about 71% of the microbial community present in the mud spring belonged to domain Archaea of which 63% were *Crenarchaeota* and 8% were *Euryarchaeota*. Seventeen percent (17%) of the population consisted of bacteria as indicated by the positive hybridization with the BACT338 probe, and the remaining 12% are unidentified.

4.2 Whole Community Analysis of Thermophilic, Anaerobic Bacteria by Molecular Methods

Whole-genome sequencing sometimes also called full genome or complete genome sequencing is a process of determining the entire DNA sequence of an organism. Several investigations have been on whole-genome sequencing using thermophilic bacteria although not as much as compared with mesophilic bacteria, especially pathogens. The whole genome of several thermophilic, anaerobic bacteria has been sequenced, especially strains with biotechnological relevance. Many *Thermoanaerobacter* strains have been WGS because of their potential for bioethanol production from lignocellulose (Verbeke et al. 2013). Other examples include the facultative thermophilic *Paenibacillus* strain DA-C8 capable of xylan degradation (Chhe et al. 2021) and the extremophile *Caldicellulosiruptor saccharolyticus* (Van De Werken et al. 2008). For recent developments on whole-genome sequencing, we refer to the recent overview of Verma and coworkers (Verma et al. 2022).

In amplicon-based community analysis, only the target DNA is sequenced. However, in shotgun metagenomics "all" the isolated eDNA is sequenced to collect genomic information from microbes without cultivation. A study on the taxonomic diversity and functional potential of two hot springs in northwestern Spain (Burgas and Muino da Veiga) was done by using metagenomic sequence-based analysis (DeCastro et al. 2021). Using these methods, the investigators could determine a significant difference in the abundance of various phyla between the two hot springs. Another example of SBM is in a recent study where metagenomic analysis was made from a large composting operation (Antunes et al. 2016). GC fractionation is used when total community DNA is analyzed in an environmental sample and is independent of PCR amplification and thus provides a sense of the relative abundance of bacterial populations (Holben et al. 2004). In a study where authors investigated the competition of acetate for the acetoclastic methanogens and syntrophic acetate-oxidizing bacteria in thermophilic anaerobic digestion, metagenome-assembled genomes were used (Dyksma et al. 2020). Firmicutes was the most abundant phylum in the metagenome covering 53% of species that were responsible for various functions, e.g., polymer hydrolysis and syntrophic acetate oxidation. The Wood–Ljungdahl pathway for syntrophic acetate oxidation and corresponding genes for energy conservation were identified in *Dethiobacteraceae* metagenome-assembled genomes. 16S rRNA gene amplicon sequencing and enrichment cultivation identified the uncultured *Dethiobacteraceae* together with *Syntrophaceticus, Tepdianaerobacter,* and *Clostridia*. Thus, this study gave new insight into a complex anaerobic digestion ecosystem where acetate catabolism was mainly performed by *Bacteria*.

Functional metagenomics is a method for studying gene function, starting from extracted DNA of mixed microbial consortium. The method relies on the construction of metagenomic libraries by cloning environmental DNA into expression vectors and propagating them into appropriate hosts, followed by activity-based screening. When the active clone has been identified, the sequence of it is determined, and the gene of interest is amplified and cloned with a subsequent expression and characterization of the product. The thermophile *T. thermophilus* has been used to detect thermophilic enzymes, such as esterases using this method (Leis et al. 2015). This approach was used to investigate the microbial diversity of two hot springs in the Himalayas with the main outcome showing that both hot springs possess a diverse set of genes analogous to various (osmotic, heat, and acid) stress factors (Najar et al. 2020).

Metatranscriptomics studies retrieve information about the gene expression of microbes within natural environments of complex microbial communities. A study by Chen and coworkers where metagenomics and metatranscriptomics methods were used to compare mesophilic and thermophilic propionate degradation communities showed that microbial interactions, metabolic pathways, and niche diversity are distinct between mesophilic and thermophilic microbial communities responsible for syntrophic propionate degradation (Chen et al. 2020).

Finally, metaproteomics is a term used for experimental approaches to study all proteins in a microbial community and microbiomes from environmental sources. Its main use is to classify experimental data where all proteins are identified and quantified from complex microbial communities. In a recent study, proteomic approach was used to define thermophilic communities in the anaerobic digestion of cellulose (Lu et al. 2014). The investigators found more than 500 non-redundant protein functions and the taxonomic community structure as inferred from the metaproteomic data set was in overall agreement with 16S rRNA gene identification. Numerous protein functions were related to cellulose and hemicellulose degradation catalyzed by *Caldicellulosiruptor* and *Clostridium* species.

5 Conclusions

Although the methodology of cultivation of anaerobic thermophilic bacteria has not changed dramatically since the era of Hungate in the 1960s our understanding of oxygen sensitivity of these microbes has increased to a large extent. Also, our understanding of extremophilic thermophiles has emerged to increase the time needed for the preparation of an anaerobic medium because of the high temperature tolerance of the microbes involved. New methods using molecular microbiology identifying methods have increased our understanding of the role of thermophilic anaerobes in various natural and manmade environments although much more emphasis has been on mesophilic bacteria, especially pathogens.

References

Akinosho H, Yee K, Close D et al (2014) The emergence of *Clostridium thermocellum* as a high utility candidate for consolidated bioprocessing applications. Front Chem 2:66. https://doi.org/10.3389/fchem.2014.00066

Antunes LP, Martins LF, Pereira RV et al (2016) Microbial community structure and dynamics in thermophilic composting viewed through metagenomics and metatranscriptomics. Life below the gum line. https://doi.org/10.1038/srep38915

Auten RL, Davis JM (2009) Oxygen toxicity and reactive oxygen species: the devil is in the details. Pediatr Res 66(2):121–127. https://doi.org/10.1203/PDR.0b013e3181a9eafb

Avire NJ, Whiley H, Ross K (2021) A review of *Streptococcus pyogenes*: public health risk factors, prevention and control. Pathogens 10:1–18. https://doi.org/10.3390/pathogens10020248

Balch WE, Fox GE, Magrum LJ et al (1979) Methanogens: re-evaluation of a unique biological group. Microbiol Rev 43(2):260–296. https://doi.org/10.1128/MMBR.43.2.260-296.1979

Baughn AD, Malamy MH (2004) The strict anaerobe *Bacteroides fragilis* grows in and benefits from nanomolar concentrations of oxygen. Nature 427(6973):441–444. https://doi.org/10.1038/nature02285

Bolhuis H, Cretoiu MS, Stal LJ (2014) Molecular ecology of microbial mats. FEMS Microbiol Ecol 90:335–350. https://doi.org/10.1111/1574-6941.12408

Breznak JA, Costilow RN (2007) Physiochemical factors in growth. In: Reddy CA (ed) Methods for general and molecular microbiology, 3rd edn. ASM Press, Washington, DC, pp 309–329

Buerger S, Spoering A, Gavrish E et al (2012) Microbial scout hypothesis and microbial discovery. Appl Environ Microbiol 78(9):3229–3233. https://doi.org/10.1128/AEM.07308-11

Chellapandi P, Bharathi M, Sangavai C et al (2018) *Methanobacterium formicicum* as a target rumen methanogen for the development of new methane mitigation interventions: a review. Vet Anim Sci 6:86–94. https://doi.org/10.1016/j.vas.2018.09.001

Chen YT, Zeng Y, Wang HZ et al (2020) Different interspecies electron transfer patterns during mesophilic and thermophilic syntrophic propionate degradation in chemostats. Microb Ecol 80: 120–132. https://doi.org/10.1007/s00248-020-01485-x

Chhe C, Uke A, Baramee S et al (2021) Draft genome sequence data of the facultative, thermophilic, xylanolytic bacterium *Paenibacillus* sp. strain DA-C8. Data Brief 35:106787. https://doi.org/10.1016/j.dib.2021.106784

Choe E, Min DB (2006) Chemistry and reactions of reactive oxygen species in foods. Crit Rev Food Sci Nutr 46(1):1–22. https://doi.org/10.1080/10408390500455474

Collins MD, Lawson PA, Willems A et al (1994) The phylogeny of the genus *Clostridium*: proposal of five new genera and eleven new species combinations. Int J Syst Bacteriol 44(4):812–826. https://doi.org/10.1099/00207713-44-4-812

Daffonchio D, De Biase A, Rizzi A et al (1998) Interspecific, intraspecific and introperonic variability in the 16S rRNA gene of methanogens revealed by length and single-strand conformation polymorphism analysis. FEMS Microbiol Lett 164:403–410. https://doi.org/10.1111/j.1574-6968.1998.tb13116.x

DeCastro ME, Escuder-Rodriguez J-J, Becerra M et al (2021) Comparative metagenomic analysis of two hot springs from Ourense (Northwest Spain) and others worldwide. Front Microbiol 12: 769065. https://doi.org/10.3389/fmicb.2021.769065

Delarouzee A, Ferreira NL, Wasels F (2023) Alleviation of carbon catabolite repression through araR and xylR inactivation in *Clostridium acetobutylicum* DSM 792. Appl Environ Microbiol 89:3. https://doi.org/10.1128/aem.02135-22

Dohrmann AB, Baumert S, Klingbiel L et al (2011) Bacterial community structure in experimental methanogenic bioreactors and search for pathogenic clostridia as community members. Appl Microbiol Biotechnol 89:1991–2004. https://doi.org/10.1007/s00253-010-2955-y

DSMZ (2012) Cultivation of anaerobes. Available at dsmz.de

Durre P (2001) From pandora's box to cornucopia: clostridia – a historical perspective. In: Bahl H, Durre P (eds) In clostridia biotechnology and medical applications. Wiley, New York, pp 1–18

Durre P (2005) Formation of solvents in clostridia. In: Durre P (ed) Handbook of clostridia. CRC Press, Boca Raton, FL, pp 671–693

Dyksma S, Jansen L, Gallert C (2020) Syntrophic acetate oxidation replaces acetoclastic methanogenesis during thermophilic digestion of biowaste. Bicrobiome 8:105. https://doi.org/10.1186/s40168-020-00862-5

Einarsson H, Eklund T, Nes IF (1988) Inhibitory mechanisms of Maillard reaction products. Microbios 53:27–36

Engelkirk PG, Duben-Engelkirk J, Dowel VR (1992) Clinical anaerobic bacteriology. Star Publishing Company, Atlanta

Ezraty B, Gennaris A, Barras F et al (2017) Oxidative stress, protein damage and repair in bacteria. Nat Rev Microbiol 15:385–396. https://doi.org/10.1038/nrmicro.2017.26

Handelsman J, Rondon MR, Brady SF et al (1998) Molecular biological access to the chemistry of unknown soil microbes: a new frontier for natural products. Chem Biol 5:R24–R25. https://doi.org/10.1016/S1074-5521(98)90108-9

Holben WE, Feris KP, Kettunen A et al (2004) GC fractionation enriches microbial community diversity assessment and detection of minority populations of bacteria by denaturing gradient gel electrophoresis. Appl Environ Microbiol 70:2263–2270. https://doi.org/10.1128/AEM.70.4.2263-2270.2004

Holdeman LV, Cato EP, Moore WEC (1977) Anaerobe laboratory manual, 4th edn. Virginia Polytechnic Institute Anaerobe Laboratory, Blacksberg

Hsia CCW, Schmitz A, Lambertz M et al (2013) Evolution of air breathing: oxygen homeostasis and the transitions from water to land and sky. Compr Physiol 3:849–915. https://doi.org/10.1002/cphy.c120003

Hungate RE (1947) Studies on cellulose fermentation: III. The culture and isolation for cellulose-decomposing bacteria from the rumen of cattle. J Bacteriol 53(5):631–645. https://doi.org/10.1128/jb.53.5.631-645.1947

Hungate RE (1969) A roll tube method for cultivation of strict anaerobes. In: Norris JR, Ribbons DW (eds) Methods in microbiology. Academic, New York, pp 117–132

Hyun HH, Zeikus JG, Longin R et al (1983) Ultrastructure and extreme heat resistance of spores from thermophilic *Clostridium* species. J Bacteriol 156(3):1332–1337. https://doi.org/10.1128/JB.156.3.1332-1337

Jacob HE (1970) Redox potential. Methods Microbiol 2:91–129

Jeon JB, Kim SY, Park YK et al (2009) Enrichment of hydrogenotrophic methanogens in coupling with methane production using electrochemical bioreactor. J Microbiol Biotechnol 19:1665–1671. https://doi.org/10.4014/jmb.0904.04002

Jessen JE, Orlygsson J (2012) Production of ethanol from sugars and lignocellulosic biomass by *Thermoanaerobacter* J1 isolated from a hot spring in Iceland. J Biomed Biotechnol 186982: 1869–1882. https://doi.org/10.1155/2012/186982

Kaeberlein T, Lewis K, Epstein SS (2002) Isolating "uncultivable" microorganisms in pure culture in a simulated natural environment. Science 296:1127–1129. https://doi.org/10.1126/science.1070633

Krishna KV, Bharathi N, Malaviya A (2022) An updated review on advancement in fermentative production strategies for biobutanol using *Clostridium* species. Environ Sci Pollut Res 29: 47988–48019. https://doi.org/10.1007/s11356-022-20637-9

Lamont RJ, Jenkinson HF (1998) Life below the gum line: pathogenic mechanisms of *Porphyromonas gingivalis*. Microbiol Mol Biol Rev 62:1244–1263. https://doi.org/10.1128/MMBR.62.4.1244-1263

Lantican NB, Diaz Ma GQ, Cantera JJL et al (2011) Microbial community of a volcanic mudspring in The Philippines as revealed by 16S rDNA sequence analysis and fluorescence in situ hybridization. World J Microb Biotechnol 27:859–867. https://doi.org/10.1007/s11274-010-0528-y

Leis B, Angelov A, Mientus M et al (2015) Identification of novel esterase active enzymes from hot environments by use of the host bacterium *Thermus thermophilus*. Front Microbiol 6:275. https://doi.org/10.3389/frricb.2015.00275

Liu CG, Xue C, Lin YH et al (2013) Redox potential control and applications in microaerobic and anaerobic fermentations. Biotechnol Adv 31(2):257–265. https://doi.org/10.1016/j.biotechadv.2012.11.005

Lu F, Bize A, Guillot A et al (2014) Metaproteomics of cellulose methanisation under thermophilic conditions reveals a surprisingly high proteolytic activity. ISME J 8(1):88–102. https://doi.org/10.1038/ismej.2013.120

McClung LS (1935) Studies on anaerobic bacteria VI. J Bacteriol 29(2):189–203

Mitchell WJ (2001) Biology and physiology. In: Bahl H, Durre P (eds) Clostridia biotechnology and medical applications. Wiley, New York, pp 49–104

Mohammadali M, Davies J (2018) Antimicrobial resistance genes and wastewater treatment. In: Keen PL, Fugere R (eds) Antimicrobial resistance in wastewater treatment processes, 1st edn. Wiley, Hoboken, pp 1–14

Najar IN, Sherpa MT, Das S et al (2020) Diversity analysis and metagenomic insights into the antibiotic resistance and metal resistances among Himalayan hot spring bacteriobiome-insinuating inherent environmental baseline levels of antibiotic and metal tolerance. J Glob Antimicrob Res 21:342–352. https://doi.org/10.1016/j.jgar.2020.03.026

Padden AN, Dillon VM, Edmonds J et al (1999) An indigo-reducing moderate thermophile from a woad vat, *clostridium isatidis* sp. nov. Int J Syst Bacteriol 49(3):1025–1031. https://doi.org/10.1099/00207713-49-3-1025

Padgett PJ, Cover WH, Krieg NR (1982) The microaerophile *Spirillum volutans*: cultivation on complex liquid and solid media. Appl Environ Microbiol 43(2):469–477. https://doi.org/10.1128/AEM.43.2.469-477

Pedersen MB, Gaudu P, Lechardeur D et al (2012) Aerobic respiration metabolism in lactic acid bacteria and uses in biotechnology. Annu Rev Food Sci Technol 3(1):37–58. https://doi.org/10.1146/annurev-food-022811-101255

Saghatelyan A, Margaryan A, Panosyan H et al (2021) Microbial diversity of terrestrial Geothermal Springs in Armenia and Nagorno-Karabakh: a review. Microorganisms 9:1473. https://doi.org/10.3390/microorganisms9071473

Schmidt TM, DeLong EF, Pace NR (1991) Analysis of marine picoplankton community by 16S rRNA gene cloning and sequencing. J Bacteriol 173:4371–4378. https://doi.org/10.1128/JB.173.14.4371-4378.1991

Scully SM, Orlygsson J (2020) Branched-chain amino acid catabolism of *Thermoanaerobacter pseudoethanolicus* reveals potential route to branched-chain alcohol formation. Extremophiles 24(1):121–133. https://doi.org/10.1007/s00792-019-01140-5

Selmer T (2005) Anoxic testing and purification of enzymes. In: Durre P (ed) Handbook of clostridia. CRC Press, Boca Raton, FL, pp 71–98

Shahin M, Jamalo W, Verghese T et al (2003) Comparative evaluation of anoxomat and conventional anaerobic GasPak jar systems for the isolation of anaerobic bacteria. Med Princ Pract 12: 81–86. https://doi.org/10.1159/000069116

Shu WS, Huang LN (2021) Microbial diversity in extreme environments. Nat Rev Microbiol 20: 219–235. https://doi.org/10.1038/s41579-021-00648-y

Srinivas SP, Rao G, Mutharasan R (1988) Redox potential in anaerobic and microaerobic fermentation. In: Fung DYC (ed) Handbook on anaerobic fermentations. Marcel-Dekker, New York, pp 147–186

Takada K, Hirasawa M (1998) Isolation and characterization of non-pigmented rough colony of *Porphyromonas gingivalis* from periodontitis. Anaerobe 4(5):233–240. https://doi.org/10.1006/anae.1998.0168

Ueda Y, Yamamoto M, Urasaki T et al (2007) Sequencing and reverse transcription-polymerase chain reaction (RT-PCR) analysis of four hydrogenase gene clusters from an obligately autotrophic hydrogen-oxidizing bacterium *Hydrogenobacter thermophilus* T-6. J Biosci Bioeng 104:470–475. https://doi.org/10.1263/jbb.104.470

Van de Werken HJG, Verhaart MRA, VanFossen AL et al (2008) Hydrogenomics of the extremely thermophilic bacterium *Caldicellulosiruptor saccharolyticus*. Appl Environ Microbiol 74: 6720–6729. https://doi.org/10.1128/AEM.00968-08

Van Gylswyk NO, Van der Toorn JJTK (1987) *Clostridium aerotolerans* sp. nov., a xylanolytic bacterium from corn Stover and from the rumina of sheep fed corn Stover. Int J Syst Bacteriol 37(2):102–105. https://doi.org/10.1099/00207713-37-2-102

Verbeke T, Zhang X, Henrissat B et al (2013) Genomic evaluation of *Thermoanaerobacter* spp. for the construction of designer co-cultures to improve lignocellulosic biofuel production. PLoS One 8:e59362. https://doi.org/10.1371/journal.pone.0059362

Verma G, Kumar V, Satyanarayana T (2022) Genomic attributes of thermophilic and hyperthermophilic bacteria and archaea. World J Microbiol Biotechnol 38:8. https://doi.org/10.1007/s11274-022-03327-z

Wiegel J, Ljungdahl LG (1981) *Thermoanaerobacter ethanolicus* gen. nov., spec. nov., a new, extreme thermophilic, anaerobic bacterium. Arch Microbiol 128:343–348. https://doi.org/10.1007/BF00405910

Williams E, Lowe TM, Savas J et al (2007) Microarray analysis of the hyperthermophilic archaeon *Pyrococcus furiosus* exposed to gamma irradiation. Extremophiles 11:19–29. https://doi.org/10.1007/s00792-006-0002-9

Willis AT (1969) Chapter III Techniques for the study of anaerobic, spore-forming bacteria. Methods Microbiol 4:79–115

Willquist K, Pawar SS, Van Niel EWJ (2011) Reassessment of hydrogen tolerance in *Caldicellulosiruptor saccharolyticus*. Microb Cell Fact 10(1):111. https://doi.org/10.1186/1475-2859-10-111

Wilson CM, Rodriguez M, Johnson CM et al (2013) Global transcriptome analysis of *Clostridium thermocellum* ATCC 27405 during growth on dilute acid pretreated populus and switchgrass. Biotechnol Biofuels 6:179. https://doi.org/10.1186/1754-6834-6-179

Zengler K, Toledo G, Rappe M et al (2002) Cultivating the uncultured. Proc Natl Acad Sci USA 99(24):15681–15686. https://doi.org/10.1073/pnas.252630999

Part II
Biochemistry and Physiology of Thermophiles

Physiology of Chemoheterotrophic Thermoanaerobes

Ed W. J. van Niel, Sean M. Scully, and Johann Orlygsson

Abstract The industry is increasingly discovering the potential of strict anaerobic thermophilic bacteria, as there are many applications to consider, including the production of stable thermozymes and degradation of lignocellulosic materials, proteins, and alkanes. However, the metabolism of these thermoanaerobes is still poorly understood and so far, there is comprehensive knowledge of only a few model species. Of the studies published to date, the vast majority have focused almost exclusively on the anaerobic catabolism of carbohydrates, including polysaccharides, hexoses, pentoses, and disaccharides. While research on the fermentation of proteins and amino acids has enjoyed some recent developments, the exploration of the degradation of lipids and alkanes under these conditions has only just begun. In this chapter, the different central carbon pathways are discussed along with the utilization of alternative intracellular energy carriers other than ATP, namely GTP and pyrophosphate. Subsequent steps and approaches in deeper understanding of the physiology are concisely discussed.

Abbreviations

AcCoA	Acetyl-Coenzyme A
ADP	Adenine diphosphate
ATP	Adenine triphosphate
ED	Entner-Doudoroff pathway
EMP	Embden-Meyerhof pathway
Fd	Ferredoxin
FNR	Ferredoxin:NAD(P)$^+$ oxidoreductase

E. W. J. van Niel (✉)
Division of Applied Microbiology, Lund University, Lund, Sweden
e-mail: ed.van_niel@tmb.lth.se

S. M. Scully · J. Orlygsson
Faculty of Natural Resource Sciences, School of Health, Business and Natural Sciences, University of Akureyri, Borgir, Akureyri, Iceland

S. M. Scully, J. Orlygsson (eds.), *Thermophilic Anaerobes*, Emerging Issues and Trends, https://doi.org/10.1007/978-3-031-41720-7_5

H_2ase	Hydrogenase
NADH	Nicotinamide adenine dinucleotide
NADPH	Nicotinamide adenine dinucleotide phosphate
PFL	Pyruvate formate lyase
PFOR	Pyruvate ferredoxin:oxidoreductase
pH_2	partial hydrogen pressure in the gas phase
PPi	Pyrophosphate
PPP	Pentose phosphate pathway
PTS	Phosphotransferase system
TCA	Tricarboxylic acid cycle
VFA	Volatile fatty acids

1 Introduction

In general, anaerobic microorganisms, which include also various extremophiles, are essential contributors to global elemental biogeochemical cycles. As an example, the strictly anaerobic heterotrophic thermophilic bacteria inhabiting hot springs and other warm environments take part in at least the global cycling of carbon, sulfur, and nitrogen. Amongst thermophilic anaerobes, the majority characterized thus far have their optimum pH between 5.0 and 8.5 and some can be spore forming (Kristjansson and Stetter 1998). These anaerobic bacteria earn their energy through fermentation or anaerobic respiration, such as denitrification or sulfate reduction (Martinez-Spinosa 2020), with a number of organisms combining these two strategies to varying degrees. Some thermophiles, such as *Thermotoga* species, can use thiosulfate and elemental sulfur as electron acceptors (Fardeau et al. 1997). Consequently, the utilization of these alternative electron acceptors improves growth, and their fermentation product profiles are altered such that lower quantities of reduced end product such as hydrogen and alanine are formed. These heterotrophic thermophiles reside both in naturally niches, such as geothermal and volcanic areas, and in man-made thermophilic habitats, such as biological waste treatment plants, mines, and compost piles (Mehti and Satyanarayana 2013).

Like their mesophilic counterparts, there are examples of thermophilic anaerobes using all of the major biomolecule categories including carbohydrates, lipids, proteins, and nucleic acids as substrates, although the latter has received scant attention in the literature. Among the substrates that are fermented by these organisms, carbohydrates represent the majority, which include polysaccharides, oligosaccharides, pectin, disaccharides, and monosaccharides. Typically, the end products produced from carbohydrates include a mixture of organic acids, such as acetic and lactic acids, and ethanol as well as hydrogen, CO_2, and alanine. In addition, several thermophiles are also proteolytic, often using both exo- and endoproteases, and fermenting select amino acids to varying degrees depending on the environmental conditions, which similarly yield mixtures of the aforementioned gases and acids. Therefore, these bacteria are of commercial interest for their production of carbohydrate hydrolases and proteolytic enzymes. The variety of metabolic pathways that convert these various substrates are described in this chapter.

2 Central Carbon Pathways

In microbial metabolism, the central carbon pathways or catabolic pathways possess the highest fluxes and foresee the cell with reducing equivalents (NADH and NADPH), energy carriers (ATP), and precursors for biosynthesis of new cell material. The majority of the chemoheterotrophic thermoanaerobes can convert pentoses, hexoses, disaccharides, oligosaccharides, and polysaccharides to various end products. Depending on the species, other compounds such as methanol, glycerol, pectin, a polysaccharide containing D-galacturonic acid residues, and various other organic compounds can be degraded. Unfortunately, the utilization of carbon compounds is not standardized leaving many knowledge gaps when novel strains are characterized. Despite the prevalent nature of carbohydrate metabolism among thermoanaerobes, very few have been studied in depth although *Acetivibrio thermocellus* (formerly *Clostridium thermocellum)*, and species of *Thermotoga* and *Caldicellulosiruptor* have received great scrutiny in this regard. The sugars are taken up by ABC transporters in most of these thermophiles and many own a phosphotransferase system for fructose (Conners et al. 2005; Van de Werken et al. 2008; Lin et al. 2011).

2.1 Catabolite Repression and Transporters

In contrast to *Thermotoga maritima*, *Caldicellulosiruptor* species have been tested on different monosaccharide combinations and displayed no catabolite repression, allowing for simultaneous uptake and fermentation of multiple monosaccharides (Van de Werken et al. 2008; Van Fossen et al. 2009; Vongkampang et al. 2021). In a rich sugar mixture, *C. saccharolyticus* showed the following preference: fructose > xylose/arabinose > mannose/glucose/galactose (Van Fossen et al. 2009). Fructose is transported with a phosphotransferase system (PTS), which was expressed in the presence of fructose in the medium (Van Fossen et al. 2009). A similar phenomenon and transporter were observed for *Thermotoga* strain RQ2, the only *Thermotoga* strain possessing a PTS for fructose (Frock et al. 2012). *Thermotoga* species possess a kind of catabolite repression, as in a mixture of fructose, galactose, glucose, xylose, and arabinose, only glucose and xylose are first taken up simultaneously. Both these sugars were taken up by the cell with the same transporter (Frock et al. 2012). Later studies revealed that the xylose transporter of *C. saccharolyticus* also transported glucose, but a dedicated glucose transporter was only expressed in the absence of xylose (Björkmalm et al. 2018). Remarkably, even though *Thermoanaerobacterium thermosaccharolyticum* strain W16 possesses a transporter for glucose and for xylose, in mixtures of glucose and xylose high concentrations of either sugar inhibited the uptake of the other (Zhao et al. 2019, 2020). In this case, the glucose is transported by a PTS and xylose with an ABC transporter, as has been described also for *Thermoanaerobacter* sp. X514 (Lin et al. 2011). On the other hand, *Thermoanaerobacter thermohydrosulfuricus* contains two xylose transporters, of which one is induced at relatively high xylose concentrations (>50 mM) that does not transport, nor is inhibited by, glucose (Cook et al. 1994). The second transporter operates at low xylose concentrations (~5–10 mM), and transports glucose as well. A transcriptional regulation mechanism for expressing coutilization of hexoses and pentoses has been presented for *Thermoanaerobacter* (Lin et al. 2011), which has no relationship to the known catabolite repression mechanism. Interestingly, *C. kronotskyensis* relied on glucose uptake via the xylose transporter only, as a dedicated glucose transporter appears to be lacking (Vongkampang et al. 2021). In the presence of both glucose and xylose, these xylose/glucose transporters are prone to undergo mutual competitive inhibition (Vongkampang et al. 2021; Cook et al. 1994). It was speculated that *C. kronotskyensis* has adapted to glucose uptake in the form of cellobiose instead, as hinted by its faster uptake compared to glucose. Similarly, *T. maritima* grows slower on monosaccharides than on polysaccharides (Chhabra et al. 2003), which might be due to that glucose is not stable at higher temperatures (Conners et al. 2006). The simultaneous conversion of various saccharides by one species reflects the way these types of thermophiles have adapted to environments low in free carbohydrates (Roh et al. 2002). This makes them very effective and simplifying industrial bioprocesses using complex feedstocks. This effectiveness can be further enhanced by designing appropriate cocultures of bacteria with complementary metabolic capabilities such as described for *T. saccharolyticum* and *C. thermocellum* (Jacobson et al. 2020).

2.2 Hexose Metabolism

The Embden-Meyerhof-Parnas pathway (EMP) is most commonly used central metabolic pathway among the facultative and strict anaerobic chemoheterotrophic thermophiles (Flamholz et al. 2013) (Fig. 1). In fermentation mode, these anaerobes rely mostly on substrate-level phosphorylation (SLP) for ATP production and the EMP yields 2 mol ATP/mol hexose. A small percentage of organisms are known to possess the Entner-Doudoroff pathway (ED) commonly associated with Gram-negative organisms (Fig. 1), although only a few anaerobes are known to rely on the ED only. The likely reason for this preference for EMP over ED is that the latter is inferior anaerobically as the lower energy yield is only 1 mol ATP/mol hexose, whereas under aerobic conditions the ATP loss at substrate level is well compensated by the oxidative phosphorylation. However, in many thermophiles, the pyruvate is further converted to acetate, which is accompanied by an additional production of ATP via SLP. In that case, the overall amount of ATP per hexose generated is 3 and 4 via the ED and EMP pathways, respectively, making the loss of

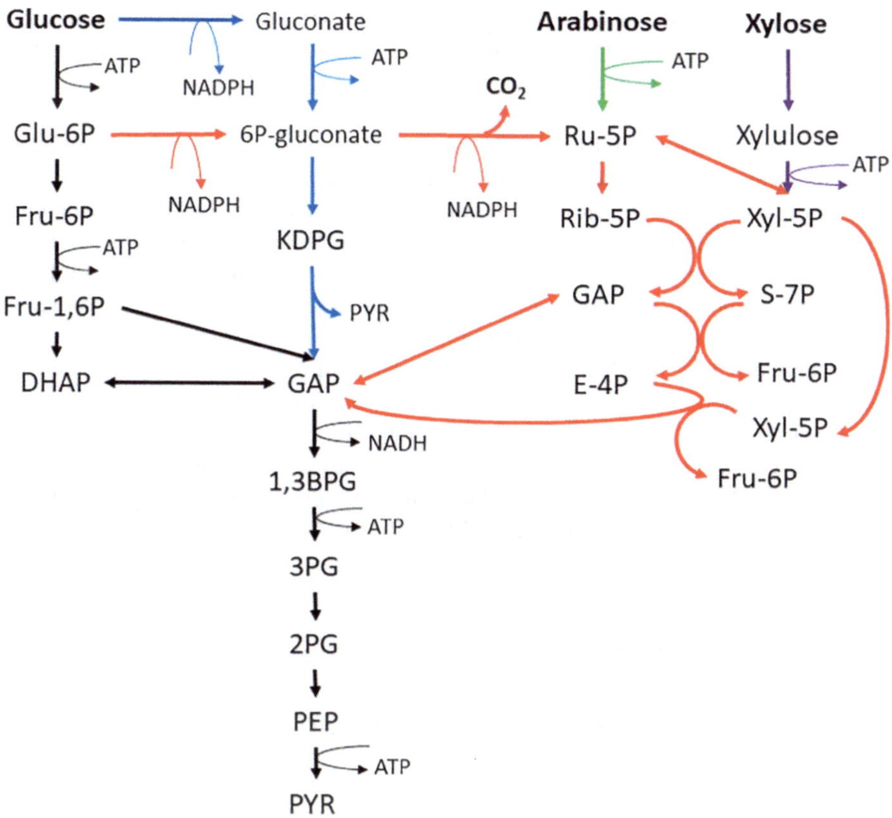

Fig. 1 Combination of the central carbon pathways in thermophilic bacteria: The Embden Meyerhof pathway (EMP) (*black* arrows); Entner-Doudoroff pathway (ED) (*blue* arrows); Pentose Phosphate Pathway (PPP) (*red* arrows); Arabinose pathway (*green* arrow); Xylose pathway (*purple* arrows). In the ED pathway, the C6-intermediate splits into two C3-intermediates, i.e., pyruvate (PYR) and glyceraldehyde 3 phosphate (GAP) and thus enters the pathway that it has in common with the EMP

1 ATP durable. Alternatively, it has been proposed that pathways such as the ED display higher redox carrier (NAD(P)H) formation fluxes relative to energy carrier (ATP) formation fluxes, thereby severely affecting the growth rate and cell yields (Van Niel et al. 2017). Therefore, these microorganisms contain a dedicated metabolic "highway" to convert more substrate to product as a means to extrapolate sufficient energy for proliferation as has been depicted in the mesophilic *Zymomonas mobilis* (Fuhrer et al. 2005). However, studies with *T. maritima* have shown that the ED and EMP are running in parallel with a flux distribution of 15% and 85%, respectively (Selig et al. 1997). This was later confirmed with an in silico metabolic model of *T. maritima* (Nogales et al. 2012). The low flux through the ED accounts for that it plays a role in biosynthesis, which may be linked to keeping an optimal ratio of reduced ferredoxin over NADH (Nogales et al. 2012). *T. maritima* also

possesses a complete pentose phosphate pathway (PPP), but an increasing flux through the oxidative part of the PPP apparently may be detrimental to growth, whereas it is the opposite for increasing fluxes through the EMP or ED.

Fewer studies have focused on the metabolism of other hexoses. The thermophilic bacteria studied so far that can ferment galactose do so via the Leloir pathway (Van de Werken et al. 2008; Fushinobu 2021; Qian et al. 2009), which include species of *Caldicellulosiruptor*, *Marinithermus*, *Thermoanaerobacter*, *Thermotoga*, and *Thermus*. Rhamnose, a deoxyhexose that is available in significant quantities in pectin and hemicelluloses, can be converted into several fermentation products, which was studied in depth in *C. saccharolyticus*. It ferments rhamnose to 1,2-propanediol, acetate, CO_2, and H_2 in a 1:1:1:1 ratio (Bielen et al. 2013). This sugar is not effective for H_2 production as all NADH are used to reduce lactaldehyde into 1,2-propanediol. Nonetheless, the latter is a commodity chemical that is of industrial interest. Many other thermophilic bacteria, including other *Caldicellulosiruptor* species, can also ferment rhamnose and fucose to equimolar amounts of 1,2-propanediol plus similar fermentation products as mentioned above (Weimer et al. 1984; Ingvadottir et al. 2017, 2018).

2.3 Pentose Metabolism

Pentoses are rich energy carriers and carbon sources in nature, especially in the form of hemicellulose; therefore, it is no surprise that many thermophiles can ferment these sugars. Moreover, preference for xylose over other sugars is quite common among anaerobic thermophilic bacteria. This is similar for arabinose as has been observed at least in *C. saccharolyticus* (Van Fossen et al. 2009; Björkmalm et al. 2018), but not so in *Thermotoga* species (Frock et al. 2012). Conversion of xylose is through the pentose phosphate pathway (PPP) (Fig. 1) in thermophilic bacteria (Gottschalk 1986), which is entered as the metabolite xylulose-5P via a xylose isomerase and xylulokinase. L-arabinose (pentose in hemicellulose) and D-arabinose are also converted to xylulose-5P with their respective isomerase and kinase (Van de Werken et al. 2008; Rodionov et al. 2021). However, the gene coding for D-arabinose isomerase has been annotated also as a L-fucose isomerase (Van de Werken et al. 2008) and the purified recombinant protein could use both L-fucose and D-arabinose, among other sugars, as substrates (Ju and Oh 2010). Therefore, D-arabinose is also funneled into the fucose pathway resulting in the production of significant amounts of ethylene glycol next to acetate (Isern et al. 2013).

Several of these thermophiles studied so far are missing the oxidative part of the PPP (Van de Werken et al. 2008; Feng et al. 2009; Cordova et al. 2016), including species of *Caldicellulosiruptor*, *Thermoanaerobacter*, and *Thermus*. Therefore, the source of NADPH production must be positioned elsewhere in the metabolic network. For *Caldicellulosiruptor* species, the source has not yet been found, but isocitrate dehydrogenase has been identified for *Thermus thermophile* (Cordova et al. 2016), which has been indicated to be a rate-limiting step in biosynthesis due to low fluxes through the TCA cycle. Otherwise, Ferredoxin-$NADP^+$ reductase has been observed as the catalyst for NADPH production in several thermophilic anaerobes (Feng et al. 2009; Lacis and Lawford 1991).

One of the exceptions among thermophiles is *Cl. thermocellum* that can take up xylose, but accumulation of the intermediate xylitol inhibits further conversion (Verbeke et al. 2017). However, recently heterologous expression of xylose isomerase and xylose kinase plus overexpression of the transketolases could turn *Cl. thermocellum* into a successful xylose-consumer (Rangel et al. 2020).

2.4 Energy Currencies

ATP is considered the universal energy carrier in the cell (Stryer 1988). It is produced in the glycolytic pathways, TCA cycle, and in aerobic microorganisms from NADH via the electron transport chain—building up a proton motive force (pmf)—and ATPase. The strict anaerobes rely on ATP production at substrate phosphorylation level in the glycolytic pathways and via acetate formation if they possess an acetate kinase. Some thermophiles, like *Thermoanaerobacter* species, might also form ATP from a bidirectional PEP carboxykinase (Verbeke et al. 2013). Finally, anaerobic thermophiles can also use the pmf to produce ATP thereby mediated by three types of ATP synthases, i.e., the F-type (most common), the V-type (linked to vacuoles), and A-type (Archaeal) (Kuhlbrandt and Davies 2016).

However, among the anaerobic thermophilic bacteria, pyrophosphate (PPi) and GTP have been notified as additional energy carriers in several, mainly catabolic, reactions (Verbeke et al. 2017; Bielen et al. 2010; Zhou et al. 2013) (Fig. 2). GTP can be generated from ATP mediated by an NDP-kinase and in the catabolic enzymes, phosphoglycerate kinase and PEP carboxykinase, as depicted in *Cl. thermocellum* (Zhou et al. 2013). In the latter organism, GTP is then used to phosphorylate glucose and PPi to phosphorylate fructose-6P in the EMP.

PPi in the role of energy carrier occurs mainly in plants, several unicellular eukaryotes, and thermophilic prokaryotes (Mertens 1991). The source of PPi is the nucleotide polymerization and protein synthesis metabolism that is especially active under growth conditions. In those reactions, PPi is a by-product that normally is hydrolyzed to two inorganic phosphates with a cytosolic pyrophosphatase. This is an essential reaction that makes the formation of these macromolecules thermodynamically possible. However, if we focus on the thermophilic bacteria only, PPi can be removed by the kinases in the central carbon pathways (pyrophosphate-dependent 6-phosphofructokinase and pyruvate phosphate dikinase). So far, PPi as an energy carrier has been studied only for a few thermophiles. It has been revealed that *C. saccharolyticus* is relying on PPi as main energy carrier, being present at higher concentrations (approx. 4 ± 2 mM) than ATP (approx. 0.43 ± 0.07 mM), at least during growth conditions (Bielen et al. 2010). During exponential growth, the ATP concentration is even lower than the ADP concentration, which is normally a sign of starvation. Relatively high PPi levels (0.8–1.5 mM) have also been observed in the

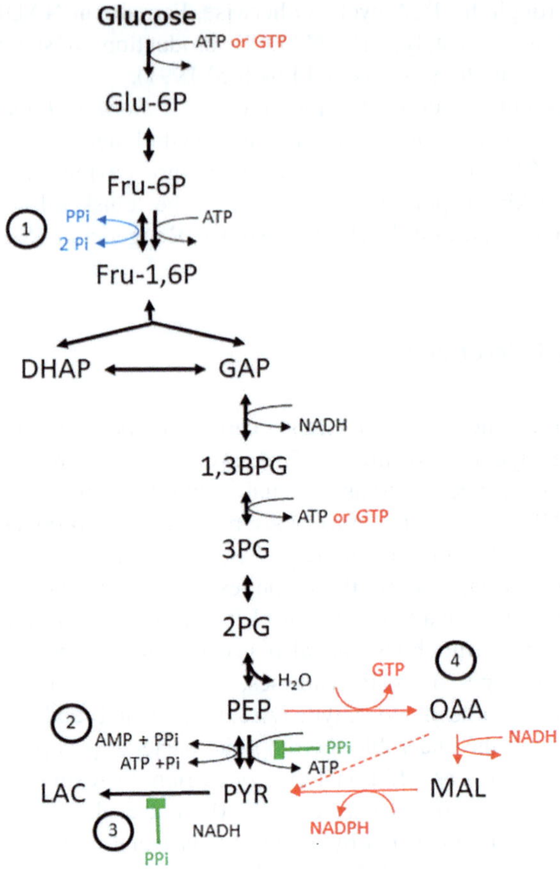

Fig. 2 Combination of Embden-Meyerhof pathways of several thermophilic bacteria. *Black*: common backbone but specialized for *C. saccharolyticus*; *Red*: special metabolism of *Cl. thermocellum*. Specific differences in thermophilic bacteria: (1) ATP-dependent phosphofructokinase and pyrophosphate-dependent phosphofructokinase of *C. saccharolyticus* and *Cl thermocellum* only possessing the latter (*blue*); (2) pyruvate phosphodikinase and phosphokinase, of which the latter is inhibited by PPi (*green*; see main text); (3) inhibition of lactate dehydrogenase by pyrophosphate (PPi); (4) pathway from PEP to PYR due to a missing pyruvate kinase in *Cl. thermocellum*. *Glu-6P* glucose 6 phosphate, *Fru-6P* fructose 6-phosphate, *Fru-1,6P* fructose bisphosphate, *DHAP* dihydroxyacetone phosphate, *1,3BPG* 1,3-bisphosphoglycerate, *3PG* 3-phosphoglycerate, *2PG* 2-phosphoglycerate, *PEP* phospho-enolpyruvate, *PYR* pyruvate, *LAC* lactate, *OAA* oxaloacetate, *MAL* malate. Modified from Bielen et al. (2010) and Zhou et al. (2013)

mid-log phases of *Cl. pasteurianum* and *Moorella* (formerly *Clostridium*) *thermoacetica* (Heinonen and Drake 1988). Decline of these concentrations sets in at the end of the log phase until the PPi levels are similar to those in *Escherichia coli* (0.3 mM); in the latter the level remains the same, irrespective of the growth phase.

That PPi is deeply involved in the catabolism of *C. saccharolyticus*, is further reflected in its role of inhibitor of ADP-dependent pyruvate kinase[42] and lactate dehydrogenase (LDH) (Fig. 2) (Willquist and van Niel 2010). Interestingly, this bacterium also possesses a cytosolic and a membrane-bound proton-translocating pyrophosphatases. The latter may contribute to maintaining a pmf (Serrano et al. 2004). The presented studies strongly indicate that PPi plays a central role in the catabolic metabolism in several anaerobic thermophilic bacteria. Whether this exemplifies that PPi is a primitive form of energy carrier was recently debated (Wimmer et al. 2021) because its hydrolysis is a pivotal irreversible event that makes biosynthesis of macromolecules possible. The latter is evident, nevertheless several microorganisms and plants have opted to utilize PPi as a central energy carrier as demonstrated in *C. saccharolyticus* (Bielen et al. 2010).

3 Cellulolytic Fermenting Thermophiles

The interest in thermophilic enzymes degrading microbes and their enzymes has increased dramatically in the past few decades. Cellulose and hemicellulose, part of lignocellulose in plants and some macroalgae, are the most dominant renewable biomass available on Earth and is a material of intense interest for the production of both fine chemicals and biofuels. There are several interesting thermophilic bacteria capable of degradation of cellulose anaerobically. The most well-known and highly studied cellulolytic thermophile is the moderately thermophilic *Clostridium thermocellum*, especially because of their possession of a working scaffold, the so called cellulosome (Doi and Kosugi 2004; Fontes and Gilbert 2010; Akinosho et al. 2014). Other well-known cellulolytic thermophiles belong to other *Clostridium* species, *Caldicellulosiruptor* species (Rainey et al. 1994; Kadar et al. 2004; Chung et al. 2015) and *Thermotoga* species (Liebl 2001; Cheng et al. 2011; Obeng et al. 2017). In order to achieve the complete deconstruction of crystalline cellulose into glucose, multiple enzymes working synergistically are required: cellobiohydrolases (CBH), endoglucanases (EGL), and B-glucosidases (BGL) (Wood and Garcia-Campayo 1990). For further information on cellulose degradation with thermophiles, we refer to (Scully and Orlygsson 2023).

Unlike cellulose, hemicellulose tends to have random, amorphous structures that are only comprised of 500–300 residues (Glazer and Nikaido 2007). Furthermore, there is a tremendous diversity of hemicellulose that can be comprised of a mixture of hexoses, such as glucose, galactose, and mannose, pentoses (D-arabinose, D-xylose), methylpentose (L-rhamnose), and uronic acids each in addition to various modifications such as esterification and etherification. Because of the heterogeny of hemicellulose biomass, there is a wide variety of enzymes needed for its degradation. The main types of hemicellulases are endoxylanases, β-xylosidase, arabinofuranosidase, and acetyl-xylan esterase among others (Ebringerová 2005). Thermophilic enzymes have advantage over mesophilic enzymes and can withstand the extreme conditions needed in the various industries, e.g., bleaching of pulp in the paper industry. Therefore, enzymes from extremophilic bacteria have gained increased interest in the last decades. Many thermophilic and hyperthermophilic bacteria have been reported to produce endoxylanases such as *Thermotoga* (Bok et al. 1994; Winterhalter et al. 1995), *Clostridium* (Rani and Nand 2000; Heinze et al. 2017), *Caldicellulosiruptor* (Crosby et al. 2022), and various aerobic thermophiles (*Geobacillus*, *Bacillus*, and *Streptomyces* species). For further information on hemicellulose degradation with thermophiles, see Scully and Orlygsson (2023).

4 Protein and Amino Acid-Degrading Thermophiles

Studies of thermophiles have mainly focused on carbohydrate degradation, partly because of the vast availability and broad range of sugars they can ferment. Nonetheless, proteolytic thermophiles have become of increased interest in connection to improving degradation of protein-rich wastes. The underlying reason is that these microorganisms possess thermostable enzymes, such as proteases, peptidases, and keratinases, that can withstand harsh industrial conditions and/or can tackle recalcitrant animal proteins (Suzuki et al. 2006; Elleuche et al. 2014). Microorganisms need several types of proteases and peptidases for proper protein degradation. Proteases can be distinguished between highly specific to highly promiscuous. Exopeptidases mainly cleave the terminal peptide bonds releasing single amino acids at the N-terminus (aminopeptidases) or the C-terminus (carboxypeptidases) (Mótyán et al. 2013). Apart from them, there are also dipeptidyl peptidases, tripeptidyl peptidases, and peptidyl dipeptidases. Keratinases, as the name suggest, break down keratin (as found in feathers), are used in industrial sectors including feather recycling, leather, textile, biofertilizers, feed, and cosmetics (Brandelli 2008; Hassan et al. 2020). Keratin is a fibrous and tough protein and due to its recalcitrant nature cannot be degraded by common proteases. Interestingly, keratinases of *Thermoanaerobacter*, *Thermococcus,* and *Thermosipho* have potential, thanks to excellent performance properties, to degrade also recalcitrant proteins such as prions (Gupta et al. 2013). Other thermophilic anaerobic bacteria that are under investigation for their proteolytic potential include species of *Caloramator* (Tarlera and Stams 1999), *Coprothermobacter* (Sasaki et al. 2011), *Thermotoga* (Ward et al. 2002), and *Fervidobacterium* (Akram et al. 2022).

In addition to exploitation of thermozymes, there is also production of potential chemicals in amino acid fermentations. For instance, longer chain and branched-chain alcohols can be produced from anaerobic degradation of amino acids that can be of interest to the chemical industry. Organic wastes containing proteins vary from municipality waste, food industry (dairy waste) to slaughterhouse waste. As such waste treatment at mesophilic conditions may be a potential source of pathogens, it is safer to perform these processes at higher temperatures with strict anaerobic thermophiles with the benefit of value-added product formation.

In general, proteins are first hydrolyzed extracellularly to (oligo)peptides and amino acids, which are then transported into the cell and further degraded to volatile fatty acids (VFA), hydrogen, and ammonium (McInerney 1988). Genes coding for transporters of amino acids and oligopeptides have been annotated in several thermophiles, such as the ABC transporters in *C. saccharolyticus* (Van Werken et al. 2008) and *T. maritima* (Nelson et al. 1999). As a carbon source for biosynthesis, amino acids are degraded to the level of acetyl-CoA, propionyl-CoA, pyruvate, and/or α-ketoglutarate, which are metabolites on the crossroads of catabolism and anabolism.

4.1 Proteolytic Thermophiles

For the hydrolysis of the peptide bonds, proteolytic thermophiles use several types of proteases that can be peptidases with narrow or a broad specificity. These enzymes can be characterized by their active site chemistry, which include acid proteases being active at low pH, alkaline proteases being active at neutral to high pH, and metalloproteases containing a metal ion at their active site. Hydrolysis starts outside the cell with exocellular proteases cleaving the protein into oligopeptides and amino acids. Oligopeptides can be further hydrolyzed with intracellular proteases. As mesophilic proteases may be slow in degrading especially insoluble proteins (Hobson and Wallace 1982), thermophilic conditions have seen benefits of protein hydrolysis due to thermal denaturation that causes loss of secondary structure of the protein. Hardly any studies have focused on the regulation of proteases in thermophiles. However, older studies with mesophilic strict anaerobes revealed that these proteases are highly induced under energy-limited conditions, such as sugar-limited batch cultures entering the stationary phase or under low dilution rates in continuous cultures (Allison and Macfarlane 1990). Extracellular protease activity was also seen to increase three-fold in cocultures with a methanogen (Tarlera and Stams 1999). Despite *T. maritima* possessing at least 35 proteases and peptidases of which about 12 have a function outside the cell as annotated by Ward and coworkers (2002), its growth on peptides as sole carbon and energy source has still not been reported.

4.2 Amino Acid-Degrading Thermophiles

In nature, there are two types of amino acid-degrading mechanisms: (1) amino acid fermentation, and (2) conversion of pairs of amino acids via the Stickland reaction, in which one amino acid is the oxidant and the other the reductant receiving the electrons (Stickland 1934). Amino acid fermentation will be limited outside a syntrophic consortium as the deamination step is a thermodynamically constraint due to hydrogen production (Orlygsson et al. 1995). Alternatively, in the absence of hydrogenotrophic microorganisms, e.g., methanogens, chemical electron acceptors improve also amino acid fermentation, such as has been observed with *Thermoanaerobacter* species reducing thiosulfate to sulfide (Faudon et al. 1997). Fermentation products of amino acids consist of VFA, ammonium, and hydrogen, of which the latter needs to be immediately removed by the hydrogenotrophic partner to sustain the fermentation flow. In the presence of adequate amounts of thiosulfate, hydrogen is not produced as the electrons produced in the oxidation step go to this electron acceptor that is reduced to sulfide. Interestingly, several *Thermoanaerobacter* and *Caldanaerobacter* species contain promiscuous alcohol dehydrogenases (ADH) that can reduce, with both NADH and NADPH, these VFA further to their corresponding alcohol (Scully et al. 2015). Moreover, *Thermoanaerobacter* species possess a number of ADHs having different substrate specificities for either primary or secondary alcohols (Lamed and Zeikus 1981) and an aldehyde:ferredoxin oxidoreductase (Hitschler et al. 2018) making them equipped for formation of a variety of longer chain and branched-chain alcohols. Indeed, *T. pseudethanolicus* and *Thermoanaerobacter* strain AK85 can degrade the branched-chain amino acids isoleucine, leucine, threonine, and valine to their respective fatty acids (Scully and Orlygsson 2019, 2020). The corresponding alcohols were formed when thiosulfate was present, although the concentrations remain relatively low. The amino acid fermentation process needs to be further optimized to increase product yields, including temperature, pH, substrate concentrations, and removal of accumulating inhibitors. Of these inhibitors, hydrogen is the most important, as has been demonstrated by the effect of scavenging it through using thiosulfate or methanogens as described above. During active fermentation, hydrogen may accumulate very fast in the liquid creating oversaturation due to mass transfer limitations to the gas phase (Pauss et al. 1990). The study of Tarlera and Stams (1999) is one of the few studies that investigated the critical pH_2 of amino acid fermentation inhibition by thermophiles. They found that fermentation of leucine and valine was terminated in *Caloramator proteoclasticus* at partial hydrogen pressure (pH_2) values of 7.8 and 4 kPa, respectively. This is a factor 5 to 8 lower than for the maximum pH_2 of sugar fermentation by *Caldicellulosiruptor saccharolyticus* (Willquist et al. 2011), showcasing the high sensitivity of amino acid fermentation to hydrogen. As mentioned before, this issue can be prevented by either adding a hydrogenotrophic methanogen, glycine (to stimulate the Stickland reaction) or thiosulfate. The former has the advantage that methane is produced as an added-value product, but the latter has the drawback of sulfide accumulation that can lead to corrosion and precipitation of precious metal ions negatively influencing the fermentation process.

Ammonium is another potential inhibitor of that accumulates during amino acid fermentation. Although, it might be ammonia that is the effective inhibitor as it can enter the cell through diffusion, and once inside, it will become protonated and thus increase the cytoplasmic pH. This will be a bigger issue at higher pH due to the high pK_a value of ammonia ($pK_a = 9.24$). There are thermophiles that are able to alleviate ammonium inhibition by forming alanine through amination of pyruvate, such as the moderately thermophilic amino acid fermenting *Clostridium* strain P2 (Orlygsson et al. 1995). Accumulation of ammonium and VFA plus alcohols can also be inhibitory as they contribute to the increase of the osmolarity of the fermentation broth.

5 Future Perspectives and Conclusions

So far, the physiology of only a few strict anaerobic thermophiles has been investigated in more detail, including *Cl. thermocellum*, and species of *Caldicellulosiruptor*, *Thermoanaerobacter,* and *Thermotoga*, and has been captured in a metabolic model (Nogales et al. 2012; Zhang et al. 2009, 2021; Tong et al. 2013; Garcia et al. 2020; Gautam and Xu 2021). Most attention has been paid to the sugar metabolism for production of biofuels and chemicals, due to the availability of substantial sugar-rich waste streams and lignocellulosic biomass resources. There is an increasing attention in the physiology of protein and amino acid degradation as this may lead to other types of biochemicals and production of enzymes of interest to the industry. Thermophilic lipid and alkane degradation is still a young study field of interest to the oil industry and for bioremediation of oil spills. Recently, a thermophilic community with a dominant novel thermophilic bacterium within the phylum actinobacteria, "*Candidatus* Syntraliphaticia," was described able to degrade n-alkanes (Liu et al. 2020). It improves degradation in the presence of fumarate, forming an n-alkyl-succinate, which is succeeded by an n-alkyl-succinyl-CoA that will undergo β-oxidation steps forming acetate, hydrogen/formate, and CO_2. This consortium also contains methanogens that further improve the performance by removing these inhibiting fermentation products through producing methane. Anaerobic alkane degradation works better under thermophilic than under mesophilic conditions due to improved accessibility of these compounds through increased solubility and desorption (Hlihor et al. 2017). This can be an issue deeper in oil reservoirs where temperatures are high.

To further improve yields of these products, metabolic engineering might be required for commercial application of these thermophiles. However, genetic protocols have been accomplished for only a few strict anaerobic thermophiles, including *Cl. thermocellum* (Tripathi et al. 2010), *Caldicellulosiruptor bescii* (Cha et al. 2013), and species of *Thermotoga* (Han et al. 2014), *Thermoanaerobacter,* and *Thermoanaerobacterium* (Shaw et al. 2010), of which a few are promoted to become platforms for cell factories. Tools and protocols for applying CRISPR/Cas genome editing in thermophiles are becoming available as well (Le and Sun 2022). Engineering studies started with removing genes to avoid by-product formation, which led to slight increases of yields but can make the performance of the strain weaker due to lower energy yields (Scully and Orlygsson 2017). Currently, the focus is on improving yields of fermentation products already generated by these bacteria such as hydrogen, ethanol (Crosby et al. 2019), butanol (Bhandiwad et al. 2014), and lactate (Mazzoli and Olson 2020), but also for the production of thermostable enzymes (Sahoo et al. 2020). The intention is to make several of these thermoanaerobes into cell factories or "thermochassis" (Vavitas et al. 2022) through introducing foreign genes to produce a wider palette of commodities.

References

Akinosho H, Yee K, Close D et al (2014) The emergence of *Clostridium thermocellum* as a high utility candidate for consolidated bioprocessing applications. Front Chem 2:66. https://doi.org/10.3389/fchem.2014.00066

Akram F, Ageel A, Shoaib M et al (2022) Multifarious revolutionary aspects of microbial keratinases: an efficient green technology for future generation with prospective applications. Environ Sci Pollut Res. https://doi.org/10.1007/s11356-022-23638-w

Allison C, Macfarlane GT (1990) Regulation of protease production in *Clostridium sporogenes*. Appl Environ Microbiol 56:3485–3490. doi: https://doi.org/10.1128/AEM.56.11.3485-3490.1990

Bhandiwad A, Shaw AJ, Guss A et al (2014) Metabolic engineering of *Thermoanaerobacterium saccharolyticum* for n-butanol production. Metab Eng 22:83–88. https://doi.org/10.1016/j.ymben.2013.10.012

Bielen B, Willquist K, Engman J et al (2010) Pyrophosphate as a central energy carrier in the hydrogen producing extreme-thermophile *Caldicellulosiruptor saccharolyticus*. FEMS Microbiol Lett 307:48–54. https://doi.org/10.1111/j.1574-6968.2010.01957.x

Bielen AAM, Verhaart MRA, van der Oost J et al (2013) Biohydrogen production by the thermophilic bacterium *Caldicellulosiruptor saccharolyticus*: current status and perspectives. Life 3:52–85

Björkmalm J, Byrne E, van Niel EWJ et al (2018) A non-linear model of hydrogen production by *Caldicellulosiruptor saccharolyticus* for diauxic-like consumption of lignocellulosic sugar mixtures. Biotechnol Biofuels 11:1. https://doi.org/10.1186/s13068-018-1171-3

Bok JD, Goers DA, Eveleigh DE (1994) Cellulase and xylanase systems of *Thermotoga neapolitana*. ACS Symp Ser 566:54–65

Brandelli A (2008) Bacterial keratinases: useful enzymes for bioprocessing agroindustrial wastes and beyond. Food Bioprocess Technol 1:105–116. https://doi.org/10.1007/s11947-007-0025-y

Cha M, Chung D, Elkins JG et al (2013) Metabolic engineering of *Caldicellulosiruptor bescii* yields increased hydrogen production from lignocellulosic biomass. Biotechnol Biofuels 6:85. https://doi.org/10.1186/1754-6834-6-85

Cheng YS, Ko TP, Wu TH et al (2011) Crystal structure and substrate-binding mode of cellulase 12A from *Thermotoga maritima*. Proteins 79(4):1193–1204. https://doi.org/10.1002/prot.22953

Chhabra SR, Schockley KR, Conners SB et al (2003) Carbohydrate-induced differential gene expression patterns in the hyperthermophilic bacterium *Thermotoga maritima*. J Biol Chem 278:7540–7552. https://doi.org/10.1074/jbc.M211748200

Chung D, Cha M, Snyder EN et al (2015) Cellulosic ethanol production via consolidated bioprocessing at 75 °C by engineered *Caldicellulosiruptor bescii*. Biotechnol Biofuels 8(1):163. https://doi.org/10.1186/s13068-015-0346-4

Conners SB, Montero CI, Comfort DA et al (2005) An expression-driven approach to the prediction of carbohydrate transport and utilization regulons in the hyperthermophilic bacterium *Thermotoga maritima*. J Bacteriol 187:7267–7282. https://doi.org/10.1128/JB.187.21.7267-7282.2005

Conners SB, Mongodin EF, Johnson MR et al (2006) Microbial biochemistry, physiology, and biotechnology of hyperthermophilic *Thermotoga* species. FEMS Microbiol Rev 30:872–905. https://doi.org/10.1111/j.1574-6976.2006.00039.x

Cook GM, Janssen PH, Russell JB et al (1994) Dual mechanisms of xylose uptake in the thermophilic bacterium *Thermoanaerobacter thermohydrosulfuricus*. FEMS Microbiol Lett 116:257–262. https://doi.org/10.1016/0378-1097(94)90044-2

Cordova LT, Lu J, Cipolla RM et al (2016) Co-utilization of glucose and xylose by evolved *Thermus thermophiles* LC113 strain elucidated by ^{13}C metabolic flux analysis and whole genome sequencing. Metab Eng 37:63–71. https://doi.org/10.1016/j.ymben.2016.05.001

Crosby JR, Laemthong T, Lewis AM et al (2019) Extreme thermophiles as emerging metabolic engineering platforms. Curr Opin Biotechnol 59:55–64

Crosby JR, Laemthong T, Bing RG et al (2022) Biochemical and regulatory analyses of xylanolytic regulons in *Caldicellulosiruptor bescii* reveal genus-wide features of hemicellulose utilization. Appl Environ Microbiol 88. https://doi.org/10.1128/aem.01302-22, https://doi.org/10.1016/j.copbio.2019.02.006

Doi RH, Kosugi A (2004) Cellulosomes: plant-cell-wall-degrading enzyme complexes. Nat Rev Microbiol 2(7):541–551. https://doi.org/10.1038/nrmicro925

Ebringerová A (2005) Structural diversity and application potential of hemicelluloses. Macromol Symp 232(333):1–12. https://doi.org/10.1002/masy.200551401

Elleuche S, Schroder C, Sahm K et al (2014) Extremozymes – biocatalysts with unique properties from extremophilic microorganisms. Curr Microbiol 29:116–123. https://doi.org/10.1016/j.copbio.2014.04.003

Fardeau M-L, Ollivier B, Patel BKC et al (1997) *Thermotoga hypogea* sp. nov., a xylanolytic, thermophilic bacterium from an oil-producing well. Int J Syst Bacteriol 47:1013.1019. doi: https://doi.org/10.1099/00207713-47-4-1013

Faudon C, Fardeau ML, Heim J et al (1997) Peptide and amino acid oxidation in the presence of thiosulfate by members of the genus *Thermoanaerobacter*. Curr Microbiol 31:152–157. https://doi.org/10.1007/BF00293546

Feng X, Mouttaki H, Lin L, Huang R et al (2009) Characterization of the central metabolic pathways in *Thermoanaerobacter* sp. strain X514 via isotopomer-assisted metabolite analysis. Appl Environ Microbiol 75:5001–5008. https://doi.org/10.1128/AEM.00715-09

Flamholz A, Noor E, Bar-Even A et al (2013) Glycolytic strategy as a tradeoff between energy yield and protein cost. Proc Natl Acad Sci USA 110:10039–10044. https://doi.org/10.1073/pnas.1215283110

Fontes CMGA, Gilbert HJ (2010) Cellulosomes: highly efficient nanomachines designed to deconstruct plant Cell Wall complex carbohydrates. Annu Rev Biochem 79(1):655–681. https://doi.org/10.1146/annurev-biochem-091208-085603

Frock AD, Grey SR, Kelly RM (2012) Hyperthermophilic *Thermotoga* species differ with respect to specific carbohydrate transporters and glycoside hydrolases. Appl Environ Microbiol 78: 1978–1986. https://doi.org/10.1128/AEM.07069-11

Fuhrer T, Fischer E, Sauer U (2005) Experimental identification and quantification of glucose metabolism in seven bacterial species. J Bacteriol 187:1581–1590. https://doi.org/10.1128/JB.187.5.1581-1590.2005

Fushinobu S (2021) Molecular evolution and functional divergence of UDP-hexose-4-epimerases. Curr Opin Chem Biol 61:53–62. https://doi.org/10.1016/j.cbpa.2020.09.007

Garcia S, Thompson RA, Giannone RJ et al (2020) Development of a genome-scale metabolic model of *Clostridium thermocellum* and its applications for integration of multi-omics datasets and computational strain design. Front Bioeng Biotechnol 8:772. https://doi.org/10.3389/fbioe.2020.00772

Gautam J, Xu Z (2021) Construction and validation of a genome-scale metabolic network of *Thermotoga* sp. strain RQ7. Appl Biochem Biotechnol 193:896–911. https://doi.org/10.1007/s12010-020-03470-z

Glazer AN, Nikaido H (2007) Biomass. In: Glazer AN, Nikaido H (eds) Microbial biotechnology, Cambridge University Press, New York

Gottschalk G (1986) Bacterial metabolism. Springer, New York, USA

Gupta R, Sharma R, Beg QK (2013) Revisiting microbial keratinases: next generation proteases for sustainable biotechnology. Crit Rev Biotechnol 33:216–228. https://doi.org/10.3109/07388551.2012.685051

Han D, Xy H, Puranik R et al (2014) Natural transformation of *Thermotoga* sp. strain RQ7. BMC Biotechnol 14:39. https://doi.org/10.1186/1472-6750-14-39

Hassan MA, Abol-Fotouh D, Omer AM et al (2020) Comprehensive insights into microbial keratinases and their implication in various biotechnological and industrial sectors: a review. Int J Biol Macromol 154:567–583. https://doi.org/10.1016/j.ijbiomac.2020.03.116

Heinonen JK, Drake HL (1988) Comparative assessment of inorganic pyrophosphate and pyrophosphatase levels of *Escherichia coli*, *Clostridium pasteurianum*, and *Clostridium thermoaceticum*. FEMS Microbiol Lett 52:205–208. https://doi.org/10.1016/0378-1097(88)90361-8

Heinze S, Mechelke M, Kornberger P et al (2017) Identification of endoxylanase XynE from *Clostridium thermocellum* as the first xylanase of glycoside hydrolase family GH141. Sci Rep 7:11178. https://doi.org/10.1038/s41598-017-11598-y

Hitschler L, Kuntz M, Langschied F (2018) *Thermoanaerobacter* species differ in their potential to reduce organic acids to their corresponding alcohols. Appl Microbiol Biotechnol 102:8465–8476. doi: https://doi.org/10.1007/s00253-018-9210-3

Hlihor R-M, Apostol L-C, Gavrilescu M (2017) Environmental bioremediation by biosorption and bioaccumulation: principles and applications, enhancing cleanup of environmental pollutants. Springer, pp 289–315

Hobson PN, Wallace RJ (1982) Microbial ecology and activities in the rumen: Part II. Crit Rev Microbiol 9:253–320. https://doi.org/10.3109/10408418209104490

Ingvadottir EM, Scully SM, Örlygsson J (2017) Evaluation of the genus of *Caldicellulosiruptor* for production of 1,2-propanediol from methylpentoses. Anaerobe 47:86–88. https://doi.org/10.1016/j.anaerobe.2017.04.015

Ingvadottir EM, Scully SM, Örlygsson J (2018) Production of (S)-1,2-propanediol from L-rhamnose using the moderate thermophilic *Clostridium* strain AK1. Anaerobe 54:26–30. https://doi.org/10.1016/j.anaerobe.2018.07.003

Isern NG, Xue J, Rao JV et al (2013) Novel monosaccharide fermentation products in *Caldicellulosiruptor saccharolyticus* identified using NMR spectroscopy. Biotechnol Biofuels 6:47. https://doi.org/10.1186/1754-6834-6-47

Jacobson TB, Korosh TK, Stevenson DM et al (2020) In vivo thermodynamic analysis of glycolysis in *clostridium thermocellum* and *Thermoanaerobacterium saccharolyticum* using ^{13}C and ^{2}H tracers. mSystems 5:e00736–e00719. https://doi.org/10.1128/mSystems.00736-19

Ju YH, Oh DK (2010) Characterization of a recombinant L-fucose isomerase from *Caldicellulosiruptor saccharolyticus* that isomerizes L-fucose, D-arabinose, D-altrose, and L-galactose. Biotechnol Lett 32:299–304. https://doi.org/10.1007/s10529-009-0154-7

Kadar Z, De Vrijek T, van Noorden GE et al (2004) Yields from glucose, xylose, and paper sludge hydrolysate during hydrogen production by the extreme thermophile *Caldicellulosiruptor saccharolyticus*. Appl Biochem Biotechnol 113–116:497–508

Kristjansson JK, Stetter KO (1998) Thermophilic bacteria. In: Kristjansson JK (ed) Thermophilic bacteria. CRC Press, Boca Raton, FL, pp 1–18

Kühlbrandt W, Davies KM (2016) Rotary ATPases: a new twist to an ancient machine. Trends Biochem Sci 41:106–116. https://doi.org/10.1016/j.tibs.2015.10.006

Lacis LS, Lawford HG (1991) *Thermoanaerobacter ethanolicus* growth and product yield from elevated levels of xylose or glucose in continuous cultures. Appl Environ Microbiol 57:579–585. doi: https://doi.org/10.1128/AEM.57.2.579-585.1991

Lamed RJ, Zeikus JG (1981) Novel NADP-linked alcohol-aldehyde/ketone oxidoreductase in thermophilic ethanologenic bacteria. Biochem J 195:183–190. https://doi.org/10.1042/bj1950183

Le Y, Sun J (2022) CRISPR/Cas genome editing systems in thermophiles: current status, associated challenges, and future perspectives. In: Gadd GM, Sariaslani S (eds) Advances in applied microbiology, vol 118. Elsevier, pp 1–30

Liebl W (2001) Cellulolytic enzymes from *Thermotoga* species. Methods Enzymol 330(1997): 290–300

Lin L, Song H, Tu Q et al (2011) The *Thermoanaerobacter* glycobiome reveals mechanisms of pentose and hexose co-utilization in bacteria. PLos Genet 7:e1002318. https://doi.org/10.1371/journal.pgen.1002318

Liu Y-F, Chen J, Liu Z-L et al (2020) Anaerobic degradation of paraffins by thermophilic actinobacteria under methanogenic conditions. Environ Sci Technol 54:10610–10620. https://doi.org/10.1021/acs.est.0c02071

Martínez-Spinosa RM (2020) Microorganisms and their metabolic capabilities in the context of the biogeochemical nitrogen cycle at extreme environments. Int J Mol Sci 21:4228. https://doi.org/10.3390/ijms21124228

Mazzoli R, Olson DG (2020) *Clostridium thermocellum*: a microbial platform for high-value chemical production from lignocellulose. In: Gadd GM, Sariaslani S (eds) Advances in applied microbiology, vol 113. Elsevier, pp 111–161

McInerney MJ (1988) Anaerobic hydrolysis and fermentation of fats and proteins. In: Zehnder AJB (ed) Biology of anaerobic microorganisms. Wiley, New York, pp 373–415

Mehti D, Satyanarayana T (2013) Diversity in hot environments and thermophilic microbes. In: Satayanaryana T, Littlechild J, Kawarabayasi Y (eds) Thermophilic microbes in environmental and industrial biotechnology. Springer, pp 3–60

Mertens E (1991) Pyrophosphate-dependent phosphofructokinase, an anaerobic glycolytic enzyme? FEBS Lett 285:2–5. https://doi.org/10.1016/0014-5793(91)80711-B

Mótyán JA, Tóth F, Tözser J (2013) Research applications of proteolytic enzymes in molecular biology. Biomol Ther 3:923–942

Nelson KE, Clayton RA, Gill SR et al (1999) Evidence for lateral gene transfer between archaea and bacteria from genome sequence of *Thermotoga maritima*. Nature 399:323–329. https://doi.org/10.1038/20601

Nogales J, Gudmundsson S, Thiele I (2012) An in silico re-design of the metabolism in *Thermotoga maritima* for increased hydrogen production. Int J Hydro Energy 37:12205–12218. https://doi.org/10.1016/j.ijhydene.2012.06.032

Obeng EM, Adam SNN, Budiman C et al (2017) Lignocellulases: a review of emerging and developing enzymes, systems and practices. Bioresour Bioprocess 4(1)

Orlygsson J, Houwen FP, Svensson BH (1995) Thermophilic anaerobic amino acid degradation: deamination rates and end product formation. Appl Microbiol Biotechnol 43:235–241

Pauss A, Andre G, Perrier M et al (1990) Liquid-to-gas mass transfer in anaerobic processes: inevitable transfer limitations of methane and hydrogen in the biomethanation process. Appl Environ Microbiol 56:1636–1644. https://doi.org/10.1128/AEM.56.6.1636-1644.1990

Qian Z, Meng B, Wang Q et al (2009) Systematic characterization of a novel *gal* operon in *Thermoanaerobacter tengcongensis*. Microbiology 155:1717–1725. https://doi.org/10.1099/mic.0.025536-0

Rainey FA, Donnison AM, Janssen PH et al (1994) Description of *Caldicellulosiruptor saccharolyticus* gen. nov., sp. nov: an obligately anaerobic, extremely thermophilic, cellulolytic bacterium. FEMS Microbiol Lett 120:263–266. https://doi.org/10.1016/0378-1097(94)90481-2

Rangel AET, Croft T, Barrios AFG et al (2020) Transcriptomic analysis of a *Clostridium thermocellum* strain engineered to utilize xylose: responses to xylose versus cellobiose feeding. Sci Rep 10:14517. https://doi.org/10.1038/s41598-020-71428-6

Rani DS, Nand K (2000) Production of thermostable cellulase free xylanase by *Clostridium absonum* CFR-701. Process Biochem 36:355–362. https://doi.org/10.1016/S0032-9592(00)00224-7

Rodionov DA, Rodionova IA, Rodionov VA et al (2021) Transcriptional regulation of plant biomass degradation and carbohydrate utilization genes in the extreme thermophile *Caldicellulosiruptor bescii*. mSystems 6:e01345–e01320. https://doi.org/10.1128/mSystems.01351-20

Roh Y, Liu SV, Li G et al (2002) Isolation and characterization of metal-reducing *Thermoanaerobacter* strains from deep subsurface environments of the Piceance Basin, Colorado. Appl Environ Microbiol 68:6013–6020. https://doi.org/10.1128/AEM.68.12.6013-6020.2002

Sahoo K, Sahoo RK, Gaur M et al (2020) Cellulolytic thermophilic microorganisms in white biotechnology: a review. Folia Microbiol 65:25–43. https://doi.org/10.1007/s12223-019-00710-6

Sasaki K, Morita M, Sasaki D et al (2011) Syntrophic degradation of proteinaceous materials by the thermophilic strains *Coprothermobacter proteolyticus* and *Methanothermobacter thermautotrophicus*. J Biosci Bioeng 112:469–472. https://doi.org/10.1016/j.jbiosc.2011.07.003

Scully SM, Örlygsson J (2017) Recent advances in genetic engineering of thermophilic ethanol producing bacteria. In: Gosset G (ed) Engineering of microorganisms for the production of chemicals and biofuels from renewable resources. Springer, Cham, pp 1–29

Scully SM, Örlygsson J (2019) Branched-chain amino acid catabolism of *Thermoanaerobacter* strain AK85 and the influence of culture conditions on branched-chain alcohol formation. Amino Acids 51:1039–1054. https://doi.org/10.1016/j.dib.2019.103938

Scully SM, Örlygsson J (2020) Branched-chain amino acid catabolism of *Thermoanaerobacter pseudethanolicus* reveals potential route to branched-chain alcohol formation. Extremophiles 24:121–133. https://doi.org/10.1007/s00792-019-01140-5

Scully SM, Orlygsson J (2023) Thermostable enzymes and their applications. In: Scully SM, Orlygsson J (eds) Thermophilic anaerobes – phylogeny, physiology and biotechnological applications. Springer, Berlin

Scully SM, Iloranta P, Myllymaki P et al (2015) Branched-chain alcohol formation by thermophilic bacteria within the genera of *Thermoanaerobacter* and *Caldanaerobacter*. Extremophiles 19: 809–818. https://doi.org/10.1007/s00792-015-0756-z

Selig M, Xavier KB, Santos H et al (1997) Comparative analysis of Embden-Meyerhof and Entner-Doudoroff glycolytic pathways in hyperthermophilic archaea and the bacterium *Thermotoga*. Arch Microbiol 167:217–232. https://doi.org/10.1007/BF03356097

Serrano A, Perez-Castiñeira JR, Baltscheffsky H et al (2004) Proton-pumping inorganic pyrophosphatases in some archaea and other extremophilic prokaryotes. J Bioenerg Biomembr 36(127):133. https://doi.org/10.1023/B:JOBB.0000019604.49875.b3

Shaw AJ, Hogsett DA, Lynd LR (2010) Natural competence in *Thermoanaerobacter* and *Thermoanaerobacterium* species. Appl Environ Microbiol 76:4713–4719. https://doi.org/10.1128/AEM.00402-10

Stickland LH (1934) Studies in the metabolism of the strict anaerobes (genus *clostridium*). Biochem J 28:1746–1759

Stryer L (1988) Biochemistry, 3rd edn. Freeman & Company, New York, p 316

Suzuki Y, Tsujimoto Y, Matsui H et al (2006) Decomposition of extremely hard-to-degrade animal proteins by thermophilic bacteria. J Biosci Bioeng 102:73–81. https://doi.org/10.1263/jbb.102.73

Tarlera S, Stams AJM (1999) Degradation of proteins and amino acids by *Caloramator proteoclasticus* in pure culture and in coculture with *Methanobacterium thermoformicicum* Z245. Appl Microbiol Biotechnol 53:133–138. https://doi.org/10.1007/s002530051626

Tong W, Chen Z, Cao Z et al (2013) Robustness analysis of a constraint-based metabolic model links cell growth and proteomics of *Thermoanaerobacter tengcongensis* under temperature perturbation. Mol Biosyst 9:713–722. https://doi.org/10.1039/c3mb25278g

Tripathi SA, Ohlson DG, Argyros DA et al (2010) Development of pyrF-based genetic system for targeted gene deletion in *Clostridium thermocellum* and creation of a pta mutant. Appl Environ Microbiol 76:6591–6599. https://doi.org/10.1128/AEM.01484-10

Van de Werken HJG, Verhaart MRA, VanFossen AL et al (2008) Hydrogenomics of the extremely thermophilic bacterium *Caldicellulosiruptor saccharolyticus*. Appl Environ Microbiol 74: 6720–6729. https://doi.org/10.1128/AEM.00968-08

Van Fossen AL, Verhaart MRA, Kengen SMW et al (2009) Carbohydrate utilization patterns for the extremely thermophilic bacterium *Caldicellulosiruptor saccharolyticus* reveal broad growth substrate preferences. Appl Environ Microbiol 75:7718–7724. https://doi.org/10.1128/AEM.01959-09

Van Niel EWJ, Bergdahl B, Hahn-Hägerdal B (2017) Close to the edge: growth restrained by the NAD(P)H/ATP formation flux ratio. Front Microbiol 8:1149. https://doi.org/10.3389/fmicb.2017.01149

Vavitas K, Glekas PD, Hatzinikolaou DG (2022) Synthetic biology of thermophiles: taking bioengineering to the extremes? Appl Microbiol 2:165–174

Verbeke TJ, Zhan X, Henrissat B et al (2013) Genomic evaluation of *Thermoanaerobacter* spp. for the construction of designer co-cultures to improve lignocellulosic biofuel production. PLoS One 8:e59362. https://doi.org/10.1371/journal.pone.0059362

Verbeke TJ, Giannone RJ, Klingeman DM et al (2017) Pentose sugars inhibit metabolism and increase expression of an AgrD-type cyclic pentapeptide in *Clostridium thermocellum*. Sci Rep 7:43355. https://doi.org/10.1038/srep43355

Vongkampang T, Sreenivas K, Engvall J et al (2021) Characterization of simultaneous uptake of xylose and glucose in *Caldicellulosiruptor kronotskyensis* for optimal hydrogen production. Biotechnol Biofuels 14:1. https://doi.org/10.1186/s13068-021-01938-6

Ward DE, Shockley KR, Chang LS et al (2002) Proteolysis in hyperthermophilic microorganisms. Archaea 1:63–74

Weimer PJ, Wagner W, Knowlton S et al (1984) Thermophilic anaerobic bacteria which ferment hemicellulose: characterization of organisms and identification of plasmids. Arch Microbiol 138:31–36. https://doi.org/10.1007/BF00425403

Willquist K, van Niel EWJ (2010) Lactate formation in *Caldicellulosiruptor saccharolyticus* is regulated by the energy carriers pyrophosphate and ATP. Metab Eng 12:282–290. https://doi.org/10.1016/j.ymben.2010.01.001

Willquist K, Pawar SS, van Niel EWJ (2011) Reassessment of hydrogen tolerance in *Caldicellulosiruptor saccharolyticus*. Microb Cell Fact 10:111. https://doi.org/10.1186/1475-2859-10-111

Wimmer JLE, Kleinermanns K, Martin WF (2021) Pyrophosphate and irreversibility in evolution, or why PPi is not an energy currency and why nature chose triphosphates. Front Microbiol 12: 759359. https://doi.org/10.3389/fmicb.2021.759359

Winterhalter C, Heinrich P, Candussion A et al (1995) Identification of novel cellulose binding domain within the multidomain 120 kDA xylanase xyna of hyperthermophilic bacterium *Thermotoga maritima*. Mol Microbiol 15:431–444. https://doi.org/10.1111/j.1365-2958.1995.tb02257.x

Wood TM, Garcia-Campayo V (1990) Enzymology of cellulose degradation. Biodegradation 1: 147–161. https://doi.org/10.1007/BF00058833

Zhang Y, Thiele I, Weekes D et al (2009) Three-dimensional structural view of the central metabolic network of *Thermotoga maritima*. Science 325:1544–1549. https://doi.org/10.1126/science.1174671

Zhang K, Zhao W, Rodionov DA et al (2021) Genome-scale metabolic model of *Caldicellulosiruptor bescii* reveals optimal metabolic engineering strategies for bio-based chemical production. mSystems 6:e01351–e01320

Zhao L, Wang Z-H, Wu J-T et al (2019) Co-fermentation of a mixture of glucose and xylose to hydrogen by *Thermoanaerobacter thermosaccharolyticum* W16: characteristics and kinetics. Int J Hydro Energ 44:9248–9255. https://doi.org/10.1016/j.ijhydene.2019.02.125

Zhao L, Chen C, Ren H-Y et al (2020) Unravelling hydrogen production potential by glucose and xylose co-fermentation of *Thermoanaerobacterium thermosaccharolyticum* W16 and its metabolisms through transcriptomic sequencing. Int J Energy Res 44:9617–9628. https://doi.org/10.1002/er.5468

Zhou J, Olson DG, Argyros DA et al (2013) Atypical glycolysis in *Clostridium thermocellum*. Appl Environ Microbiol 79:3000–3009. https://doi.org/10.1128/AEM.04037-12

Part III
Biotechnological Applications

Thermostable Enzymes and Their Applications

Sean Michael Scully and Johann Orlygsson

Abstract Microbial enzymes are of great importance for the degradation of organic material with relevance in the bioremediation of organic waste, bioenergy generation, large-scale industrial bioprocesses, and more. The current chapter deals with the most important enzymes thermophilic bacteria produce, starch-degrading enzymes, celluloses, hemicelluloses, proteases, and more. Thermophilic bacteria are ubiquitous in nature and grow on wide variety of substrates due to their secretion of extracellular enzymes. In present study, the most relevant thermophilic, anaerobic bacteria degrading various polymeric substrates are described.

1 Introduction

Enzymes are classified into six groups; ligases, isomerases, oxidoreductase, esterases, lyases, transferases, and hydrolases (Rigoldi et al. 2017). It has been estimated that 85% of the enzymes of biotechnological potential today are hydrolases with around 30% of all bulk enzyme sales being enzymes involved in the degradation of starch (Elleuche and Antranikian 2013).

The catalytic properties of enzymes produced by thermophiles have been exploited to carry out efficient and cost-effective degradation of various compounds in research, and in various industrial sectors like food, health, cosmetics, agriculture, chemistry, and energy (Rigoldi et al. 2017; Barzkar et al. 2018; Wu et al. 2021; Wang et al. 2019; Sahoo et al. 2020). Enzymes have long been used as detergents and for food production, mainly in cheese, sourdough, beer, and wine together with the production of leather, indigo, and linen (Prakash et al. 2013). The development of large-scale production, isolation, and purification of enzymes from thermophilic bacteria has provided biocatalysts to be used as molecular tools of actual industry processes (Leisola et al. 2017).

S. M. Scully · J. Orlygsson (✉)
Faculty of Natural Resource Sciences, University of Akyreyri, Akureyri, Iceland
e-mail: jorlygs@unak.is

S. M. Scully, J. Orlygsson (eds.), *Thermophilic Anaerobes*, Emerging Issues and Trends, https://doi.org/10.1007/978-3-031-41720-7_6

The use of enzymes can be broadly categorized as either biocatalysis or biotransformation; biocatalysis is generally understood to use specific enzymes while biotransformation typically involves the use of whole cells. As with any technology, the use of enzymes, whether as purified enzymes or as a part of a whole-cell system, presents several advantages as well as potential drawbacks. First and foremost, enzymes are incredibly efficient catalytic agents. A typical traditional chemical catalyst is often used at a concentration between 0.1 and 1% (on a mole basis) while enzymes are often employed at concentrations between of 10^{-3} to 10^{-4} mole). By their nature, enzymes are both renewable and biodegradable. Enzymes frequently exhibit a high degree of reaction selectivity, often being chemoselective, stereoselective, and regioselective with some enzymes displaying a degree of catalytic promiscuity catalyzing reactions beyond their "native" selectivity. A common selling point of the use of biocatalysis is the use of mild reaction conditions. While many thermophilic enzymes have temperature optima close to the boiling point of water, some of these enzymes retain activity at lower temperatures (Sahoo et al. 2020). Many enzymes are also compatible with organic solvents. Enzymes and whole cells can often be fixed to a solid surface or encapsulated in a solid matrix allowing their facile reuse.

The use of biocatalysts is not without its potential drawbacks. Often-cited concern is the cost associated with using purified enzymes or bioreactors as well as the range of conditions in which they can operate. In the case of some classes of enzymes, such as oxidoreductases, cofactors such as NAD(H), NADP(H), pyridoxal phosphate, Vitamin B12, and FADH2, are often required. While purified cofactors are often expensive, they are often only required in small quantities and can often be recycled by coupling the desired reaction with a secondary reaction that will recycle the cofactor needed (Sharma et al. 2022; Bachosz et al. 2023). Alternately, whole cell systems may not require the addition of exogenous cofactors. Also, the use of purified enzymes or whole cells may complicate reactions although this can often be mitigated by immobilization or the bulk separation of cells (Li et al. 2022). While the stereoselectivity of many enzymes is often cited as a tremendous advantage, it can also be a limitation (Raczynska et al. 2021). Another potential drawback to enzyme catalysis is various inhibition phenomena as well as the prospect of denaturation over time, limiting the usable life of a given catalyst. That said, there are reports of some enzyme systems being used for dozens if not hundreds of cycles in industrial settings (Zheng et al. 2017).

Enzymes have numerous applications. Thermophilic microorganisms have shown to be a good source of novel enzymes which are suitable for industrial applications. They produce a broad variety of enzymes, e.g., amylases, phosphatases, cellulases, hemicellulases, proteases, lipases, laccases, and other more specific enzymes usable in food, textile, dairy, pharmaceutical, and more (Atalah et al. 2019; Ajeje et al. 2021). The main reason for the importance of thermophilic enzymes for industrial applications is economical both in value-added product and biofuel production. The use of enzymes for biomass utilization at high temperatures improves enzyme penetration through the cell wall of lignocellulosic biomass and helps to disrupt crystalline elements within its structure.

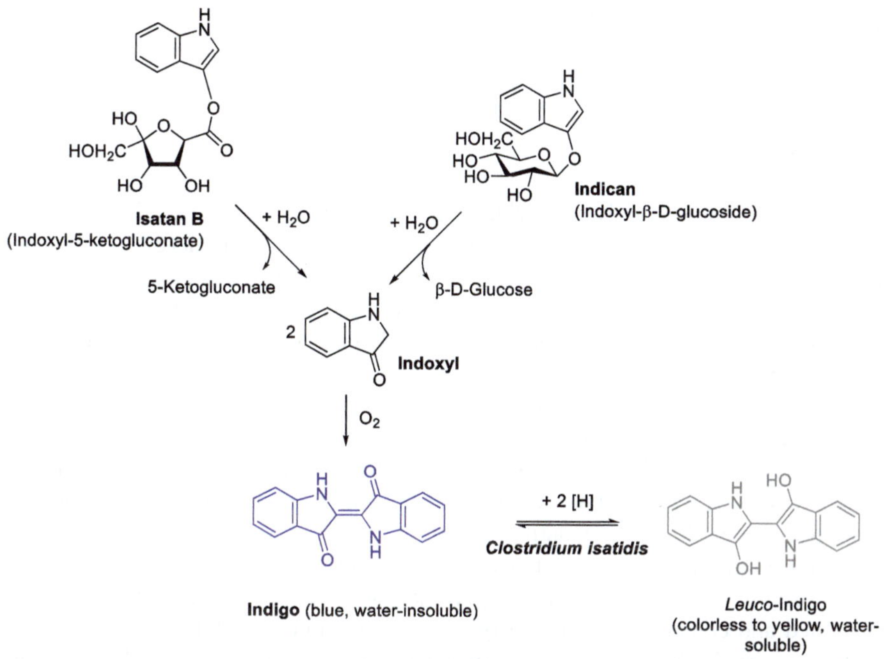

Fig. 1 Indigo is the result of the reduction of the *yellow*, water soluble form of indigo (Leuco-indigo) by a moderately thermophilic anaerobe, *Clostridium isatidis*

One of the early applications of microorganisms, albeit unknown to early users, was the biotransformation of glycosides in wood (*Isatis tinctoria*) for the production of a truly high-value compound included the formation of indigo, a blue dye used for coloring cotton fabrics since antiquity. Historically, the preparation of indigo from balled woad was performed in wood vats with the addition of wood ash to maintain an alkaline pH and by keeping the vat's temperature around 50–60 °C (Clark et al. 1993; Andreesen 2005). The organism responsible for the biotransformation is the moderately thermophilic and somewhat aerotolerant *Clostridium isatidis* which was isolated from woad vats used in Europe during medieval times (Padden et al. 1998, 2000). Subsequent work with isolates from other historic woad preparation concluded that *C. istatidis* is the responsible organism for the biotransformation (Padden et al. 2000). Figure 1 shows the production of indigo by *Clostridium isatidis*.

When evaluating the use of an enzyme for a specific purpose, understanding its operating parameters, such as temperature, pH, and salinity ranges, is critical. An often-overlooked parameter is the tolerance of an enzyme to oxygen, something of particular importance with enzymes from strictly anaerobic bacteria. Other important parameters are the use of activators, inhibitors, cofactors, thermal stability, solvent stability, and half-life. In this chapter, we will give an overview of the main types of enzymes produced by thermophiles and the wide variety of applications.

2 Hydrolases from Thermophilic Anaerobes

2.1 *Amylases and Pullulanases*

2.1.1 Structure of Starch

Starch is a ubiquitous polysaccharide composed of glucose residues. In plants and some algae species, starch serves as a storage of carbon and energy in a manner analogous to glycogen in higher animals. Starch is widely associated with edible terrestrial plants such as wheat, potatoes, rice, oats, and corn and typically forms granules within the cell (Wang et al. 2022; Apriyanto et al. 2022). Green algae such as *Ulva* species also contain starch which can make up to 45% of their dry weight (Wang et al. 2022). Starch is water soluble, depending upon the molecular weight and degree of branching, and is easily hydrolyzed in the presence of dilute acid, particularly at elevated temperatures. As such, starch is an abundant raw material for bioprocessing as well as other noteworthy applications related to its gelling properties (Wang et al. 2022; Apriyanto et al. 2022).

Starch is not a singular substance but can mainly be divided into amylose and amylopectin. Amylose is a linear polysaccharide while amylopectin is highly branched and constitutes the majority of starches found in nature (Fig. 2). In both instances, the fundamental unit of this glycan is maltose, a glucose disaccharide. Amylose is a linear homopolysaccharide composed of repeating glucose residues linked via a α-1,4-*O*-glycosidic bonds. The number of repeating glucose residues are typically between 500 and 20,000 units (Bergthaller and Hollmann 2014), although this is highly dependent upon the source of the starch. In practice, even amylose contains some degree of branching although this is typically less than 0.1% of the linkages present (Bergthaller and Hollmann 2014). Due to its hydrophilic nature, amylose is somewhat soluble in water although it tends to be quite viscous and forms a gel-like slurry. Similar to amylose, amylopectin has a linear backbone composed of the aforementioned α-1,4-*O*-glycosidic bonds with a high degree of branching at C6

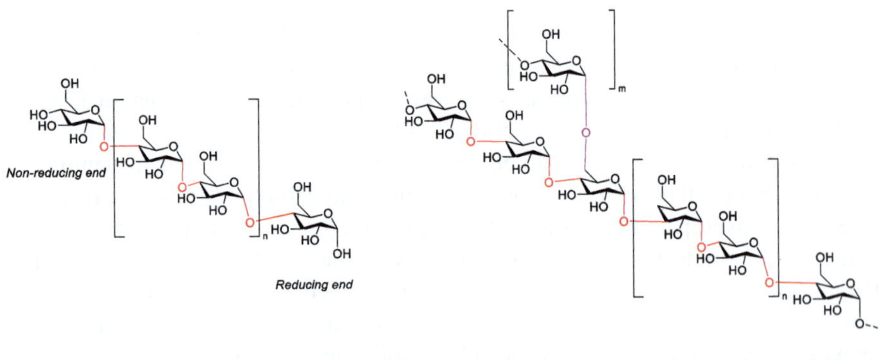

Fig. 2 The structure of starch. Amylose (left) and amylopectin (right); α-1,4-O-glycosidic bonds are shown in *red*, α-1,6-O-glycosidic bonds are shown in *magenta*; $n = 500–20,000$, $m = 20–30$

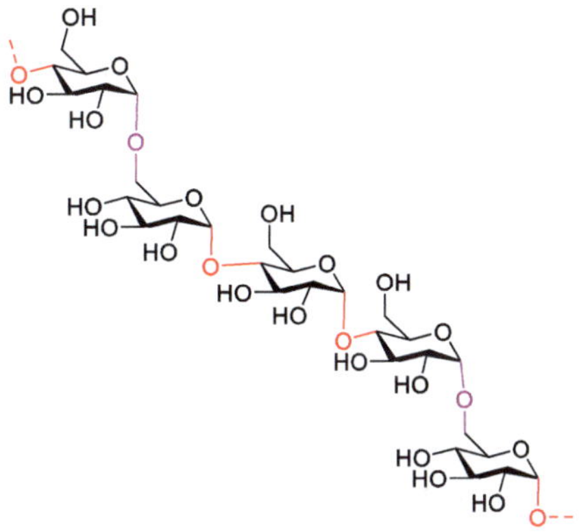

Fig. 3 The structure of pullalan which is composed of a maltotriose with two internal α-1,4-*O*-glycosidic bonds with the first and third glucose residues liking to other maltotriose residudes via an α-1,6- *O*-glycosidic linkages

position via an α-1,6-*O*-glycosidic bonds with the sidechains composed of up to 30 glucose residues. Amylopectin molecules are insoluble and can contain up to $6x10^6$ glucose residues (Naguleswaran et al. 2014), highlighting their utility as storage of carbon and energy.

Similar to starch, pullulan is a polysaccharide consisting of maltotriose units, also known as glucan composed of α-1,4-*O*- and α-1,6-*O*-glycosidic bonds which is synthesized by a number of microorganisms such as the black yeast-like fungi *Aureobasidium pullulans* (Cheng et al. 2011). Unlike starch, pullulan is a highly structured glucan with a 2:1 ratio of α-1,4- to α-1,6 bonds; the fundamental unit is maltotriose in which individual glucose residues bear an α-1,4-linkage while maltotrioses units are connected by an adjoining α-1,6-linkage (Fig. 3). The molecular weight range of pullulans can range from 4.5×10^4 to 6×10^5 Da (Lee and Yoo 1993). Pullulan is a tasteless, odorless, nonhygroscopic powder, transparent films, viscous, adhesive solution (Domań-Pytka and Bardowski 2004).

2.1.2 Enzymes

The enzymatic deconstruction of starch to its constituent glucose units is a multistep process requiring the synergistic action of multiple enzymes. Fortunately, a wide number of organisms across all three domains of life possess amylases. Even a large number of thermophilic anaerobes have the native ability to degrade amylose, amylopectin, and maltose.

A wide range of enzymes acting on starch and derivatives thereof have been described including, α-amylases, β-amylase, glucoamylase, α-glucosidases, pullulanases, isoamylases, and α-glucan lyase. Each of these enzymes differ in its substrate specificity and mechanism of action (Table 1).

The first step of starch degradation is liquefaction which requires an endo-acting α-amylases to hydrolyze the internal glycosidic linkages resulting in the liberation of maltodextrins. The subsequent step (saccharification) employs two enzymes: a pullulanase which debranches the maltodextrins by hydrolyzing any α-1,6 linkages

Table 1 Specificity of starch-acting enzymes

Enzyme	Substrate	Other names	Specificity	Notes
α-Amylase	Starch	1,4-α-D-glucan glucanohydrolase (EC 3.2.1.1)	α-1,4-*O*-glycosidic	Endo-acting
β-Amylase	Starch	1,4-α-D-glucan maltohydrolase (EC 3.2.1.2)	α-1,4-*O*-glycosidic	Exo-acting; cleaves second linkage from nonreducing end, release maltose
Glucoamylase	Starch	1,4-α-D-glucan glucohydrolase (EC 3.2.1.3)	α-1,4-*O*-	Exo-acting; requires acidic conditions
α-Glucosidases	Maltodextrins	1,4-α-D-glucoside glucanohydrolase, maltase (EC 3.2.1.20)	α-1,4-*O*-	Exo-acting; removes single glucose resides from the end of the maltodextrin
Pullulanse	Amylopectin, glycogen, pullulan	Pullulan α-1,6-glucanohydrolase (EC 3.2.1.41)	α-1,6-*O*-	Endo-acting, debranching
Isoamylase	Amylopectin, glycogen	Glycogen α-1,6-glucanohydrolase (EC 3.2.1.68)	α-1,6-*O*-	Debranching
α-Glucan lyase	Starch, maltodextrin	1,4-α-D-glucan exo-4-lyase (EC 4.2.2.13)	α-1,4-*O*-glycosidic	Yields 1,5-anhydro-D-fructose

and exo-acting glucoamylase which releases individual glucose residues from the nonreducing end of the polymer (Rodriguez-Sanoja et al. 2005).

Given the ubiquity of starch-rich crops, the demand for amylolytic enzymes by the starch processing industry represents one of the largest enzyme markets in the world. The modern industrial hydrolysis of starch into high-fructose corn syrup uses a number of enzymes at temperatures well above 50 °C, although none are sources from thermophilic anaerobes. *Bacillus licheniformis*, a thermotolerant facultative anaerobe, is widely used for enzyme production used in starch processing today (Hussain et al. 2013).

For an amylase to be useful for industrial starch processing, it should have an operating temperature near 100 °C, have high thermal stability, and not require the presence of a divalent metal ion for activity (due in part to gelatinization). The rationale for selecting enzymes for this application that operate at high temperatures is twofold: to decrease the viscosity of the starch solution and to prevent microbial contamination.

2.1.3 Thermophilic Starch-Degrading Enzymes: Microorganisms

A number of thermophilic anaerobes are known to produce various starch-acting enzymes. The ability to ferment starch as carbon source is a common feature of many genera including *Clostridium thermosulfurogenes* (Hyun and Zeikus 1985a, b),

Thermoanaerobacter (Zheng et al. 2010), and more. That said, relatively few amylases from these organisms have been characterized. Table 2 summarizes the main thermophilic bacteria producing α-amylases, glucoamylases, β-glucosidases, and pullulanases showing temperature and pH optimum and some other properties.

2.2 *Cellulases*

2.2.1 Structure of Celluose

Cellulose is a renewable carbon source that is available in abundance on Earth and is produced by a large number of organisms, although it is primarily associated with terrestrial plants and seaweeds (Urbina et al. 2021). It has been estimated that globally 100 billion tons of cellulose are produced annually (Varshney and Naithani 2011). The utilization of cellulose for bioprocessing is of intense interest, particularly in the context of producing biofuels such as ethanol from renewable lignocellulosic biomass (Ajeje et al. 2017).

Cellulose is one of the main components of lignocellulose where it serves as the primary structural component of the cell wall of plants. Conversely, bacterial celluloses serve as extracellular material. Regardless of its origin, cellulose is a linear homopolysaccharide comprised of β-D-glucopyranose units that are linked together by β-1,4-*O*-glycosidic bonds (Fig. 4). The fundamental unit of cellulose is cellobiose, a disaccharide of glucose connected by the aforementioned β-1,4-*O*-glycosidic bond, as the enzymatic deconstruction of cellulose often removes glucose residues in increments of two glucose residues (Chen et al. 2023).

Cellulose is one of the most recalcitrant polymers among all polysaccharides present in nature due to its highly stable glycosidic bond. Unlike starch, the ether linkage between the C1 and C4 carbons of adjoining glucose residues is difficult to attack as it is sequestered between adjacent heterocyclic rings. Furthermore, due to the presence of a large number of free hydroxyl groups, cellulose exhibits a high degree of inter- and intramolecular hydrogen bonding within individual cellulose strands and between strands. For this reason, cellulose often exists as a semicrystalline solid with large crystalline regions where there is a high degree of hydrogen bonding interspersed with amorphous sections (Fig. 4). Very long arrays of parallel chains aggregate into microfibrils.

The vast majority of the attention on the deconstruction of cellulose has focused on the action of fungal enzymes due in no small part to organisms such as *Trichoderma reesei* being able to secrete large quantities of cellulases. Most often these enzymes have been cloned to rapidly growing bacterial hosts. In the past few years though, more attention has been on bacterial cellulases because of their fast growth rates and survival in harsh environments and their ability to produce multiple enzyme complexes which provide increased function and synergy.

Table 2 Starch-acting enzymes from thermophilic organisms

Organism	Activity	T_{opt} (°C)	pH_{opt}	MW (kDa)	Refs
Alicyclobacillus acidocaldarius	α-Amylase	75	7.5	160	Schwermann et al. (2004)
Thermotoga neapolitana	α-Amylase	75	6.5	48	Park et al. (2010)
Thermococcus profundus	α-Amylase	80	5.0–6.0	43	Chung et al. (1995)
Thermotoga maritima	α-Amylase	85–90	7.0	61	Liebl et al. (1997)
Dictyoglomus thermophilum	α-Amylase	90	5.5	81	Fukusumi et al. (1988)
Thermotoga maritima	α-Amylase	90	8.5	241	Ballschmiter (2006)
Pyrococcus furiosus	α-Amylase	90	4.5	76	Yang et al. (2004)
Thermococcus sp.	α-Amylase	95	5.0	51	Wang et al. (2008)
Pyrococcus furiosus	α-Amylase	100	5.5–6.0	100	Koch et al. (1990)
Pyrococcus woesei	α-Amylase	100	5.5	68	Koch et al. (1991)
Clostridium thermosaccharolyticum	Glucoamylase	70	5.0	75	Specka et al. (1991)
Thermoanaerobacter tengcongensis	Glucoamylase	75	5.0	°77	Zheng (2010)
Clostridium thermohydrosulfuricum	α-Glucosidase	75	5.0–5.5	–	Saha and Zeikus (1991)
Thermotoga maritima	α-Glucosidase	90	7.5	110	Raasch et al. (2000)
Thermococcus sp. *AN1*	α-Glucosidase	98	7.0	60	Piller et al. (1996)
Pyrococcus furiosus	α-Glucosidase	105	5.0–6.0	125	Costantino et al. (1990)
Pyrococcus furiosus	α-Glucosidase	105–115	5.5	125	Chang et al. (2001)
Anaerobranca gottschalkii	Pullulanase	70	8.0	96	Bertoldo (2004)
Thermoactinomyces thalpophilus	Pullulanase	70	7.0	79	Odibo and Obi (1988)
Thermoanaerobacter ethanolicus	Pullulanase	80	6.0	109	Lin and Leu (2002)
Thermotoga neapolitana		80	5.0–7.0	93	Kang et al. (2011)
Clostridium thermohydrosulfuricum	Pullulanase	85	5.5–6.0	–	Hyun and Zeikus (1985a, b)
Clostridium thermohydrosulfuricum	Pullulanase	90	5.0–5.5	136.5	Saha et al. (1988)
Thermococcus celer	Pullulanase	90	5.5	–	Canganella et al. (1994)
Thermotoga maritima	Pullulanase	90	6.0	96.3	Bibel et al. (1998)
Thermococcus siculi	Pullulanase	95	6.0	148.6	Jiao et al. (2011)
Pyrococcus furiosus	Pullulanase	98	5.5	110	Brown and Kelly (1993)

(continued)

Table 2 (continued)

Organism	Activity	T_{opt} (°C)	pH_{opt}	MW (kDa)	Refs
Thermococcus litoralis	Pullulanase	98	5.5	119	Brown and Kelly (1993)
Desulfurococcus mucosus	Pullulanase	100	5.0	–	Canganella et al. (1994)
Pyrococcus woesei	Pullulanase	100	6.0	90	Rudiger et al. (1995)
Thermococcus aggregans	Pullulanase	100	6.5	–	Canganella et al. (1994)

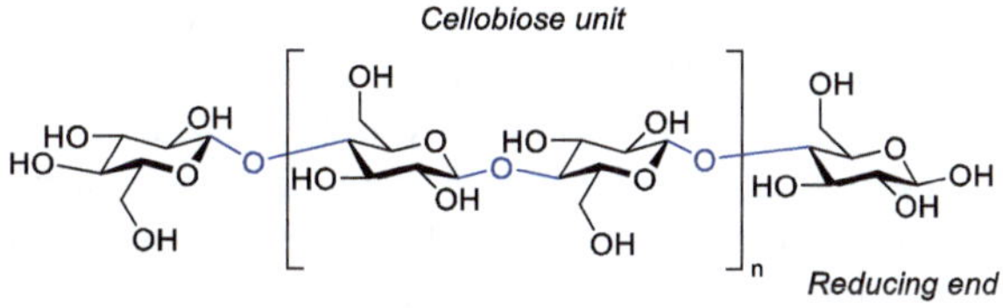

Fig. 4 Cellulose is a common carbohydrate composed disaccharide glucose residues (cellobiose) linked by β-1,4-*O*-glycosidic bonds

Table 3 Specificity of enzymes acting on cellulose

Enzyme	Substrate	Other names	Specificity	Notes
Cellobiohydrolase (CBH)	Cellulose	Cellulose 1,4-β-cellobiosidase, exo-1,4-β-D-glucanase, Avicelase (EC 3.2.1.91)	β-1,4-*O*	Exo-acting, yields oligosaccharides
Endo-1,4-β-D-glucanase (EG)	Cellulose, cellodextrin	4-β-D-glucan 4-glucanohydrolase, CMCase (EC 3.2.1.4)	β-1,4-*O*	Endo-acting, random attacks internal 1,4-linkage
β-Glucosidase (BG)	Cellobiose, other soluble cellodextrins	β-D-glucoside glucohydrolase, cellobiase	β-1,4-*O*	Exo-acting; liberates individual glucose residues

2.2.2 Enzymes

To achieve the complete deconstruction of crystalline cellulose into glucose, multiple enzymes working synergistically are required. Table 3 summarizes the main enzymes responsible for cellulose degradation.

Cellulases are grouped into glycoside hydrolyse families (Park et al. 2017) and dominate the enzyme market worldwide (Linares-Pasten et al. 2014). Hydrolysis of cellulose requires the collaboration of three classes of enzymes: cellobiohydrolases (CBH), endoglucanases (EGL), and β-glucosidases (BGL) (Wood and Garcia-Campayo 1990).

Degradation of cellulose to glucose includes various EGLs to cleave cellulose chains internally, releasing shorter fragments and also cleave of the smaller fragments to 2–4 carbon saccharides. The CBHs further degrade the smaller fragments to release the disaccharide cellobiose units. The CBHs are further classified as BCH I that acts on the reducing end and BCH II that acts on the nonreducing end. Further degradation of cellobiose is accomplished by β-glucosidase (BGL) releasing glucose and by cellobiose phosphorylase releasing glucose-1-phosphate and glucose. For an effective cellulose degradation, synergistic action of these three cellulase components is needed. EG forms nicks in the polymer opening the reducing and nonreducing end which are further used by CBH to release the cellobiose and oligosaccharide units. Finally, the BGL releases glucose from cellobiose (Dadwal et al. 2021).

Enzymatic hydrolysis of cellulose most often occurs at 40–50 °C which is often performed at slow rate with low yields of sugars and incomplete hydrolysis. Thermostable or thermotolerant enzymes are, therefore, often suitable for various use in industries, e.g., textile, detergent, animal feed, pharmaceutical, and paper and pulp industry. The ability to hydrolyze the polymer at high temperatures is currently in focus because of the increased demand for bioethanol (Zuliani et al. 2021). The use of high temperatures for cellulose degradation is also prerequisite for various other industrial processes. Additional factors favoring high temperatures for cellulose degradation are decreased risk of contamination, low cost of cooling, and recovery of end products like ethanol (Scully and Orlygsson 2015).

2.2.3 Thermophilic Cellulose-Degrading Enzymes: Microorganisms

Mainly there are two types of cellulases existing in thermophilic microorganisms. Firstly, extracellular enzymes mainly produced by filamentous fungi and aerobic bacteria, and secondly, via secretion of a multienzyme complex, the so-called cellulosome complex in anaerobic bacteria like the thermophile *Clostridium thermocellum,* but also by mesophiles like *Ruminococcus, Acetovibrio, Bacterioides,* and in anaerobic fungi like *Neocallimastigo mycota* (Bayer et al. 2008; Doi 2008; Haitjema et al. 2017). The former is more common but anaerobic bacterial cellulases are often more stable and have increased specific activity.

Microorganisms capable of cellulose degradation have been isolated from various environments, such as soil (Schut 2022) hot springs (Peng et al. 2015), and compost systems (Munir et al. 2021). Among thermophilic bacteria producing cellulases are the aerobic *Bacillus, Geobacillus, Caldibacillus, Acidothermus, Caldocellum* and the anaerobic genera of *Clostridium, Caldicellulosiruptor,* and more. Several investigations have been conducted on *Thermotoga* species concerning their cellulolytic activity. *Thermotoga neapolitiana* CelA and Cel B were investigated for their cellulose production (Bok et al. 1998). The Cel A protein was optimally active at pH 6.0 and 95 °C, but the Cel B protein had a broader pH range (6.0–6.5) at 106 °C. Both enzymes showed a high level of activity, or 1.219 and 1.536 U/mg. *Thermotoga maritima* MSB8 was investigated by fusing cellulase-β-glucosidase by gene fusion (Hong et al. 2007). The fusion protein showed both cellulase (Cel5C) and β-glucosidase (BglB) activity. The former at pH 8.0, 70 °C and the latter at pH 8.0 and 80 °C. Exo-1,4-β-cellobiohydrolase was isolated from

Thermotoga sp. FjSS3-B1 (Ruttersmith and Daniel 1991). This enzyme showed a half-life of 70 min at 108 °C and was active against amorphous cellulose and CM-cellulose but had only limited effects on filter paper and Sigmacell 20 hydrolysates. The pH optimum was around neutral.

Several *Caldicellulosiruptor* have also been investigated for the cellulase acidity at extreme temperatures. Cellulolytic *Caldicellulosiruptor* uses unique approaches to degrade cellulosic substrates. Many thermophilies in general freely secrete individual cellulases and others assemble large multiprotein complexes; cellulosomes (Bayer et al. 2008) with binding and cellulolytic properties, *Caldicellulosiruptor* species deploy a novel system with properties of both. For a recent review on the subject, see Lee et al. (2018). *C. kristjanssoni* was isolated from Iceland and showed an optimum growth around 78 °C (Bredholt et al. 1999) and was later investigated for cellulolytic activity on avicel (Bredholt et al. 1995). *Caldicellulosiruptor obsidiansis* was isolated from Yellowstone and grows optimally at 78 °C on avicel, filter paper, and processed cardboard (Hamilton-Brehm et al. 2010). Other known thermophiles that have been reported to be cellulolytic are, e.g., *Aneurinibacillus thermoaerophilus* (Acharya and Chaudhary 2012), *Anoxybacillus flavithermus* EHP2 (Salah et al. 2007), *Clostridium thermocellum* (Lv and Yu 2012; Otajevwo and Aluyi 2011), and more.

2.3 Hemicellulases

2.3.1 Structure

Hemicellulose is highly diverse family of heteropolysaccharides that represent the most abundant renewable terrestrial energy source after cellulose and make up about a third of annual biomass production (Spiridon and Popa 2008). As one of the components of lignocellulose, hemicelluloses are closely associated with the cellulose and lignin found in plants where they act as a matrix substance within the cell wall (Spiridon and Popa 2004). However, in addition to being found in terrestrial plants, hemicelluloses can also be found in a number of seaweeds (Misurcova 2012) while others still are of microbial origin.

Unlike cellulose, hemicellulose tends to have random, amorphous structures that are only comprised of 500–3000 residues (Microbial Biotech textbook) which can be easily hydrolyzed by solutions of dilute acid or base (Fig. 5). Furthermore, there is a tremendous diversity of hemicellulose that can be comprised of a mixture of hexoses (such as glucose, galactose, and mannose), pentoses (D-arabinose, D-xylose), methylpentose (L-rhamnose), and uronic acids each in addition to various modifications such as esterification and etherification. While the "backbone" of a hemicellulose is commonly composed of β-1,4-linkages, other configurations are not uncommon, and branching is also a common occurrence. The physical properties of specific hemicelluloses are dependent upon its composition, nature of the glycosidic bonds, molecular weight, and modifications, all of which vary greatly contingent upon the plant source (Ebringerová 2005).

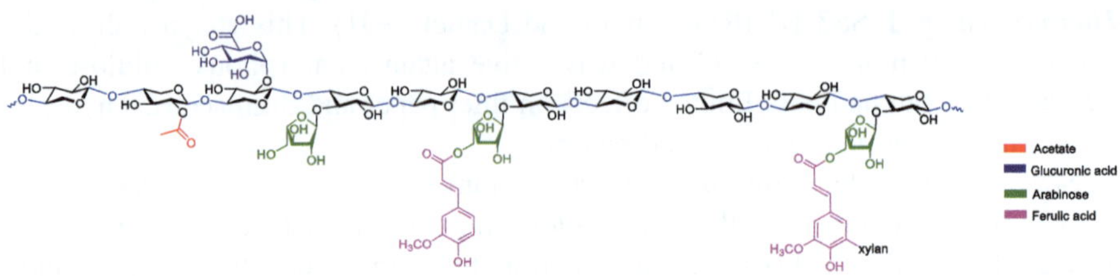

Fig. 5 Xylan is among the most common of the hemicelluloses. Here xylose residues are linked by β-1,4-*O*-glycosidic bonds with individual residues bearing modifications such as linking to uronic acids, acyl groups such as acetate, or lignols such as ferulic acid which can serve as a linker between strands of xylan

Given the large number of possible constituents and modifications, there is a tremendous diversity of hemicellulases present in nature. Four categories are used to describe hemicellulases based upon their composition: xyloglycans (xylans), mannoglycans (mannans), mixed β-glucan, and xyloglucans (Ebringerová 2005). The most common of the hemicelluloses is xylan. Other common hemicelluloses include β-glucan (a glucose polymer consisting of β-1,3-*O* glycosidic bonds), xyloglucan, arabinoxylan, mannan, galactomannan, arabinan, galactan, and polygalacturonan, although these glucans require specific enzymes systems to hydrolyze their specific linkages.

Xylans usually consist of a backbone of β-(1 → 4)-linked xylose residues and can be further divided into homoxylans and heteroxylans. Homoxylans have a backbone of D-xylopyranose residues linked by β(1 → 3) or mixed β(1 → 3, 1 → 4)-glycosidic linkages. Homoxylans mainly have structural functions. Heteroxylans such as glucuronoxylans, glucuronoarabinoxylans, and complex heteroxylans, have a backbone of D-xylopyranose and short carbohydrate branches. For example, glucuronoxylan has a substitution with α-(1 → 2)-linked glucuronosyl and 4-O-methyl glucuronosyl residues. Arabinoxylans and glucuronoarabinoxylans contain arabinose residues attached to the backbone (Salmen 2022).

The mannan-type hemicellulose can be classified into two types based on their main chain difference, galactomannans and glucomannans. Galactomannans have only β-(1 → 4) linked D-mannopyranose residues in linear chains. Glucomannans consist of both β-(1 → 4) linked D-mannopyranose and β-(1 → 4) linked D-glucopyranose residues in the main chains. As for the side chains, D-galactopyranose residues tend to be 6-linked to both types as the single side chains with various amount (Voiniciuc 2022).

The conformation of the mixed linkage glucan chains usually contains blocks of β-(1 → 4) D-Glucopyranose separated by single β-(1 → 3) D-Glucopyranose. The ratio of β-(1 → 4) and β-(1 → 3) are about 70% and 30%. These glucans primarily consist of cellotriosyl ($C_{18}H_{32}O_{16}$) and cellotetraosyl ($C_{24}H_{42}O_{21}$) segments in random order. There are some studies that show the molar ratio of cellotriosyl/cellotetraosyl for oat (2.1–2.4), barley (2.8–3.3), and wheat (4.2–4.5) (Salem 2022; Voiniciuc 2022; Hemati et al. 2022).

Xyloglucans have a backbone similar to cellulose with α-D-xylopyranose residues at position 6. To better describe different side chains, a single letter code notation is used for each side chain type. *G* unbranched Glc residue; *X* α-d-Xyl-(1 → 6)-Glc. *L* β-Gal, *S* α-l-Araf, *F* α-l-Fuc. These are the most common side chains (Hemati et al. 2022).

2.3.2 Enzymes

Given the tremendous diversity of hemicelluloses, the enzymatic deconstruction of these polysaccharides needs a much more complex set of enzymes for degradation as compared with cellulose degradation. Hemicellulases can be divided into two major classes. Firstly, enzymes with depolymerization action, which hydrolyze the main chain glycosidic bonds (xylanases, glucanases, and mannanases), and secondly, accessory enzymes, which break the ester bonds and glycosidic bonds of hemicellulose side chains (α-L-arabinofuranosidase, acetyl xylan esterase, β-glucuronidase, glucuronyl esterase, and ferulic acid esterase (Shallom and Shoham 2003). Xylan is the main constituent of the hemicellulosic compounds that account for one-third of the total organic carbon on Earth (Agger et al. 2022). Industrial processes to remove xylan components from lignocellulose include high temperatures and pressure under alkaline conditions (Kumar and Satyanarayana 2011). These processes are energy demanding and environmentally unfriendly. Therefore, mild enzyme methods using various hemicellulases are preferred today, especially enzymes that are active at high temperatures. The main types of hemicellulases are endoxylanases, β-xylosidase, arabinofuranosidase, acetyl-xylan esterase, and more. The endoxylanases (EC: 3.2.1.8) mainly cleave the β-glycosidic bonds of the xylan backbone. The β-xylosidases (EC: 3.2.1.37) degrade xylobiose and other xylooligosaccharides to yield xylose. Arabinofuranosidases (EC: 3.2.1.55) and acetyl-xylan esterases (EC:3.1.1.72) attack side chains of heterogenous xylan substrates and help xylanases and β-xylosidases to degrade xylan completely (Collins et al. 2005). Additionally, the synergistic action of these enzymes facilitates xylan and lignin removal from cellulose without affecting the cellulose structure (Collins et al. 2005).

2.3.3 Thermophilic Hemicellulose-Degrading Enzymes: Microorganisms

Hemicellulose-degrading enzymes were first discovered from mesophilic bacteria and fungi. Such enzymes do not withstand the extreme conditions needed in the various industries, e.g., bleaching of pulp in the paper industry. Therefore, enzymes from thermophilic bacteria have gained increased interest in the last decades. There are several reasons for considering bacterial over fungi enzymes (Zhu et al. 2011; Verma et al. 2019). Bacteria are easily cultivated and grow fast and harvesting methods have been well documented. Thus, bacteria are regarded as attractive factories for enzyme production (Verma et al. 2019). The majority of fungal xylanases are either acidic or neutral, whereas many bacterial enzymes work at alkaline pH range, often an advantage in the pulp and paper industry (Verma et al. 2019). Finally, bacterial hemicellulose enzymes are often smaller than enzymes of fungal origin, which increases their diffusion into the rigid lignocellulosic biomass (Breccia et al. 1998; Verma et al. 2019).

Below are examples of the various literature on thermophilic anaerobes degrading various portions of hemicelluoses. *Geobacillus* species are perhaps the best-known thermophiles known to produce xylanases and other hemicellulose-degrading components (Chadha et al. 2019). These bacteria are however aerobic and thus not a topic in this chapter. While only a few *Caldicellulosiruptor* species can degrade cellulose, all known species are hemicellulolytic and among good producers of xylanases are species within the genus, such as *C. bescii, C. lactoaceticus,* and *C. owenensis* (Crosby et al. 2022; An et al. 2015; Jia et al. 2014; Liu et al. 2017). *C. bescii* GH10 xylanase was overexpressed in *E. coli* and showed high activity with a half-life of 7.7 h at 60 °C. The enzyme was also capable of cellulose degradation and may therefore be a good choice for biomass degradation (An et al. 2015). *C. lactoaceticus* also possesses GH10 xylanase activity at 80 °C and pH 4.5 (Jia et al. 2014). The enzymatic degradation of xylan resulted in liberation of XOS with methyl-glucouronic acid sub-chains. *C. owensis* produces thermostable xylanases which are optimally active at 90 °C and exhibited a half-life of 1 h at 80 °C and is capable of degrading hemicellulose of corn stover and cob (Liu et al. 2017). Other known thermophilic anaerobic xylanase-producing bacteria are, e.g., *Herbivorax saccincola* (Aikawa et al. 2018), *Caldicoprobacter algertensis* (Amel et al. 2016), and *Thermoanaerobacterium aotearoense* (Huang et al. 2015). Many thermophilic and hyperthermophilic bacteria have been reported to produce endoxylanases, including *Thermotoga* (Bok et al. 1998; Winterhalter et al. 1995), *Clostridium* (Rani and Nand 2000; Heinze et al. 2017), *Caldicellulosiruptor* (Crosby et al. 2022), and various aerobic thermophiles (*Geobacillus, Bacillus, Streptomyces* sp).

Degradation of other hemicellulose types like β-glucan, xyloglucan, arabinoxylan, mannan, galactomannan, arabinan, galactan, and polygalacturonan is much less studied as compared with xylan. The most thermoactive α-galactosidase was isolated from *Thermotoga neapolitana* 5068 possessing a temperature optimum at 100–103 °C (Duffaud et al. 1997). Almost all thermostable mannanases have been isolated from fungi (Aulitto et al. 2019), except for a description of β-mannanase from *Thermotoga* species (Chhabra et al. 2001) and *Dictyglomus thermophilum* (Gibbs et al. 1999).

2.4 *Proteases and Amino Acid-Degrading Enzymes*

2.4.1 Enzymes

Proteases (EC 3.4) are a diverse class of hydrolyses that are among the most fundamental and most flexible family due to their role in the hydrolysis of proteins via the hydrolysis of peptide bonds making them critical to the core physiological activities of all organisms. Proteases have found an expansive range of applications ranging from leather processing, silver recovery, medical purposes, food and feed processing, chemical synthesis, and waste treatment (Homaei et al. 2010, 2014; Homaei 2015; Homaei and Etemadipour 2015). According to some sources, the market value for proteases in 2016 is more than 65% of all enzymes produced (Shanmugavel et al. 2016) highlighting their widespread utility. Additionally, many proteases can hydrolyze carboxylic esters and often exhibit a high degree of catalytic promiscuity while retaining enantioselectivity, often toward the enantiomer most similar to their natural substrate, making them useful agents in organic synthesis (Bornscheuer and Kazlauskas 2006; Faber 2011).

Proteases are complex groups of enzymes that differ by their site of cleavage along a peptide, the nature of the active site, and associated mechanism of action, as well as their optimum pH and temperature. As is the case with other hydrolases, proteases can be classified as either endo- or exo-acting based on the location of peptide hydrolysis. The former cleaves at the terminal amino acid residues near the end of a polypeptide while the latter hydrolyzes internal peptide bonds (Souza et al. 2005). The specific location of peptide hydrolysis is often determined by the neighboring steric factors dictated by neighboring sides chains; as an example, two serine proteases, trypsin and chymotrypsin, differ in their location of the hydrolysis with trypsin hydrolyzing peptide bonds. Another classification may be done based on their location of cleavage, i.e., aminoprotease acts on free amino-terminal of the polypeptide chain and carboxylprotease on the carboxyl-terminal of the polypeptide chain. Further classification of proteases may be done according to their catalytic active site into four groups, serine-, cysteine-, aspartic-, and metallo-proteases (Gupta et al. 2002).

2.4.2 Thermophilic Protein-Degrading Bacteria

Thermophilic, anaerobic degradation of proteins and amino acids are much less studied as compared with mesophilic bacteria (Orlygsson 1994). Early work on mesophilic bacteria mainly focused upon degradation of rumen microorganisms to understand the classification of protein-degrading organisms (Mead 1971; Elsden and Hilton 1979) and later on physiology of the rumen and soil consortia (Firkins et al. 2007). Of particular interest were *Clostridium* species of clinical relevance, such as *Clostridium botulinum* (Elsden and Hilton 1978). During these studies, it became clear that the importance of hydrogenotrophic methanogenesis in systems where protein degradation occurs was of great importance (Orlygsson et al. 1994). This was later investigated further leading to the observation that many amino acids, especially reduced amino acids (e.g., alanine, branched-chain amino acids) cannot be degraded anaerobically as single substrates unless the hydrogen produced in the initial oxidative step is scavenged (Faudon et al. 1995). These amino acids are therefore not degraded as single substrates unless the electrons produced are scavenged, either by coculturing the amino acid-degrading organism in a coculture with hydrogen scavenging organisms, or by adding a chemical such as thiosulfate to scavenge the hydrogen produced (Fardeau et al. 1997; Scully and Orlygsson 2014). Thus, the branched-chain amino acids, leucine, isoleucine, and valine were degraded to their corresponding, one carbon shorter, branched-chain fatty acids, isovalerate, 2-methylbutyrate, and isobutyrate (2-methylpropionate) under such conditions (Scully and Orlygsson 2014; Scully and Orlygsson 2019). Interestingly, recent investigations revealed that some species within the genera of *Thermoanaerobacter* and *Caldanaerobacter*, did not only produce their corresponding branched-chain fatty acids from branched-chain amino acids, but also to their corresponding alcohol when cultivated with thiosulfate as an electron acceptor (Scully and Orlygsson 2019). This is an interesting mechanism of producing valuable high-carbon alcohols from amino acids.

Among thermophilic anaerobic bacteria that have been reported as proteolytic are members within the genera of *Coprothermobacter* (Gagliano et al. 2015), and the archaea *Thermococcus* (Koga et al. 2014) and *Pyrococcus* (Borissenko and Groll 2005). Several *Thermoanaerobacter* and *Caldanaerobacter* species also exhibit proteolytic features (Scully 2019).

2.5 Esterases and Lipases

Lipases and esterases belong to the hydrolase family of enzymes and the class of serine hydrolases (Akram et al. 2023). They are classified into eight families (I–VIII) based on their differences in their physiological properties, conserved motifs, and amino acid sequences (Ramnath et al. 2017). Their activity involves synthesis, hydrolysis of fat, esterification, transesterification, interesterification, acidolysis, alcoholysis, and aminolysis (Verissimo et al. 2018; Nawal et al. 2019; Monteiro et al. 2020; Qiu et al. 2021; Adetunji and Olaniran 2021). These catalytic activities, however, have often limitations in most synthetic and chemical processes (Li et al. 2012). Lipases do not require any cofactors to function and are highly soluble in water and can catalyze insoluble substrates like long-chain triacylglycerols. Many lipases have, however, high thermostability often necessary in various industrial processes; detergent, food, oleochemical, pulp and paper, resolution of drugs, wastewater treatment, peptide synthesis, and biodiesel (Bornscheuer 2002; Hasan et al. 2006; Sharma et al. 2001). Thermostable or thermotolerant lipases are widespread research fields because of the desired properties of thermophilic enzymes compared with their mesophilic equivalents. A very significant feature of thermophilic lipases is their thermostability, which is an excellent property in most reaction conditions correlated with optimum enzymatic activity and growth. The thermostability of lipases has pushed them to the frontiers of being the most suitable catalyst in biodiesel production (Milasinovic et al. 2012). One of the main benefits of using lipases is the fact that these enzymes often work in solvent-free reactions. Important requirements for thermostability of lipases are a small hydrophobic surface, exposed N- and C-terminal loops, strong ion-pairing with arginine residues, hydrogen and disulfide bonds, hydrophobic interactions, and interactions between aromatic pairs (Jaenicke and Böhm 1998). One of the main obstacles of using microbial lipases is the fact that they do not fulfill all the requirements for efficient biocatalysis. Therefore, in the past few years there has been extensive research to use genetic engineering to increase their thermostability (Li and Zhang 2005).

2.5.1 Thermophilic, Anaerobic Bacteria Producing Lipases

There are very few thermophilic fermentative anaerobic bacteria known to produce lipases. However, there are numerous reports on thermophilic and thermostable lipases from aerobic bacteria and we refer to an excellent review from 2021 on this subject (Hamdam et al. 2021).

In a study by Royter et al. (2009), two novel genes responsible for lipase production from *Thermoanarobacter thermohydrosulfuricus* strain (DSM) and *Caldanaerobacter subterraneus* subsp. *tencongensis* (DSM 15242, formerly *Thermoanaerobacter tencogensis*) were successfully cloned to *E. coli* (Royter et al. 2009). These enzymes were robust proteins and resistant against a large number of organic solvents and detergents and showed activity (10.90–12.15 U/mg) toward a broad spectrum of substrates, including triacylglycerols, monoacylglycerols, ester of secondary alcohols, and p-nitrophenyl esters. Some studies have been performed on the moderate thermophile species within the genus *Anoxybacillus* (Bakir and Metin 2016). Esterases from *Anoxybacillus gonensis* (Colak et al. 2005), a lipase from *Anoxybacillus kamchatkensis* (Olusesan et al. 2009), a lipase from various *Anoxybacillus* species (Pinzon-Martinez et al. 2010), a carboxylesterase from Anoxybacillus sp. PDF1 (Ay et al. 2011), and an esterase/lipase from *Anoxybacillus flavithermus* (Chis et al. 2013). An intracellular lipase from *Anoxybacillus flavithermus* was isolated and purified by Bakir and Metin (2016). The enzyme from the strain was stable at pH between 6.0 and 11.0 and at temperatures between 25 and 50 °C for 24 h. The enzyme was highly stable against glycerol, sorbitol, and mannitol and preferred long-chain triacylglycerol and saturated fatty acids. Two lipases from the thermophilic, anaerobic bacterium, *Thermosyntropha lipolytica* were described by Salameh and Wiegel (2009). The bacterium retained 50% activity after incubation for 6 h at 100 °C. The enzymes, LipA and LipB are the most alkali-thermophilic lipases (optimal pH at 25 °C = 9.4–9.6; optimal T = 96 °C) known today. It catalyzes the synthesis of diacylglycerols and various alcohol fatty acids. Finally, *Fervidobacterium nodusum* was investigated for lipase production capacity (Yu et al. 2010). The strain hydrolyzes triacylglycerols with long chains and has maximum activity at 70 °C and pH 9.0. The enzyme was activated by treating it with polar organic solvents like propanol and acetone.

2.6 Pectinases

The vast majority of biomass used as a feedstock for fuels, chemicals, and other bioproducts has been focusing on crystalline cellulose, lignin, and xylan (Taylor et al. 2009; Scully and Orlygsson 2015). Pectin is, however, overlooked although it is produced in secondary walls of most plant species. It is also known that saccharification of plant biomass can be improved by modifying the structure of pectin. Also, some types of biomass, such as apple pomace, citrus processing waste, and sugar beet pulp may contain relatively high concentrations of pectin (Lin et al. 2022). The first commercial use of pectinases was in 1930 where the enzyme was used for clarification of apple juice (Kertesz 1930). The applications of pectinases vary to a great extent depending on physical conditions and microbes used. These enzymes have been used in various industrial processes, such as tea, coffee, textile, wastewater treatment, oil extraction, and more. Perhaps the most important use of pectinases is their use in fruit processing industry as juice clarifiers, color and yields enhancers, and in fruit mash treatment (Zhang et al. 2010).

2.6.1 Structure

Pectin is a complex and heterogenous polymer formed by various substructures like homogalacturonan (HG), xylogalacturonan (XGA), rhamnogalacturonan I (RG-I), and rhamnogalacturonan II (RG-II) (Fig. 6) (Harholt et al. 2010).

The ratio of the various substructures of pectin vary to a great deal depending on the source but in most cases, HG is in highest concentrations with 65% of the molecule, followed by RG-1 (25–30%) (Harholt et al. 2010). HG forms a backbone of α-1,4-D-galacuronic acid made up of galacturonic acid or methyl galacturonic acids residues linked with a-1,4 glycosidic linkage. In XGA, these residues are substituted with xylose and clusters of side chains of glycosyl residues. For further information on structure and function of pectin, we refer to Harholt et al. (2010) and Caffall and Mohnen (2009).

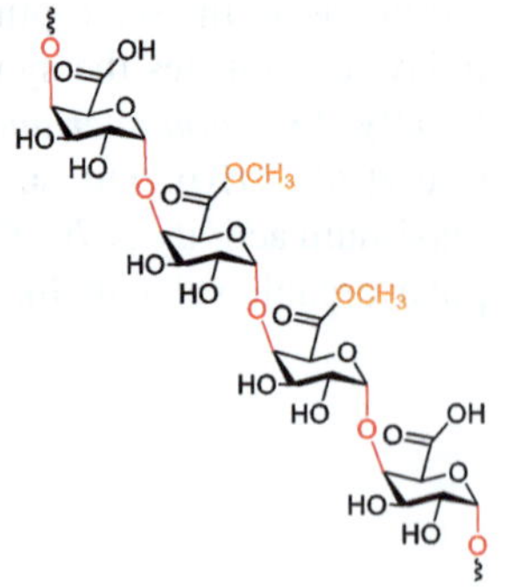

Fig. 6 Pectins (galacturonans)are a class consisting of α-(1–4)-linked D-galacturonic acid residues with varying degrees of esterification of the carboxylic acids group. The example shown is a homogalacturonan, although heterogalcturonans contain other carbohydrates and may have branching that incorporates neutral sugars

2.6.2 Enzymes

Pectinases are a heterogenous group of related enzymes capable of pectin degradation, mostly present in plants. They are classified into three groups, hydrolyses (EC 3.2.1.15), lyase/transeliminases (EC 4.2.2.10), and pectate lyases (EC 3.1.1.11). Pectinases are primarily used for processes involving plant material degradation and have also been used in wine production and are of great industrial importance. Pectin hydrolases (PG) catalyze the hydrolysis of a-1,4-glycosidic bond to depolymerize polygalacturonate. They can be divided in endo- and exo-PG with the former hydrolyzing polygalacturonate and liberates oligogalacturonate but the latter hydrolyzes pectic acids to monogalacturonate (Patidar et al. 2018). Pectin methylesterase de-esterifies methoxyl groups of pectin-forming pectic acid (Kashyap et al. 2001). Pectinolytic lyases (pectic transeliminases) degrade pectin through cleavage reaction with β-elimination producing 4,5-unsaturated oligogalacturonate products (Gummadi and Kumer 2005).

2.6.3 Thermophilic Bacteria Degradation of Pectin

Majority of pectinases are produced from mesophilic facultative aerobic bacteria like Yersinia (Abbott and Boraston 2007), *Klebsiella* (Walker and Pemberton 1987), *Pseudomonas* (Liao et al. 1992), *Bacillus* (Bekli et al. 2019), *Paenibacillus* (Li et al. 2014), *Bacteroides* (McGarthy et al. 1985), and more. Additionally, some filamentous fungi are known to produce pectinases (Damak et al. 2013; Yadav et al. 2009). Thermoactive pectinases of anaerobic origin have though been reported in literature. *Caldicellulosiruptor, Clostridia,* and *Thermoanaerobacter* species are among thermophilic, anaerobic genera that have been mentioned (Bredholt et al. 1999; Schink and Zeikus 1983; Wiegel et al. 1979; Kozianowski et al. 1997). Recently a study on *Caldicellulosiruptor kronotskyensis* showed that a thermostable pectin lyase was highly conserved and exhibited an optimal activity at 70 °C and pH 9.0 with Ca^{2+} as cofactor. This bacterium degraded polygalacturonic acid, methylated pectin through endo-cleaving action (Su et al. 2015). *Thermotoga maritima* has also been shown to grow on pectin as a sole carbon source and produces pectin lyase A in the medium (Kluskens et al. 2003). A recent investigation on genetic engineering work where exo-polygalacturonases from *T. maritima* were cloned and expressed in *E. coli* showed optimum activity at temperatures above 50 °C and good stability at high temperatures (40–90 °C) for up to 24 h (Flores-Fernandez et al. 2022).

3 Other Enzymes

Among other thermophilic enzymes used in various industries are oxidoreductases, nitrilases, transaminases, glutamate dehydrogenases, and laccases (Atalah et al. 2019). Thermostable alcohol dehydrogenases in particular have received a lot of coverage in the literature due in no small part to their ability to enantioselectivity reduce ketones to their corresponding (*S*)-enantiomer which is of particular importance in the preparation of pharmaceutical intermediates. A noteworthy feature of the ADHs produced by *Thermoanaerobacter* species is their high selectivity for secondary alcohol dehydrogenases (Fig. 7) making them useful tools in their regard.

The native TS alcohol dehydrogenases produced by *Thermoanaerobacter* often have a relatively limited preference for small substrates and also require NADP as a cofactor. While beyond the scope of this chapter, these secondary ADHs have been extensively used as targets for the expansion of the activity site to allow an expanded range of substrate specificity as well as the substitution of NAD for NADP (An et al. 2019).

Another enzyme class of synthetic value are thermostable nitrilases. Nitrilases are enzymes that contain a cyano group in their structure and are used for the production of polymers, pharmaceuticals, and pesticides (Nigam et al. 2017). Many of these enzymes are toxic and carcinogenic and it is not surprising that microorganisms have developed a way to degrade them in nature. These enzymes are enantioselective, thiolytic, and convert nitriles to their corresponding carboxylic acid and ammonia. The catalytic place is a residue of Glu-Lys-Cys and they catalyze a wide spectrum of substrates (Nigam et al. 2017). The enzymes do not require metal ions or other cofactors for activity. Many bacteria have been shown to possess nitrilases, most of mesophilic origin (Nigam et al. 2017). Several thermophiles (Almatawah and Cowan 1999; Chen et al. 2019) and extremophiles (Chen et al. 2019; Mueller et al. 2006) have thermostable nitrilases. The archaea *Pyrococcus* sp. M2D13 produces nitrilase with highest activity at 85 °C, most likely the most thermostable enzyme of its class known (Dennett and Blamey 2016).

Transaminases are a group of enzymes that use pyridoxal-5′-phosphate (PLP) as a cofactor and catalyze the reversible transfer of an amine group from a donor (often amino acid) to the carbonyl of an acceptor substrate (Mehta et al. 1993). The mechanism of these enzymes is a two-step catalysis; first the amine or the amino acid is deaminated releasing the amino donor product, whereafter the amination of a keto, ketone or aldehyde (amino acceptor) is performed, with the release of a new amino acid or amine (Guo and Berglund 2017). The active site of these enzymes

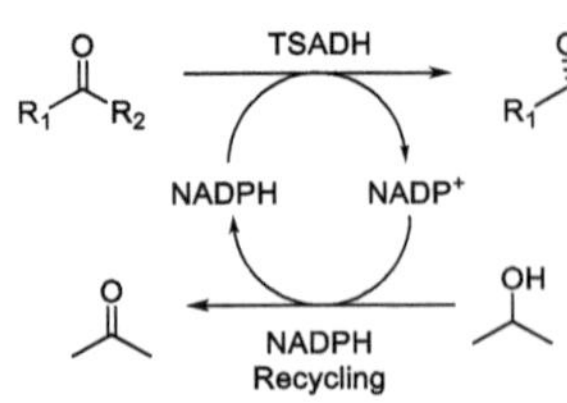

Fig. 7 The reduction of ketones to a chiral secondary alcohol is readily facilitated by *Thermoanaerobacter* secondary alcohol dehydrogenase (TSADH)

comprises a lysine residue which binds transiently to the donated amino group and after binding to the amino accepting compound, the enzyme returns back to its initial state (Steffen-Munsberg et al. 2015). The fact that these enzymes do not need addition of any more cofactors is appealing for industrial use.

Transaminases are ubiquitous enzymes since they participate in the central metabolism of all living things. However, there are relatively few transaminases from thermophiles that have been described and all are from aerobic bacteria (Atalah et al. 2019). Another enzyme that has an important role in carbon and nitrogen metabolism is glutamate dehydrogenase but it catalyzes the reverse oxidative deamination of glutamate to α-ketoglutarate and ammonia. The enzyme is dependent on NAD^+ or $NADP^+$ and most of them are homo-oligomers. There are several thermostable glutamate dehydrogenases but few of anaerobic origin. *Pyrococcus furiosus* has though been described to produce one of the most stable glutamate dehydrogenases with an impressive half-life of 10.5 h at 100 °C (Diruggiero and Robb 1995).

Other hydrolases of note have been found among thermophilic bacteria The thermophilic L-aminoacylase (esterase) was cloned and overexpressed from archaeon *Thermococcus litoralis* (Toogood et al. 2002). This esterase gene codes for pyroglutamyl carboxyl peptidase which is a cysteine protease that cleaves the pyroglutamyl group from the N-terminus of biologically important peptides. The commercial use of this enzyme is to cleave the pyroglutamyl group from blocked peptides. Other examples of commercial enzymes that have been identified are carboxyl esterase from *Thermogutta terrifontis* (Sayer et al. 2015), μ-lactamase from *Sulfolobus solfataricus* (Toogood et al. 2004), an α-carbonic anhydrase from *Thermovibrio ammonificans* (James et al. 2014), and transaminase from *Sulfolobus sulfotaricus* (Sayer et al. 2012).

4 Conclusions

The enzymes from thermophilic anaerobes are an important and exciting field of biotechnology used in various industries. Their importance is mostly due to their tolerance toward extremes, not only because of their tolerance to high temperatures but also of other extreme environmental factors used in industrial production of various products. The urge for renewable biomass for the production of biofuels and other compounds, at present mostly produced by ecologically unfriendly methods, is increasing and thus the importance of understanding both the basic microbiology of microbes involved in the production of these compounds is increasing as ever. Thermophilic bacteria possessing the enzymatic factory of degrading the various portions of lignocellulosic biomass are thus getting more and more important and at escalating rate. An example of this is the number of articles with the use of cellulose and hemicellulose at Web of Science at present date are 197,496 and 22,773, respectively. The degradation of proteins and amino acids remains an unexplored field of study, but is essential especially because of the possibility to produce higher chain alcohols from branched-chain amino acids, a route to produce high-energy dense alcohols from protein-containing substrates. Thermostable enzymes from thermophilic anaerobes are also an excellent source of synthetic tools such as nitrilases, transaminases, and alcohol dehydrogenases.

References

Abbott DW, Boraston AB (2007) A family 2 pectate lyase displays a rare fold and transition metal-assisted β-elimination. J Biol Chem 282:35328–35336. https://doi.org/10.1074/jbc.M705511200

Acharya S, Chaudhary A (2012) Alkaline cellulase produced by a newly isolated thermophile *Aneurinibacillus thermoaerophilus* WBS2 from hot spring, India. Afr J Microbiol Res 6(26): 5453–5458. doi: https://doi.org/10.5897/AJMR11.1491

Adetunji AI, Olaniran AO (2021) Production strategies and biotechnological relevance of microbial lipases: a review. Braz J Microbiol 52:1257–1269. https://doi.org/10.1007/s42770-021-00503-5

Agger JW, Jane W, Isaksen T (2022) Discovery of LPMO activity on hemicelluloses shows the importance of oxidative processes in plant cell wall degradation. Proc Natl Acad Sci USA:6287–6292. https://doi.org/10.1073/pnas.1323629111

Aikawa S, Baramee S, Sermsathanaswadi J et al (2018) Characterization and high-quality draft genome sequence of *Herbivorax saccincola* A7, an anaerobic, alkaliphilic, thermophilic, cellulolytic and xylanolytic bacterium. Syst Appl Microbiol 41:261–269. https://doi.org/10.1016/j.syapm.2018.01.010

Ajeje SB, Hu Y, Song GJ et al (2021) Thermostable cellulases/xylanases from thermophilic and hyperthermophilic microorganisms: current perspective. Front Bioeng Biotechnol 9:794304. https://doi.org/10.3389/fbioe.2021.794304

Akram F, Mir AS, Ul Hag I et al (2023) An appraisal on prominent industrial and biotechnological applications of bacterial lipases. Mol Biotechnol 65:521–543. https://doi.org/10.1007/s12033-022-00592-z

Almatawah QA, Cowan DA (1999) Thermostable nitrilase catalysed production of nicotinic acid from 3-cyanopyridine. Enzym Microb Technol 25:718–724. https://doi.org/10.1016/S0141-0229(99)00104-0

Amel BD, Nawel B, Khelifa B et al (2016) Characterization of a purified thermostable xylanase from *Caldicoprobacter algeriensis* sp. nov. strain TH7C1T. Carbohydr Res 419:60–68. https://doi.org/10.1016/j.carres.2015.10.013

An J, Xie Y, Zhang Y et al (2015) Characterization of a thermostable, specific GH10 xylanase from *Caldicellulosiruptor bescii* with high catalytic activity. J Mol Catal B Enzym 117:13–20. https://doi.org/10.1016/j.molcatb.2015.04.003

An JH, Nei Y, Xu Y (2019) Structural insights into alcohol dehydrogenases catalyzing asymmetric reductions. Crit Rev Biotechnol 39:366–379. https://doi.org/10.1080/07388551.2019.1566205

Andreesen JR (2005) Degradation of heterocyclic compounds. In: Durre P (ed) Handbook of clostridia. Taylor and Francis, Boca Raton, FL, pp 221–237

Apriyanto A, Compart J, Fettke J (2022) A review of starch, a unique biopolymer - structure, metabolism and in planta modifications. Plant Sci 318:111223. https://doi.org/10.1016/j.plantsci.2022.111223

Atalah J, Caceres-Moreno P, Espina G et al (2019) Thermophiles and the applications of their enzymes as new biocatalysts. Bioresour Technol 280:478–488. https://doi.org/10.1016/j.biortech.2019.02.008

Aulitto M, Fusco S, Limauro D et al (2019) Galactomannan degradation by thermophilic enzymes: a hot topic for biotechnological applications. World J Microbiol Biotechnol 35:32. https://doi.org/10.1007/s11274-019-2591-3

Ay F, Karaoglu H, Inan K et al (2011) Cloning, purification and characterization of a thermostable carboxylesterase from *Anoxybacillus* sp PDF. Protein Expr Purif 80:74–79. https://doi.org/10.1016/j.pep.2011.06.019

Bachosz K, Zdarta J, Bilal M et al (2023) Enzymatic cofactor regeneration systems: a new perspective on efficiency assessment. Sci Total Environ 868:161630. https://doi.org/10.1016/j.scitotenv.2023.161630

Bakir ZB, Metin K (2016) Purification and characterization of an alkali-thermostable lipase from thermophilic *Anoxybacillus flavithermus* HBB134. J Microbiol Biotechnol 26:1087–1097. https://doi.org/10.4014/jmb.1512.12056

Ballschmiter M, Futterer O, Liebl W (2006) Identification and characterization of a novel intracellular alkaline alpha-amylase from the hyperthermophilic bacterium *Thermotoga maritima* MSB8. Appl Environ Microbiol 72:2206–2211. https://doi.org/10.1128/AEM.72.3.2206-2211.2006

Barzkar N, Homaei A, Hemmati R et al (2018) Thermostable marine microbial proteases for industrial applications: scopes and risks. Extremophiles 22:335–346. https://doi.org/10.1007/s00792-018-1009-8

Bayer E, Lamed R, White B, Flint H (2008) From cellulosomes to cellulosomics. Chem Rec 8:364–377. https://doi.org/10.1002/tcr.2016

Bekli S, Aktas B, Gencer D et al (2019) Biochemical and molecular characterization of a novel pH- and temperature-stable pectate lyase from *Bacillus amyloliquefaciens* S6 for industrial application. Mol Biotechnol 61:681–693. https://doi.org/10.1007/s12033-019-00194-2

Bergthaller W, Hollmann J (2014) Starch. In: Rreference module in chemistry, molecular sciences and chemical engineering. Elsevier, pp 1–29. https://doi.org/10.1016/B978-0-12-409547-2.11374-

Bertoldo C, Armbrecht M, Becker F et al (2004) Cloning, sequencing, and characterization of thermoalkalistable type I pullulanase from *Anaerobrancha gottschalkii*. Appl Environ Microbiol 70:3407–3416. https://doi.org/10.1128/AEM.70.6.3407-3416.2004

Bibel M, Brettl C, Gosslar U et al (1998) Isolation and analysis of genes for amylolytic enzymes of the hyperthermophilic bacterium *Thermotoga maritima*. FEMS Microbiol Lett 158:9–15. https://doi.org/10.1016/S0378-1097(97)00420-5

Bok JD, Yernool DA, Eveleigh DE (1998) Purification, characterization, and molecular analysis of thermostable cellulases CelA and CelB from *Thermotoga neapolitana*. Appl Environ Microbiol 64:4774–4781

Borissenko L, Groll M (2005) Crystal structure of TET protease reveals complementary protein degradation pathways in prokaryotes. J Mol Biol 346:1207–1219. https://doi.org/10.1016/j.jmb.2004.12.056

Bornscheuer UT (2002) Microbial carboxyl esterases: classification, properties and application in biocatalysis. FEMS Microbiol Rev 26:73–81. https://doi.org/10.1111/j.1574-6976.2002.tb00599.x

Bornscheuer UT, Kazlauskas RJ (2006) Hydrolases in organic synthesis: regio- and stereoselective biotransformations. Viley-VCH Verlag GmbH & Co. KgaA. https://doi.org/10.1002/3527607544

Breccia JD, Sineriz F, Baigori MD et al (1998) Purification and characterization of thermostable xylanase from *Bacillus amyloliquefaciens*. Enzym Microb Technol 22:42–49. https://doi.org/10.1016/S0141-0229(97)00102-6

Bredholt S, Mathrani IM, Ahring BK (1995) Extremely thermophilic cellulolytic anaerobes from Icelandic hot springs. Antonie Van Leeuwenhoek 68:263–271. https://doi.org/10.1007/BF00874135

Bredholt S, Sonne-Hansen J, Nielsen P et al (1999) *Caldicellulosiruptor kristjanssonii* sp. nov., a cellulolytic, extremely thermophilic, anaerobic bacterium. Int J Syst Bacteriol 49:991–996. https://doi.org/10.1099/00207713-49-3-991

Brown SH, Kelly RM (1993) Characterization of amylolytic enzymes, having both alpha-1,4 and alpha 1,6 hydrolytic activity, from the thermophilic archaea *Pyrococcus furiosus* and

Thermococcus litoralis. Appl Environ Microbiol 59:2614–2621. https://doi.org/10.1128/AEM.59.8.2614-2621.1993

Caffall KH, Mohnen D (2009) The structure, function, and biosynthesis of plant cell wall pectic polysaccharides. Carbohydr Res 344:1879–1900. https://doi.org/10.1016/j.carres.2009.05.021

Canganella F, Andrade CM, Antrankian G (1994) Characterization of amylolytic and pullulytic enzymes from thermophilic archaea and from a new *Fervidobacterium* species. Appl Microbiol Biotechnol 42:239–245

Chadha BS, Kaur B, Basotra N et al (2019) Thermostable xylanases from thermophilic fungi and bacteria: current perspective. Bioresour Technol 277:195–203. https://doi.org/10.1016/j.biortech.2019.01.044

Chang ST, Parker KN, Bauer MW et al (2001) Alpha glucosidase from *Pyrococcus furiosus*. Hyperthermophilic enzymes. Methods Enzymol 330:260–269

Cheng KC, Demirci A, Catchmark JM (2011) Pullulan: biosynthesis, production, and applications. Appl Microbiol Biotechnol 92(1):29–44. https://doi.org/10.1007/s00253-011-3477-y

Chen Z, Aziz T, Sun HY et al (2023) Advances and applications of cellulose bio-composites in biodegradable materials. J Polym Environ 31:2273–2284. https://doi.org/10.1007/s10924-022-02561-8

Chen Z, Chen HY, Ni Z et al (2019) Expression and characterization of a novel nitrilase from hyperthermophilic bacterium *Thermotoga maritima* MSB8. J Microbiol Biotechnol 25:1660–1669. https://doi.org/10.4014/jmb.1502.02032

Chhabra S, Parker KN, Lam D et al (2001) β-Mannanases from *Thermotoga* species. In: MWW A, Kelly RM (eds) Hyperthermophilic enzymes, vol 330. Academic, pp 224–238

Chis L, Hriscu M, Bica A et al (2013) Molecular cloning and characterization of a thermostable esterase/lipase produced by a novel *Anoxybacillus flavithermus* strain. J Gen Appl Microbiol 59:119–134

Chung YC, Kobayashi T, Kanal H et al (1995) Purification and properties of extracellular amylase from the hyperthermophilic archaeon *Thermococcus profundus* DT5432. Appl Environ Microbiol 61:1502–1506. https://doi.org/10.1128/AEM.61.4.1502-1506.1995

Clark RJH, Cooksey CJ, Daniels MAM et al (1993) Indigo, woad, and Tyrian purple: important vat dyes from antiquity to the present. Endeavour 17(4):191–199. https://doi.org/10.1016/0160-9327(93)90062-8

Colak A, Sisik D, Saglam N et al (2005) Characterization of thermoalkalophilic esterase from a novel thermophilic bacterium, *Anoxybacillus gonensis* G2. Bioresour Technol 96:625–6313. https://doi.org/10.1016/j.biortech.2004.06.003

Collins T, Gerday C, Feller G (2005) Xylanases, xylanase families and extremophilic xylanases. FEMS Microbiol Rev 29:3–23. https://doi.org/10.1016/j.femsre.2004.06.005

Constantio HR, Brown SH, Kelly RM (1990) Purification and characterization of an alpha glucosidase from a hyperthermophilic archaebacterium, *Pyrococcus furiosus*, exhibiting a temperature optimum of 105°C. J Bacteriol 172:3654–3660. https://doi.org/10.1128/jb.172.7.3654-3660.1990

Crosby JR, Laemthong T, Bing RG et al (2022) Biochemical and regulatory analyses of xylanolytic regulons in *Caldicellulosiruptor bescii* reveal genus-wide features of hemicellulose utilization. Appl Environ Microbiol 88. https://doi.org/10.1128/aem.01302-22

Dadwal A, Sharma S, Satyanarayana T (2021) Thermostable cellulose saccharifying microbial enzymes: characteristics, recent advances and biotechnological applications. Int J Biol Macromol 188:226–244. https://doi.org/10.1016/j.ijbiomac.2021.08.024

Damak N, Abdeljalil S, Koubaa A et al (2013) Cloning and heterologous expression of a thermostable pectate lyase from *Penicillium occitanis* in *Escherichia coli*. Int J Biol Macromol 62:549–556. https://doi.org/10.1016/j.ijbiomac.2013.10.013

Dennet GV, Blamey JM (2016) A new thermophilic nitrilase from an antarctic hyperthermophilic microorganism. Front Bioeng Biotechnol 4:5. https://doi.org/10.3389/fbioe.2016.00005

Diruggiero J, Robb FT (1995) Expression and in-vitro assembly of recombinant glutamate-dehydrogenase from the hyperthermophilic archaeon Pyrococcus furiosus. J Environ Microbiol 61:159–164. https://doi.org/10.1128/AEM.61.1.159-164.1995
Doi RH (2008) Cellulases of mesophilic microorganisms - Cellulosome and noncellulosome producers. Ann NY Acad Sci 1125:267–279. https://doi.org/10.1196/annals.1419.002
Domań-Pytka M, Bardowski J (2004) Pullulan-degrading enzymes of bacterial origin. Crit Rev Microbiol 30(2):107–121. https://doi.org/10.1080/10408410490435115
Duffaud GD, McCutchen CM, Leduc P et al (1997) Purification and characterization of extremely thermostable beta-mannanase, beta-mannosidase, and alpha-galactosidase from the hyperthermophilic eubacterium *Thermotoga neapolitana* 5068. Appl Environ Microbiol 63:169–177. https://doi.org/10.1128/AEM.63.1.169-177.1997
Ebringerová A (2005) Structural diversity and application potential of hemicelluloses. Macromol Symp 232(333):1–12. https://doi.org/10.1002/masy.200551401
Elleuche S, Antranikian G (2013) Starch-Hydrolyzing enzymes from thermophiles. In: Satyanarayana T (ed) Thermophilic microbes in environmental and industrial biotechnology: biotechnology of thermophiles, 2nd edn. Springer, Hoboken, pp 509–533. https://doi.org/10.1007/978-94-007-5899-5_20
Elsden SR, Hilton MG (1978) Volatile fatty acid production from threonine, valine, leucine and isoleucine by clostridia. Arch Microbiol 117:165–172. https://doi.org/10.1007/BF00402304
Elsden SR, Hilton MG (1979) Amino acid utilization patterns in clostridial taxonomy. Arch Microbiol 123:137–141. https://doi.org/10.1007/BF00446812
Faber K (2011) Biotransformations in organic chemistry: a textbook, 5th edn. Springer, New York
Fardeau ML, Patel BKC, Magot M et al (1997) Utilization of serine, leucine, isoleucine and valine by *Thermoanaerobacter brockii* in the presence of thiosulfate or Methanobacterium sp. as electron acceptors. Anaerobe 3:405–410. https://doi.org/10.1006/anae.1997.0126
Faudon C, Fardeau ML, Heim J et al (1995) Peptide and amino acid oxidation in the presence of thiosulfate by members of the genus *Thermoanaerobacter*. Curr Microbiol 31:152–157. https://doi.org/10.1007/BF00293546
Firkins JL, Yu Z, Morrison M (2007) Ruminal nitrogen metabolism: perspectives for integration of microbiology and nutrition for dairy. J Dairy Sci 90:E1–E16. https://doi.org/10.3168/jds.2006-518
Flores-Fernandez CN, Cardenas-Fernandez M, Lye GJ, Ward JM (2022) Synergistic action of thermophilic pectinases for pectin bioconversion into D-galacturonic acid. Enzym Microb Technol 160:110071. https://doi.org/10.1016/j.enzmictec.2022.110071
Fukusumi S, Kamizono A, Horinouchi S et al (1988) Cloning and nucleotide-sequence of a heat stable amylase gene from an anaerobic thermophile, *Dictyoglomus thermophilum*. Eur J Biochem 174:15–21. https://doi.org/10.1111/j.1432-1033.1988.tb14056.x
Gagliano MC, Braguglia CM, Petruccioli M et al (2015) Ecology and biotechnological potential of the thermophilic fermentative *Coprothermobacter* spp. FEMS Microbiol Ecol 91:fiv018. https://doi.org/10.1093/femsec/fiv018
Gibbs MD, Reeves RA, Sunna A et al (1999) Sequencing and expression of a beta-mannanase gene from the extreme thermophile *Dictyoglomus thermophilum* Rt46B.1, and characteristics of the recombinant enzyme. Curr Microbiol 39:351–357. https://doi.org/10.1007/s002849900471
Gummadi SN, Kumar DS (2005) Microbial pectic transeliminases. Biotechnol Lett 27:451–458. https://doi.org/10.1007/s10529-005-2197-8
Guo F, Berglund P (2017) Transaminase biocatalysis: optimization and application. Green Chem 19:333–360. https://doi.org/10.1039/c6gc02328b
Gupta R, Beg QK, Lorenz P (2002) Bacterial alkaline proteases: molecular approaches and industrial applications. Appl Microbiol Biotechnol 59:15–32. https://doi.org/10.1007/s00253-002-0975-y
Haitjema CH, Gilmore SP, Henske JK et al (2017) A part list for fungal cellulosomes revealed by comparative genomics. Nat Microbiol 2(8):17087. https://doi.org/10.1038/nmicrobiol.2017.87

Hamdam SH, Majangwa J, Ali MSM et al (2021) Thermostable lipases and their dynamics of improved enzymatic properties. Appl Microbiol Biotechnol 105:7069–7094. https://doi.org/10.1007/s00253-021-11520-7

Hamilton-Brehm SD, Mosher JJ, Vishnivetskaya T et al (2010) *Caldicellulosiruptor obsidiansis* sp nov., an anaerobic, extremely thermophilic, cellulolytic bacterium isolated from obsidian pool, Yellowstone National Park . Appl Environ Microbiol 76:1014–1020. doi: https://doi.org/10.1128/AEM.01903-09

Harholt J, Suttangkakul A, Scheller HV (2010) Biosynthesis of pectin. Plant Physiol 153:384–395. https://doi.org/10.1104/pp.110.156588

Hasan F, Shah AA, Hameed A (2006) Industrial applications of microbial lipases. Enzym Microb Technol 39:235–251. https://doi.org/10.1016/j.enzmictec.2005.10.016

Heinze S, Mechelke M, Kornberger P et al (2017) Identification of endoxylanase XynE from *Clostridium thermocellum* as the first xylanase of glycoside hydrolase family GH141. Sci Rep 7:11178. https://doi.org/10.1038/s41598-017-11598-y

Hemati A, Nazari M, Lajayer BA et al (2022) Lignocellulosics in plant cell wall and their potential biological degradation. Folia Microbiol 67:671–681. https://doi.org/10.1007/s12223-022-00974-5

Homaei A (2015) Immobilization and its applications in the food industry. In: Homaei A (ed) Advances in food biotechnology. Wiley, p 145

Homaei A, Etemadipour R (2015) Improving the activity and stability of actinidin by immobilization on gold nanorods. Int J Biol Macromol 72:11767–11181. https://doi.org/10.1016/j.ijbiomac.2014.10.029

Homaei A, Sajedi RH, Aariri R et al (2010) Cysteine enhances activity and stability of immobilized papain. Amino Acids 38:937–942. https://doi.org/10.1007/s00726-009-0302-3

Homaei A, Barkheh H, Sariri R et al (2014) Immobilized papain on gold nanorods as heterogenous biocatalysts. Amino Acids 46:1649–1657. https://doi.org/10.1007/s00726-014-1724-0

Hong SY, Lee JS, Cho KM et al (2007) Construction of the bifuntional enzyme cellulase beta-glucosidase from hyperthermophilic bacterium *Thermotoga maritima*. Biotechnol Lett 29:931–936. https://doi.org/10.1007/s10529-007-9334-5

Huang X, Li Z, Du C et al (2015) Improved expression and characterization of a multidomain xylanase from *Thermoanaerobacterium aotearoense* SCUT27 in *Bacillus subtilis*. J Agric Food Chem 63:6430–6643. https://doi.org/10.1021/acs.jafc.5b01259

Hussain I, Siddique F, Mahmood MS et al (2013) A review of the microbiological aspects of alpha amylase production. Int J Agric Biol 15:1029–1034

Hyun HH, Zeikus JG (1985a) General biochemical characterization of thermostable extracellular beta-amylase from *Clostridium thermosulfurogenes*. Appl Environ Microbiol 49:1162–1167. https://doi.org/10.1128/AEM.49.5.1162-1167.1985

Hyun HH, Zeikus JG (1985b) Regulation and genetic enhancement of glucoamylase and pullulanase production in *Clostridium thermohydrosulfuricum*. J Bacteriol 164:1146–1152. https://doi.org/10.1128/JB.164.3.1146-1152.1985

Jaenicke R, Böhm G (1998) The stability of proteins in extreme environments. Curr Opin Struct Biol 8:738–748. https://doi.org/10.1016/S0959-440X(98)80094-8

James P, Isupov MN, Sayer C (2014) The structure of a tetrameric α-carbonic anhydrase from *Thermovibrio ammonificans* reveals a core formed around intermolecular disulfides that contribute to its thermostability. Acta Crystallogr D Biol Crystallogr 70:2607–2618. https://doi.org/10.1107/S1399004714016526

Jia Z, Mi S, Wang J et al (2014) Insight into glycoside hydrolases for debranched xylan degradation from extremely thermophilic bacterium *Caldicellulosiruptor lactoaceticus*. PLoS One 9:e106482. https://doi.org/10.1371/journal.pone.0106482

Jiao YL, Wang SJ, Lv MS et al (2011) A GH57 family amylopullulanase from deep sea *Thermococcus siculi*: expression of the gene and characterization of the recombinant enzyme. Curr Microbiol 62:222–228. https://doi.org/10.1007/s00284-010-9690-6

Kang J, Park KM, Choi KH et al (2011) Molecular cloning and biochemical characterization of a heat stable type I pullulanase from *Thermotoga neapolitana*. Enzym Microb Technol 48:260–266. https://doi.org/10.1016/j.enzmictec.2010.11.006

Kashyap DR, Vohra PK, Chopra S et al (2001) Application of pectinases in the commercial sector: a review. Bioresour Technol 77:215–227. https://doi.org/10.1016/S0960-8524(00)00118-8

Kertesz Z (1930) A new method for enzymic clarification of unfermented apple juice. US Patent No. 1.932.833, New York State Agricultural Experimentation Station (Geneva) Bill. No. 689

Kluskens LD, Alebeck G-JWM, Voragen AGJ et al (2003) Molecular and biochemical characterization of the thermoactive family I pectate lyase from the hyperthermophilic bacterium *Thermotoga maritima*. Biochem J 370:651–659. https://doi.org/10.1042/BJ20021595

Koch R, Zablowski P, Spreinat A et al (1990) Extremely thermostable amylolytic enzyme from the archaebacterium *Pyrococcus furiosis*. FEMS Microbiol Lett 71:21–26

Koch R, Spreinat A, Lemke K et al (1991) Purification and properties of a hyperthermoactive alpha amylase from the archaeobacterium *Pyrococcus woesei*. Arch Microbiol 155:572–578. https://doi.org/10.1007/BF00245352

Koga Y, Tanaka S, Sakudo A et al (2014) Proteolysis of abnormal prion protein with a thermostable protease from *Thermococcus kodakarensis* KOD1. Appl Microbiol Biotechnol 98:2113–2120. https://doi.org/10.1007/s00253-013-5091-7

Kozianowski G, Canganella F, Rainey FA et al (1997) Purification and characterization of thermostable pectate-lyases from a newly isolated thermophilic bacterium, *Thermoanaerobacter italicus* sp nov. Extremophiles 4:171–182. https://doi.org/10.1007/s007920050031

Kumar V, Satyanarayana T (2011) Applicability of thermoalkali stable and cellulase free xylanase from a novel thermohalophilic *Bacillus halodurans* in producing xylooligosaccharides. Biotechnol Lett 33:2279–2285. https://doi.org/10.1007/s10529-011-0698-1

Lee KY, Yoo YJ (1993) Optimization of pH for high molecular weight pullulan. Biotechnol Lett 15:1021–1024. https://doi.org/10.1007/BF00129930

Lee LL, Blumer-Schuette SM, Izquierdo JA et al (2018) Genus-wide assessment of lignocellulose utilization in the extremely thermophilic genus *Caldicellulosiruptor* by genomic, pangenomic, and metagenomic analyses. Appl Environ Microbiol:e02694-17. https://doi.org/10.1128/AEM.02694-17

Leisola M, Jokela J, Pastinen O et al (2017) Encyclopedia of life support systems (EOLSS). Physiolgoy and maintainance. Vol. II. Industrial use of enzymes (UNESCO-EOLSS)

Li H, Zhang X (2005) Characterization of thermostable lipase from thermophilic *Geobacillus* sp. TW1. Protein Expr Purif 42:153–159. https://doi.org/10.1186/1472-6750-14-18

Li X, Wang H, Zhou C et al (2014) Cloning, expression and characterization of pectate lyase from *Paenibacillus* sp. 0602 in recombinant *Escherichia coli*. BMC Biotechnol 14:18. https://doi.org/10.1186/1472-6750-14-18

Liao CH, Sasaki K, Nagahashi G et al (1992) Cloning and characterization of pectate lyase gene from the soft-rotting bacterium *Pseudomonas viridiflava*. Mol Plant-Microbe Interact 5:301–308. https://doi.org/10.1094/MPMI-5-301

Liebl W, Stemplinger I, Ruile P (1997) Properties and gene structure of the *Thermotoga maritima* alpha-amylase AmyA, a putative lipoprotein of a hyperthermophilic bacterium. J Bacteriol 179:941–948. https://doi.org/10.1128/jb.179.3.941-948.1997

Li P, Jing RX, Zhou MP (2022) Whole-cell biosynthesis of cytarabine by an unnecessary protein reduced *Escherichia coli* that coexpresses purine and uracil phosphorylase. Biotechnol Bioeng 119:1768–1780. https://doi.org/10.1002/bit.28098

Li Y, Du W, Liu DH (2012) Effect of phospholipids on the catalytic performance of lipase. Chinese J Cat 33:389–395. https://doi.org/10.3724/SP.J.1088.2012.11043

Lin DR, Ma Y, Qin W et al (2022) The structure, properties and potential probiotic properties of starch—pectin blend: a review. Food Hydrocoll 129:107644. https://doi.org/10.1016/j.foodhyd.2022.107644

Lin FP, Leu KL (2002) Cloning, expression, and characterization of thermostable region of amylopullulase gene from *Thermoanaerobacter ethanolicus* 39E. Appl Biochem Biotechnol 97:33–44. https://doi.org/10.1385/ABAB:97:1:33

Linares-Pasten J, Andersson M, Karlsson E (2014) Thermostable glycoside hydrolases in biorefinery technologies. Curr Biotechnol 3:26–44

Liu X, Liu T, Zhang Y et al (2017) Structural insights into the thermophilic adaption mechanism of endo-1,4-B-xylanase from *Caldicellulosiruptor owenensis*. J Agric Food Chem 66:187–193. https://doi.org/10.1021/acs.jafc.7b03607

Lv W, Yu Z (2012) Isolation and characterization of two thermophilic cellulolytic strains of *Clostridium thermocellum* from a compost sample. J Appl Microbiol 114:1001–1007. https://doi.org/10.1111/jam.12112

McGarthy RE, Kotarski SF, Salyers AA (1985) Location and characterization of enzymes involved in the breakdown of polygalacturonic acids by *Bacteroides thetaiotaomicron*. J Bacteriol 161: 493–499. https://doi.org/10.1128/JB.174.17.5609-5616.1992

Mead GC (1971) The amino acid fermenting Clostridia. J Gen Microbiol 67:47–56

Mehta R, Campbell JS, Laver GW (1999) Acute hepatic response to aflatoxin B(1) in rats fed a methyl-deficient, amino acid defined diet. Cancer Lett 69:93–106. https://doi.org/10.1016/0304-3835(93)90161-2

Milasinovic N, Milosavljevic N, Filipovic J et al (2012) Efficient immobilization of lipase from *Candida rugosa* by entrapment into poly(N-isopropylacrylamide-coitaconic acid) hydrogels under mild conditions. Polym Bull 69:347–361. https://doi.org/10.1007/s00289-012-0737-7

Misurcova L (2012) Chemical composition of seaweeds. In: Kim SK (ed) Handbook of marine macroalgae: biotechnology and applied phycology. Wiley, New York, pp 173–192

Monteiro RRC, de Arana-Pena S, Rocha TN et al (2020) Liquid lipase preparations designed for industrial production of biodiesel. is it really an optimal solution? Renew Energy 164:1566–1587. https://doi.org/10.1016/j.renene.2020.10.071

Mueller P, Egorova K, Vorgias CE et al (2006) Cloning, overexpression, and characterization of a thermoactive nitrilase from the hyperthermophilic archaeon *Pyrococcus abyssi*. Protein Expr Purif 47:672–681. https://doi.org/10.1016/j.pep.2006.01.006

Munir N, Malik K, Bashir R et al (2021) Isolation and characterization of polysaccharide-degrading microbes from compost samples. J Environ Sci Manag 24:10–16

Naguleswaran S, Vasanthan T, Hoover R et al (2014) Amylolysis of amylopectin and amylose isolated from wheat, triticale, corn and barley starches. Food Hydrocoll 35:686–693. https://doi.org/10.1016/j.foodhyd.2013.08.018

Nawal G, Irfan M, Ashfaq S et al (2019) An overview of bacterial lipases and their enormous applications. Punjab Univ J Zoology 34:61–71

Nigam VK, Arfi T, Kumar V et al (2017) Bioengineering of nitrilases towards its use as green catalyst: applications and perspectives. Indian J Microbiol 57:131–138. https://doi.org/10.1007/s12088-017-0645-5

Odibo FJC, Obi SKC (1988) Purification and characterization of a thermostable pullulanase from *Thermoactinomyces thalpophilus*. J Ind Microbiol 3:343–350. https://doi.org/10.1007/BF01569555

Olusesan AT, Azura LK, Abubakar F et al (2009) Phenotypic and molecular identification of a novel thermophilic *Anoxybacillus* species: a lipase-producing bacterium isolated from a Malaysian hotspring. Word J Microbiol Biotechnol 25:1981–1988. https://doi.org/10.1007/s11274-009-0097-0

Orlygsson J (1994) The role of interspecies hydrogen transfer on thermophilic protein and amino acid metabolism. PhD thesis. Swedish University of Agricultural Sciences

Orlygsson J, Houwen FP, Svensson BH (1994) Influence of hydrogenotrophic methane formation on the thermophilic anaerobic degradation of protein and amino acids. FEMS Microbiol Ecol 13:327–334. https://doi.org/10.1016/0168-6496(94)90070-1

Otajevwo FD, Aluyi HSA (2011) Cultural conditions necessary for optimal cellulase yield by cellulolytic bacterial organisms as they relate to residual sugar released in broth medium. Mod Appl Sci 5:141–151

Padden AN, Dillon VM, John P et al (1998) *Clostridium* used in mediaeval dyeing. Nature, 396(6708):225–225. doi: https://doi.org/10.1038/24290

Padden AN, John P, Collins MD et al (2000) Indigo-reducing *Clostridium isatidis* isolated from a variety of sources, including a 10th-century Viking dye vat. J Archaeol Sci 27(10):953–956. https://doi.org/10.1006/jasc.1999.0524

Park KM, Jun SY, Choi KH et al (2010) Characterization of an exo-acting intracellular alpha-amylase from the hyperthermophilic bacterium *Thermotoga neapolitana*. Appl Microbiol Biotechnol 86:555–566. https://doi.org/10.1007/s00253-009-2284-1

Park S, Lee B, Park K (2017) Extremophilic carbohydrate active enzyme (CAZymes). J Nutr Health Food Eng 7:230–237

Patidar MK, Nighojkar S, Kumar A et al (2018) Pectinolytic enzymes solid state fermentation, assay methods and applications in fruit juice industries: a review. 3 Biotech 8:199. https://doi.org/10.1007/s13205-018-1220-4

Peng XW, Qiao W, Mi SF et al (2015) Characterization of hemicellulase and cellulase from the extremely thermophilic bacterium *Caldicellulosiruptor owensensis* and their potential application for bioconversion of lignocellulosic biomass without pretreatment. Biotechnol Biofuels 8: 131. https://doi.org/10.1186/s13068-015-0313-0

Piller K, Daniel PM, Petach HH (1996) Properties and stabilization of an extracellulsar alpha glucosidase from the extremely thermophilic archaebacteria *Thermococcus* strain AN1: enzyme activity at 130°C. Biochim Biophys Acta Protein Struct Mol Enzymol 1292:197–205

Pinzon-Martinez DL, Rodriguez-Gomez C, Minana-Galbis D et al (2010) Thermophilic bacteria from Mexican thermal environments: isolation and potential applications. Environ Technol 31: 957–966. https://doi.org/10.1080/09593331003758797

Prakash D, Nawani N, Prakash M et al (2013) Actinomycetes: a repertory of green catalysts with a potential revenue resource. Biomed Res Int 2013:264020. https://doi.org/10.1155/2013/264020

Qiu J, Han R, Wang C (2021) Microbial halophilic lipases: a review. J Basic Microbiol 61:594–602. https://doi.org/10.1002/jobm.202100107

Raczynska A, Jadczyk J, Brzezinska-Rodak M (2021) Biotransformations via the physicochemical parameters impacting the processes. Catalysts 11:7. https://doi.org/10.3390/catal11070781

Raasch C, Streit W, Schanzer J et al (2000) *Thermotoga maritima* AglA, an extremely thermostable NAD(+)- Mn2+, and thiol-dependent alpha-glucosidase. Extremophiles 4:189–200. doi: https://doi.org/10.1007/PL00010711

Ramnath L, Sithole B, Govinden R (2017) Classification of lipolytic enzymes and their biotechnological applications in the pulping industry. Can J Micriobiol 63:179–192. https://doi.org/10.1139/cjm-2016-0447

Rani DS, Nand K (2000) Production of thermostable cellulase free xylanase by *Clostridium absonum* CFR-701. Process Biochem 36:355–362. https://doi.org/10.1016/S0032-9592(00)00224-7

Rigoldi F, Donini S, Redaelli A et al (2017) Review: engineering of thermostable enzymes for industrial applications. APL Bioeng 2:011501. https://doi.org/10.1063/1.4997367

Rodriguez-Sanoja R, Oviedo N et al (2005) Microbial starch binding domain. Curr Opin Microbiol 8:260–267. https://doi.org/10.1016/j.mib.2005.04.013

Royter M, Schmidt M, Elend C et al (2009) Thermostable lipases from the extreme thermophilic anaerobic bacteria *Thermoanaerobacter thermohydrolsulfuricus* SOL1 and *Caldanaerobacter subterraneus* subsp *tengcongensis*. Extremophiles 13:769–783. https://doi.org/10.1007/s00792-009-0265-z

Rudiger A, Jorgensen PL, Antranikian G (1995) Isolation and characterization of a heat stable pullulanase from the hyperthermophilic archaeon *Pyrococcus woesei* after cloning and expression of its gene in *Escherichia coli*. Appl Environ Microbiol 61:567–575

Ruttersmith LD, Daniel RM (1991) Thermostable cellobiohydrolase from the eubacterium *Thermotoga* sp strain FJSS3-B.1 - purification and properties. Biochem J 277:887–890. https://doi.org/10.1042/bj2770887

Saha BC, Zeikus JG (1991) Characterization of thermostable alpha glucosdidase from *Clostridium thermohydrosulfuricum* 39E. Appl Microbiol Biotechnol 35:568–571. https://doi.org/10.1042/bj2520343

Saha BC, Mathupala SP, Zeikus JG (1988) Purification and characterization of a highly thermostable novel pullulanase from *Clostridium thermohydrosulfuricum*. Biochem J 252:343–348

Sahoo K, Sahoo RK, Gaur M et al (2020) Cellulolytic thermophilic microorganism in white biotechnology: a review. Folia Microbiol 65:25–43. https://doi.org/10.1007/s12223-019-00710-6

Salah A, Ibrahim S, El-diwany AI (2007) Isolation and identification of new cellulases producing thermophilic bacteria from an Egyptian hot spring and some properties of the crude enzyme. Aust J Basic Appl Sci 1:473–478

Salameh MA, Wiegel J (2009) Synthesis of fatty acid esters and diacylglycerols at elevated temperatures by alkalithermophilic lipases from *Thermosyntropha lipolytica*. J Ind Microbiol Biotechnol 36:1281–1287. https://doi.org/10.1007/s10295-009-0610-3

Salmen L (2022) On the organization of hemicelluloses in the wood cell wall. Cellulose 29:1349–1355. https://doi.org/10.1007/s10570-022-04425-9

Sayer C, Bommer M, Isupov M et al (2012) Crystal structure and substrate specificity on the thermophilic serine:pyruvate aminotransferase from *Sulfolobus solfataricus*. Acta Crystallogr D Biol Crystallogr 68:763–772. https://doi.org/10.1107/S0907444912011274

Sayer C, Szabo Z, Isupov MN et al (2015) The structure of a novel thermophilic esterase from the planctomycetes species, *Thermogutta terrifontis* reveals an open active site due to a minmal "cap" domain. Front Microbiol 6:1294. https://doi.org/10.3389/fmicb.2015.01294

Schink B, Zeikus JG (1983) Characterization of pectinolytic enzmyes of *Clostridium thermosulfurogenes*. FEMS Microbiol Lett 17:295–298. https://doi.org/10.1111/j.1432-1033.1994.00981.x

Schwermann B, Pfau K, Liliensiek B et al (2004) Purification, properties and structural aspects of a thermoacidophilic alpha-amylase from *Alicyclobacillus acidocaldarius* ATCC-27009 - insight into acidostability of proteins. Eur J Biochem 226:981–991

Schut EG (2022) Environmental degradation of cellulose under anaerobic conditions. Bioresources 17:6953–6969. https://doi.org/10.15376/biores.17.4.6953-6969

Scully SM (2019) Amino acid and related catabolism of *Thermoanaerobacter* species. PhD thesis. University of Iceland

Scully SM, Orlygsson J (2014) Branched-chain alcohol formation from branched-chain amino acids by *Thermoanaerobacter brockii* and *Thermoanaerobacter yonseiensis*. Anaerobe 30:82–84. https://doi.org/10.1016/j.anaerobe.2014.09.003

Scully SM, Orlygsson J (2015) Recent advances in second generation ethanol production by thermophilic bacteria. Energies 8:1–30. https://doi.org/10.3390/en8010001

Scully SM, Orlygsson J (2019) Branched-chain amino acid catabolism of *Thermoanaerobacter* strain AK85 and the influence of culture conditions on branched-chain alcohol formation. Amino Acids 51:1039–1054. https://doi.org/10.1007/s00726-019-02744-z

Shallom D, Shoham Y (2003) Microbial hemicellulases. Curr Opin Microbiol 6:219–228. https://doi.org/10.1016/S1369-5274(03)00056-0

Shanmugavel M, Vasantharaj S, Saathiyvimal S et al (2016) Application of an alkaline protease in biological waste processing: an eco friendly approach. Int J Biosci Nanosci 3:19–24

Sharma R, Chisti Y, Banerjee UC (2001) Production, purification, characterization and applications of lipases. Biotechol Adv 19:627–662. https://doi.org/10.1016/S0734-9750(01)00086-6

Sharma VK, Hutchison JM, Allgeier AM (2022) Redox biocatalysis: quantitative comparison of nicotinamide cofactor regeneration methods. ChemSusChem 15:22. https://doi.org/10.1002/cssc.202200888

Souza ADN, Rolland C, Motta JP et al (2005) The intestinal epithelium is a major source of proteases and protease inhibitors: physiological and pathophysiological consequences. Dig Dis Sci 60:2557–2557

Specka U, Mayer F, Antranikian G (1991) Purification and properties of a thermoactive glucoamylase from *Clostridium thermosaccharolyticum*. Appl Environ Microbiol 57:2317–2323. https://doi.org/10.1128/AEM.57.8.2317-2323.1991

Spiridon I, Popa VI (2004) Hemicelluloses: structure and properties. In: Dumitriu S (ed) Polysaccharides, 2nd edn. Marcel Dekker, New York, pp 475–490. https://doi.org/10.1201/9781420030822-22

Spiridon I, Popa VI (2008) Hemicelluloses: major sources, properties and applications. In: Monomers, polymers and composites from renewable resources. Elsevier, pp 289–304. https://doi.org/10.1016/B978-0-08-045316-3.00013-2

Steffen-Munsberg F, Vickers C, Kohls H et al (2015) Bioinformatic analysis of a PLP-dependent enzyme superfamily suitable for biocatalytic applications. Biotechnol Adv 33:566–604. https://doi.org/10.1016/j.biotechadv.2014.12.012

Su H, Qiu W, Kong Q et al (2015) Thermostable pectate lyase from *Caldicellulosiruptor kronotoskyensis* provides an efficient addition for plant biomass deconstruction. J Mol Catal B Enzym 121:104–112. https://doi.org/10.1016/j.molcatb.2015.08.013

Taylor MP, Eley KL, Martin S et al (2009) Thermophilic ethanologenesis: future prospects for second-generation bioethanol production. Trends Biotechnol 27:398–405. https://doi.org/10.1016/j.tibtech.2009.03.006

Toogood HS, Brown RC, Line K et al (2004) The use of thermostable signature amidase in the resolution of the bicyclic synthon (rac)-gamma-lactam. Tetrahedron 60:711–716. https://doi.org/10.1016/j.tet.2003.11.064

Toogood HS, Hollingsworth EJ, Brown RC et al (2002) A thermostable L-aminoacylase from *Thermococcus litoralis*: cloning, overexpresseion, characterization, and applications in biotransformations. Extremophiles 6:111–122. https://doi.org/10.1007/s007920100230

Urbina L, Corcuera MA, Gabilondo N et al (2021) A review of bacterial cellulose: sustainable production from agricultural waste and applications in various fields. Cellulose 28(13):8229–8253. https://doi.org/10.1007/s10570-021-04020-4

Varshney VK, Naithani S (2011) Chemical functionalization of cellulose derived from nonconventional sources. In: Kalia S, Kaith B, Kaur I (eds) Cellulose fibers: bio- and nano-polymer composites. Springer, Berlin. https://doi.org/10.1007/978-3-642-17370-7_2

Verissimo LAA, Mol PCG, Soares WCL et al (2018) Development of a bioreactor based on lipase entrapped in a monolithic cryogel for esterification and interesterification reactions. Rev Mex Ing Quím 17:177–187

Verma D, Kumar R, Satyanarayana T (2019) Diversity in xylan-degrading prokaryotes and xylanolytic enzymes and their bioprospects. In: Satyanarayana S, Das K, Johri BN (eds) Microbial diversity in ecosystem sustainability and biotechnological applications. Springer, London, pp 325–373

Voiniciuc C (2022) Modern mannan: a hemicellulose's journey. New Phytol 234:1175–1184. https://doi.org/10.1111/nph.18091

Walker MJ, Pemberton JM (1987) Construction of a transposon containing a gen for polygalacturonate trans-eliminase from *Klebsiella oxytoca*. Arch Microbiol 146:390–395. https://doi.org/10.1007/BF00410941

Wang J, Salem DR, Sani RK (2019) Extremophilic exopolysaccharides: a review and new perspectives on engineering strategies and applications. Carbohydr Polym 205:8–26. https://doi.org/10.1016/j.carbpol.2018.10.011

Wang SJ, Lu ZX, Lu MS et al (2008) Identification of archaeon producing hyperthermophilic alpha amylase and characterization of the alpha amylase. Appl Microbiol Biotechnol 80(605):614. https://doi.org/10.1007/s00253-008-1561-8

Wang Z, Hong Y, Gu Z et al (2022) Themes, trends, and knowledge structure in 30 years of starch research in food science and technology: a visualization review. Starch 74:2100274. https://doi.org/10.1002/star.202100274

Wiegel J, Ljungdahl LG, Rawson JR (1979) Isolation from soil and properties of the extreme thermophile *Clostridium thermohydrosulfuricum*. J Bacteriol 139:800–810. https://doi.org/10.1128/JB.139.3.800-810.1979

Winterhalter C, Heinrich P, Candussion A et al (1995) Identification of novel cellulose binding domain within the multidomain 120 kDA xylanase xyna of hyperthermophilic bacterium *Thermotoga maritima*. Mol Microbiol 15:431–444. https://doi.org/10.1111/j.1365-2958.1995.tb02257.x

Wood TM, Garcia-Campayo V (1990) Enzymology of cellulose degradation. Biodegradation 1:147–161. https://doi.org/10.1007/BF00058833

Wu YX, Wang WH, Zhang JS et al (2021) Liquid or solid? A biologically inspired concentrated suspension for protective coating. Chem Eng J 428:131793. https://doi.org/10.1016/j.cej.2021.131793

Yadav PK, Singh VK, Yadav S et al (2009) In silico analysis of pectin lyase and pectinase sequences. Biochem Mosc 74:1049–1055. https://doi.org/10.1134/S0006297909090144

Yang SJ, Lee HS, Park CS et al (2004) Enzymatic analysis of an amylolytic enzyme from the hyperthermophilic archaeon *Pyrococcus furiosus* reveals its novel catalytic properties as both an alpha amylase and cyclodextrin hydrolysing enzyme. Appl Environ Microbiol 70:5988–5995. https://doi.org/10.1128/AEM.70.10.5988-5995.2004

Yu S, Yu S, Han W et al (2010) A novel thermophilic lipase from *Fervidobacterium nodosum* Rt17-B1 representing a new subfamily of bacterial lipases. J Mol Catal B Enzym 66:81–89. https://doi.org/10.1016/j.molcatb.2010.03.007

Zhang H, Woodams EE, Hang YD (2011) Influence of pectinase treatment on fruit spirits from aplle mash, juice and pomace. Process Biochem 46:1909–1913. https://doi.org/10.1016/j.procbio.2011.06.020

Zheng MM, Xiang X, Wang S et al (2017) Lipase immobilized in ordered mesoporous silica: a powerful biocatalyst for ultrafast kinetic resolution of racemic secondary alcohols. Process Biochem 53:102–108. https://doi.org/10.1016/j.procbio.2016.12.005

Zheng YY, Xue YF, Zhang YL et al (2010) Cloning, expression, and characterization of thermostable glucoamylase from *Thermoanaerobacter tengcongensis* MB4. Appl Microbiol Biotechnol 87:225–233. https://doi.org/10.1007/s00253-010-2439-0

Zhu D, Wu Q, Wang N (2011) Industrial enzymes. In: Moo-Young M (ed) Comprehensive biotechnology, 2nd edn. Academic, New York, pp 3–13

Zuliani L, Serpico A, De Simone D et al (2021) Biorefinery gets hot: thermophilic enzymes and microorganisms for second-generation bioethanol production. Processes 9:1583. https://doi.org/10.3390/pr9091583

Production of Biofuels by Thermoanaerobic Bacteria

Ed W. J. van Niel and Johann Orlygsson

Abstract Biofuel demand is gradually rising yearly and is expected to do so more rapidly due to the current energy crisis. Bioethanol production is already at a commercial scale but is primarily made using mesophiles fermenting corn and sugar, which have obvious societal drawbacks. Therefore, new technologies should be developed. In addition, hydrogen as a non-carbon energy carrier has gained renewed interest. Thermophilic hydrogen and ethanol production from lignocellulosics and organic wastes, therefore, provide new avenues of biofuel production. This overview looks concisely into the status and remaining challenges of hydrogen and ethanol production exploiting thermophilic anaerobic bacteria.

Abbreviations

AcCoA	Acetyl-Coenzyme A
BHP	Biological hydrogen process
CEM	Cation exchange membrane
CSTR	Continuous stirred tank reactor
DF	Dark fermentation
DFE	Dark fermentation effluent
DW	Dry weight
EH	Electrohydrogenesis
EMP	Embden-Meyerhof pathway
Fd	Ferredoxin
FHL	Formate hydrogen lyase
FNR	Ferredoxin:NAD(P)$^+$ oxidoreductase
H_2ase	Hydrogenase

E. W. J. van Niel (✉)
Division of Applied Microbiology, Lund University, Lund, Sweden
e-mail: ed.van_niel@tmb.lth.se

J. Orlygsson
Faculty of Natural Resource Science, University of Akureyri, Akureyri, Iceland

S. M. Scully, J. Orlygsson (eds.), *Thermophilic Anaerobes*, Emerging Issues and Trends, https://doi.org/10.1007/978-3-031-41720-7_7

HRT	Hydraulic retention time (h)
LCA	Life cycle assessment
MEC	Microbial electrolysis cell
MFC	Microbial fuel cell
PFL	Pyruvate formate lyase
PFOR	Pyruvate ferredoxin:oxidoreductase
P_{H2}	Partial hydrogen pressure (Pa)
PHB	Polyhydroxybutyrate
PNS	Purple non-sulfur
Q_{H2}	Volumetric hydrogen productivity (mmol H_2 L^{-1} h^{-1})
TRL	Technical readiness level
UA	Up-flow anaerobic reactor
Y_{H2}	Hydrogen yield (mol H_2 mol substrate^{-1})

1 Introduction

Many governments have committed themselves to arrive at a zero-carbon society by 2050. In this scenario, various biofuels, such as biomethane, butanol, bioethanol, and biohydrogen, are interesting candidates to play a part in replacing fossil fuels. Here, we will focus on two biofuels as produced by thermophilic bacteria, i.e., biohydrogen and bioethanol. Bioethanol from complex biomass using thermophilic bacteria have gained increased interest in the past decade. The main reason for this is the broad substrate spectra that many thermophiles have as they are capable of degrading a wide variety of both pentoses and hexoses present in lignocellulose in addition to di- and oligosaccharides that result from the partial hydrolysis of carbohydrates such as starch and cellulose. Bioethanol processes have been developed for large-scale production over the last three decades, but they are primarily based on ethanologenic bioprocessing organisms such as yeast and *Zymomonas mobilis*. On the other hand, there is no industrial large-scale biohydrogen production as the technology is still at a relatively nascent technical readiness level (TRL) of 3–4 (Islam et al. 2021). Biohydrogen has become a renewed interest, due to the environmental and humanistic issues related to the manufacturing of batteries for electric cars. Currently, over 20 countries develop strategies for hydrogen applications in their existing infrastructure. The application of hydrogen technologies is present at diverse stages of readiness, from existing (fuel cell cars) to early stages, such as production of carbon-free steel, ammonia, cement, ceramics, and glass (Global hydrogen review 2021). The green hydrogen market is estimated to grow by one order of magnitude in 5-year time, with a current size of over 400 million USD (Global hydrogen market (2021–2026).

2 Background Information

Bioethanol is entirely produced via dark fermentation (DF) using either mesophilic microorganisms (*Saccharomyces cerevisiae* and *Z. mobilis*) or thermophilic microorganisms, including species of *Caldicellulosiruptor*, *Caloramator*, *Clostridium*, *Thermoanaerobacter*, *Thermoanaerobacterium*, and *Thermotoga*. Biohydrogen can be produced in more than one way including splitting water by photosynthesizing microorganisms (algae and cyanobacteria), photofermentation (by, e.g., purple non-sulfur bacteria (PNS)), DF, and using electricity in microbial fuel cells (MECs). Thermophiles are used in the latter two processes only, and thus will be discussed in this chapter.

Both biofuels are mainly produced with feedstock containing predominantly sugars. The stoichiometric reactions of the fermentation for obtaining the maximum yield of each biofuel per glucose are as according to the following two reactions:

$$C_6H_{12}O_6 \rightarrow 2\,C_2H_6O + 2\,HCO_3^- + 2\,H^+ \tag{1}$$

$$C_6H_{12}O_6 + 12\,H_2O \rightarrow 12\,H_2 + 6\,HCO_3^- + 6\,H^+ \tag{2}$$

In contrast to reaction (1), reaction (2) is endergonic (Table 1) and thus will not proceed under standard conditions. Instead, two overall reactions are needed to obtain the maximum amount of H_2 from glucose (Thauer et al. 1977):

$$C_6H_{12}O_6 + 4\,H_2O \rightarrow 4\,H_2 + 2\,C_2H_3O_2^- + 2\,HCO_3^- + 2\,H^+ \tag{3}$$

$$2C_2H_3O_2^- + 2H^+ + 8H_2O \rightarrow 8H_2 + 4HCO_3^- + 4H^+ \tag{4}$$

Reaction (3) is exergonic, but reaction (4) needs energy input for it to occur (Table 1). The latter can be accomplished with photons (photofermentation by, e.g. PNS bacteria) or with electricity (electrohydrogenesis (EH) in MECs).

At a maximum yield of bioethanol and biohydrogen, it is possible to obtain 91% and 98%, respectively, of the energy conserved in the glucose molecule (Table 1). When used as an energy carrier, H_2 and ethanol produce 122 kJ/g and 26.7 kJ/g energy, respectively, showing the significantly higher energy content of hydrogen on weight basis.

Table 1 Information about the thermodynamics of biofuel formation

Substrate	Product	Stoichiometric reaction	Gibbs free energy of the reaction ($\Delta G^{0\prime}$) (kJ mol substrate^{-1})	Energy content of the fuel ($\Delta G^{0\prime}$) (kJ mol substrate^{-1})
Glucose	Bioethanol	(1)	−234.3	−2460
	Biohydrogen	(2)	+3.2	−2670
		(3)	−206.1	−890
		(4)	+104.6	−1780
Glycerol	Biohydrogen	(5)	−9.97	−667

There are also thermophiles that can convert glycerol to hydrogen and acetate according to the following stoichiometry (Maru et al. 2013):

$$C_3H_8O_3 + 2\,H_2O \rightarrow 3\,H_2 + C_2H_3O_2^{\,-} + HCO_3^{\,-} + 2\,H^+ \quad (5)$$

The yield of 3 mol H_2 per mol glycerol has been seen with *Thermotoga* species is in stark contrast to the one H_2 per glycerol obtained with mesophiles (Maru et al. 2013). This is due to the thermodynamics of reaction (5) is more favourable at higher temperatures.

There are three subclasses amongst the thermophilic hydrogen producers: moderate thermophiles (50 °C < T_{opt} < 64 °C), extreme thermophiles (65 °C < T_{opt} < 79 ° C), and hyperthermophiles (T_{opt} > 80 °C).

Of them, significant amounts of EtOH are produced predominantly by moderate thermophiles whereas H_2 is produced with higher yields by extreme and hyperthermophiles. The latter is a consequence of the higher temperature shifting reaction (3) to the right, thus improving the thermodynamics of the conversion to more highly reduced (H_2) and oxidized end products (acetic acid, CO_2).

In general, there is a tremendous metabolic diversity among anaerobic thermophilic bacteria. Central to the ethanol production is the degradation of sugars to end products usually via the Embden-Meyerhof-Parnas (EMP) pathway but the degradation of glucose to pyruvate generates 2 moles of NADH and 2 moles of pyruvate. Pyruvate can be degraded to a wide variety of both oxidized (acetate, CO_2) and reduced (lactate, ethanol, alanine, and hydrogen) compounds. The distribution of end products is highly depended on culture conditions and microorganisms (see below).

3 Biohydrogen

3.1 Physiology

One of the perspectives of this overview is to explore the status of maximizing the volumetric productivity (Q_{H2}) and the total Y_{H2} from sugars of thermophilic hydrogen production. In principle, thermophilic DF possesses superior Y_{H2} close to the theoretical maximum of 4 moles of H_2 per mole of glucose (Thauer et al. 1977), which is only 33% of the total hydrogen that can be extracted from hexoses. Interestingly, a higher Y_{H2} is obtained via glycerol than for sugars because of glycerol being a more reduced compound: the degree of reduction is 4.67 for glycerol compared to 4 for glucose (reaction 5) (Maru et al. 2013). However, a big drawback is the very low Q_{H2} on glycerol being in the order of 1–2.3 mmol $H_2\,L^{-1}\,h^{-1}$. One reason could be the low ATP yield on glycerol fermentation, which is about 1 mol ATP mol glycerol (Maru et al. 2013). Overall, in DF processes, thermophiles are surpassed in their Q_{H2} by mesophilic hydrogen producers, i.e.,

30–160 mmol H_2 L^{-1} h^{-1} and 100–600 mmol H_2 L^{-1} h^{-1}, respectively (Van Niel 2016).

Inherent to the DF process is the formation of volatile fatty acids (VFA) and ethanol (Fig. 1). In the most optimal case, it will be only acetate being formed, which is especially seen with species of *Caldicellulosiruptor*, *Thermoanaerobacter*, and *Thermotoga* and are among the best performing hydrogen producers (Pawar and van Niel 2013). It means that the DF effluent still contains 67% of potential H_2 in the form of acetate, which can be harvested in a second process that demands input of energy (reaction 4). Of the few options, thermophiles can do the job in microbial electrolysis cells (MECs) or Microbial Fuel Cells (MFCs), with (preferably renewable) electricity as energy input to drive the conversion of acetate to H_2. In general, the EH process can reach near maximum Y_{H2} of 4 mol H_2 mol acetate^{-1} (Call and Logan 2008), but Q_{H2} remains low which is partly a technological challenge.

3.2 Feedstocks

A number of hydrogenic thermophiles produce H_2 from various sugars in a DF process (for the list, refer Kengen et al. 2009; Van Niel et al. 2011). These sugars may vary from monosaccharides, disaccharides, oligo-, and polysaccharides of both hexoses and pentoses. Many thermophilic anaerobes have the enzymes necessary to degrade starches, pectin, and some lignocellulosic biomaterials such as xylan. Therefore, thermophilic DF can be applied to convert organic streams containing these polysaccharides, including plant biomass, forestry and agricultural waste, and waste from food industry and households (for the list, refer Wang and Yin 2021). It opens the opportunity for consolidated bioprocessing (CBP), or direct microbial conversion, in which no or only a minimum of pre-treatment is applied to increase the accessibility of the material to the microorganisms (Nagarajan et al. 2019). A treatment with commercial hydrolases can thus be avoided saving operation costs. However, simultaneous saccharification and fermentation make the process slower with hydrolysis as the rate-limiting step, but also due to lower cell densities (Ren et al. 2016), resulting in low Q_{H2}. A technico-economical evaluation is needed here to determine whether this option is economically viable on a case-by-case basis. The majority of test fermentations on (pre-treated) cellulosic material have been conducted at crimp-seal flask level, which have shown yields near the theoretical value can be obtained, e.g., with *C. thermocellum* on Whatman filter and delignified wood fibers (Levin et al. 2006) and *C. saccharolyticus* on untreated switchgrass (Talluri et al. 2013). As these tests were carried out under non-optimized conditions, in the latter study and others (e.g., Cao et al. 2014; Jiang et al. 2019), Q_{H2} were obtained in the order of 0.1–0.6 mmol L^{-1} h^{-1}. Best results were obtained at neutral pH and addition of 0.15–0.3% yeast extract (Talluri et al. 2013; Sheng et al. 2015) as lignocellulosic biomass is nitrogen poor. Further optimization and cost-saving factors of the CBP lies in (1) minimum pre-treatment requirement; (2) cheap nutrient sources; (3) finding the best particle size; (4) use of the right co-cultures to obtain the

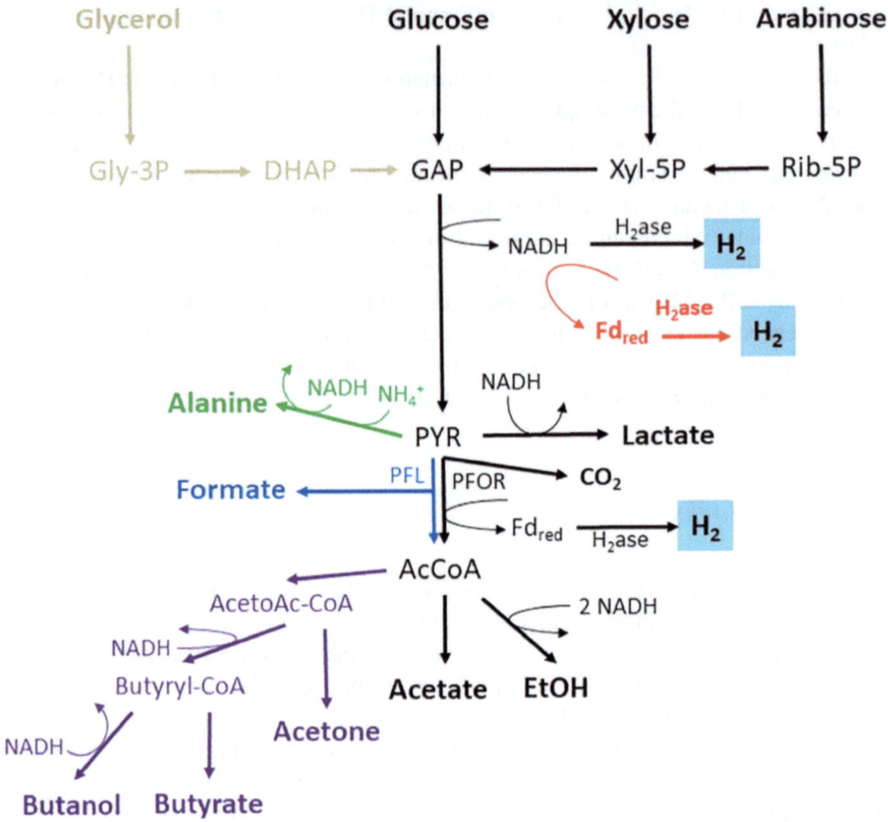

Fig. 1 Overview of the various central carbon pathways to H_2 in strict anaerobic thermophiles. The predominant route to pyruvate is via the Embden-Meyerhof pathway (EMP), although several hyperthermophiles possess both the EMP and the Entner-Douderoff (ED) pathway, such as *Thermotoga* species (Selig et al. 1997). *Black* arrows: the pathways of *Caldicellulosiruptor*, *Thermoanaerobacter*, *Thermoanaerobacterium* and *Thermotoga*, including a hydrogenase (H_2ase) reoxidizing NADH and [FeFe] and/or (NiFe] H_2ase reoxidizing reduced ferredoxin (Fd_{red}). Ferredoxin is used as electron acceptor by pyruvate ferredoxin:oxidoreductase (PFOR) that catalyzes the reaction of pyruvate to acetyl-CoA (AcCoA). A bifurcating hydrogenase (given in red) was discovered in *Thermotoga maritima* (Counts et al. 2017), and is expected to be found in other hydrogenic thermophiles (Rinker and Kelly 2000). In addition, *Thermotoga* species can also convert pyruvate to alanine as an alternative means to reoxidize NADH (Rinker and Kelly 2000) (given in *green*) and use glycerol as a substrate (Maru et al. 2013) (given in *brown-green*). In addition to PFOR, *Clostridium thermocellum* and *Caloramator celere* also possess pyruvate formate lyase (PFL) producing formate instead of H_2 and CO_2 (Ciranna et al. 2013; Lal and Levin 2016) (given in *blue*). Each of the thermophiles may use alternative ways to reoxidize NADH to avoid redox imbalances in the cell, which may lead to products such as lactate and ethanol and clostridia can also produce acetone, butyrate or butanol (given in *purple*). *DHAP* dihydroxyacetone phosphate, *EtOH* ethanol, *GAP* glyceraldehyde-3 phosphate, *Gly-3P* glycerol 3 phosphate, *PYR* pyruvate, *Rib-5P* ribulose 5 phosphate, *Xyl-5P* xylulose 5 phosphate

optimized hydrolysing enzyme cocktail; (5) minimization in water demands; and (6) dedicated bioreactor design and operation (Nagarajan et al. 2019).

An alternative substrate can be glycerol of which a vast supply is available as crude glycerol produced in the biodiesel industry (Santibanez et al. 2011). Few studies have been focusing on *T. maritima* and *T. neapolitana* showing promising conversion of glycerol to three H_2 per glycerol (reaction 5) (Maru et al. 2013; Ngo et al. 2011). However, it remains to be seen how microorganisms perform in crude glycerol knowing of the presence of impurities that can be inhibitory (Sarma et al. 2012).

3.3 Dark Fermentation

There are several process parameters that affect the performance of the thermophilic DF, i.e., the composition of the feedstock, the partial hydrogen pressure (P_{H2}), pH, substrate, and by-product concentration. Optimizing the conditions along with these parameters will maximize the Y_{H2} and the Q_{H2}.

In general, optimal operation of fermentation processes depends on the presence of all the nutrients in adequate amounts to sustain stabilized growth. H_2 is a primary product, and thus, the quality of its production is connected to the growth quality of the thermophiles. At laboratory scale, most nutrients applied are of analytical grade and thus too expensive and unsustainable to be used at large scale. Also, hydrolysates of e.g., lignocellulosic biomass need to be supplemented with minerals and vitamins, as they cover mainly the carbon and energy source. Therefore, cheap sources need to be found to provide other essential nutrients containing nitrogen, phosphorus, sulphur, trace metals, and vitamins. Examples can be manure, urine, and whey, but they need to be tested in combination with the carbon-rich feedstock. Other plant materials can be an alternative source, such as steam pre-treated Lucerne that has recently been investigated (Byrne et al. 2018).

The limitation of the P_{H2} on hydrogen formation is quite known, and being based on a thermodynamic constraint, it is important that hydrogen is removed from the liquid phase as soon as it is produced. Even with sparging, supersaturation of hydrogen may easily occur (Pauss et al. 1990; Ljunggren et al. 2011). Most studies on DF processes at lab scale are conducted in continuous stirred tank reactors (CSTR), which is not an adequate reactor for hydrogen production. Other bioreactor setups have been tested such as up-flow anaerobic reactor (O-Thong et al. 2008), packed bed reactor (Peintner et al. 2010), anaerobic sequencing blanket reactor (Prasertsan et al. 2009), membrane reactor (Oh et al. 2004; Kim et al. 2011), trickling filter (Van Groenestijn et al. 2009), and a biodisc-like reactor (Hiligsmann et al. 2014).The latter two bioreactor types possess a significant gas-liquid interphase that facilitate easy mass transfer of hydrogen to the gas phase, which makes them the most appropriate configurations for hydrogenic DF processes. Indeed, the relevance of this interphase has been demonstrated in a dedicated study (Hiligsmann et al. 2014). The membrane reactor functioned to increase the cell and solid substrate retention time, which improved the Q_{H2} and Y_{H2} by a factor 2 and 1.5, respectively (Kim et al. 2011). However, the other reactors did not seem to fare better than the CSTR (Christopher et al. 2021).

Most hydrogen producers operate optimally around neutral or slightly acidic pH. It should be kept in mind that at higher temperatures (70–80 °C), neutral pH becomes slightly acidic. As an example, *Clostridium thermocellum* operates best at 60–65 °C and pH 6.5–7. However, it reaches a Y_{H2} of 0.7–2.7 mol/mol hexose as it has several other reduced by-products such as ethanol that diminish the hydrogen yield (Wang and Yin 2021). Interestingly, *Caloramator celer* is affected by pH such that at slightly acidic pH, hydrogen is the main reduced product, but the metabolism shifts towards ethanol production at neutral to alkaline pH (Ciranna et al. 2014). This shift is due to the presence of both pyruvate ferredoxin:oxidoreductase (PFOR) and pyruvate formate lyase (PFL), which allow for metabolic flexibility.

Substrate and product inhibition has also been reported, which could be interpreted as an effect of osmolarity (Ljunggren et al. 2011; van Niel et al. 2003). It can be possible that especially polysaccharide degrading bacteria in their natural niches are exposed to only low concentrations of sugar and fermentation products and thus are never confronted with high osmolarities. Adaptation to higher sugar concentrations through Adaptive Laboratory Evolution (ALE) could be a strategy to obtain strains that can stand higher sugar concentrations, which will be more cost effective. However, even though ALE was successful for several species of *Caldicellulosiruptor*, the adapted strains could stand higher sugar concentrations but were not performing better than their parental strains (Byrne et al. 2021). Apparently, more types of adaptation are necessary to obtain improved strains, of which one is fatty acid biosynthesis and the other are local transcription factors as recently found in *C. bescii* through metabolic engineering (Sander et al. 2020).

There are several other factors to improve Y_{H2} and Q_{H2}, which include co-cultures and biofilm formation. Synthetic and undefined co-cultures have been observed to result in higher Y_{H2} and/or Q_{H2} than pure cultures (Pachapur et al. 2015). Synthetic or designed co-cultures consist of two or more species that may come from geographically different locations but together may display synergistic interactions. Synthetic co-cultures have been seen to work better than undefined co-cultures, as the latter contain microorganisms that do not contribute to improving but instead being detrimental to Y_{H2} and Q_{H2} (Zeidan and van Niel 2009). Indeed, several studies have shown improvements with synthetic co-cultures, e.g., through stimulation of one species by the other. This has been observed for two thermophilic clostridia species in CBP (Salimi et al. 2010) and two *Caldicellulosiruptor* in conventional fermentation (Zeidan et al. 2010). The latter case saw a dependent interrelationship that allowed both species to stably grow in continuous culture sharing only one substrate, thereby violating the competitive exclusion principle.

An undesired characteristic of thermophilic cultures is their low cell density (Holst et al. 1997), which limits the Q_{H2}. This could partly be due to inhibitory effects as discussed above and might be remedied by introducing growth in biofilms to create higher cell densities than in suspension cultures. Indeed, improved hydrogen production was obtained through granule formation and biofilms on material such as plastic carriers, fibrous polymers, and acrylic wool (Ahn et al. 2005; Koskinen et al. 2008; Pawar et al. 2015; Vongkampang 2021). In this way, Q_{H2} of up to 30–46 mmol H_2 L^{-1} h^{-1} could be obtained.

3.4 *Electrohydrogenesis*

The second process where hydrogenic thermophiles can be exploited consists of a microbial electrolysis cell (MEC) (Fig. 2). The working of MECs with details on materials and experiences is described elsewhere (Van Niel 2016; Bakonyi et al. 2018; Rousseau et al. 2020). In general, microorganisms are able to convert sugars,

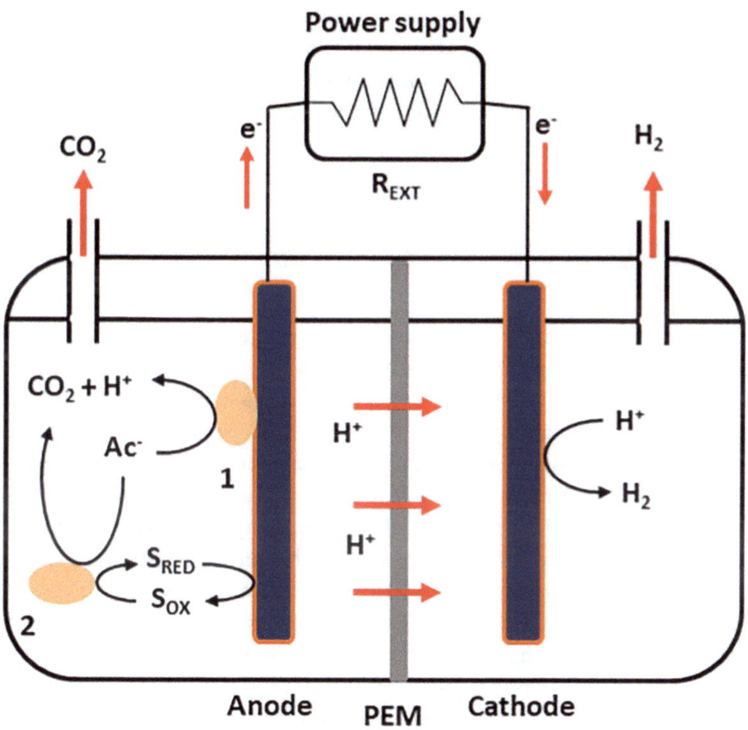

Fig. 2 Principle of the Microbial Electrolysis Cell (MEC) for biohydrogen production. In the anode chamber organic acids (e.g., DF effluent containing mainly acetate (Ac^-)) are converted to CO_2, protons and electrons. The electrons generated by the microbes can be either directly transferred to the anode electrode (1) or indirectly via an electron shuttle that becomes reduced (S_{RED}). The electron shuttle donates its electrons to the anode electrode and becomes oxidized (S_{OX}). The protons diffuse through the proton exchange membrane (PEM) to the cathode chamber. At the cathode electrons combine with the protons and oxygen to form water. As the conversion of organic acids is endergonic, a small voltage input is required to drive the reaction. This voltage is also necessary to overcome resistances in the system, including electrode overpotentials, solution conductivity and the power supply (R_{EXT}) [modified after Van Niel (2016)]

alcohols, and volatile fatty acids to produce an electrical current at the anode (e.g., Oh and Logan 2005), but acetate is one of the best substrates for this process for which electricity input is required (Cheng and Logan 2007). Therefore, electrohydrogenesis (EH) might very well integrate with the DF to configure a two-step process able to extract the maximum of hydrogen from sugars according to reactions (3) and (4) and several studies have confirmed this [refer review Bakonyi et al. 2018).

Even though MEC studies are booming, only a limited amount has focused on operating at temperatures beyond 40 °C. In principle, MECs must perform better at higher temperatures, because of higher rates of biochemical reactions, higher diffusion rates, and diffusion of protons through the proton exchange membrane (PEM) and thus increased current yield and Y_{H2} (Rathinam et al. 2019). The few studies so far have used either undefined consortia or co-cultures of various thermophiles, such as *Thermincola* and *Thermoanaerobacter* species, which reached Q_{H2} of up to 2.5 mmol $L^{-1}d^{-1}$ (Wang et al. 2017) and relatively low Y_{H2} of 1.1 mol H_2 mol $acetate^{-1}$ (Kyazze et al. 2010). Cells attached as biofilms to the cathode conducted direct electron transfer, but no electron shuttles have been found so far among these thermophiles.

3.5 Conclusions

Research in thermophilic DF processes is progressing and is on the brink of pilot scale testing. Even though Y_{H2} is between an acceptable 2–4 mol $hexose^{-1}$, it remains challenging to increase the Q_{H2} by at least a factor 5–10. Therefore, dedicated bioreactors are needed with in situ separation of fermentation by-products and maximized hydrogen mass transfer into the gas phase. Further improvements, such as minimizing undesired by-products formation, such as lactate, might be accomplished through metabolic engineering. However, genetic protocols are working for only a few hydrogen producing bacteria (Han et al. 2012; Cha et al. 2013). Further increase of hydrogen extraction from sugars can be accomplished through coupling of the DF and EH. Both processes can be run with thermophilic synthetic co-cultures, relying on the benefits of these type of bacteria as discussed in this chapter. However, there are substantial challenges to improve the EH process, including stability of electrode materials and membranes to function at elevated temperatures (Rathinam et al. 2019). Therefore, more dedicated investigations are needed before a thermophilic MEC is at the same developed stage as a mesophilic one.

4 Bioethanol

4.1 Physiology

As mentioned above, most saccharolytic thermophiles use the EMP pathway (Taylor et al. 2009) but unlike yeast and facultative anaerobic bacteria, they use pyruvate ferredoxin oxidoreductase (PFOR) for pyruvate degradation to acetyl CoA instead of pyruvate decarboxylase. A second reductive step follows via alcohol dehydrogenase rendering ethanol. Thus, these two reduction steps lead to a redox imbalance (not observed during yeast fermentation) forcing the microorganism to make produce oxidized products, most often by producing acetate via acetyl phosphate intermediate using phosphotransacetylase (PTA) and acetate kinase (AK) (Fig. 1). One of the main obstacles to achieving high ethanol titers in thermophilic bacteria is the variety of end products apart from ethanol that is produced. The theoretical yields of ethanol are 2 moles of ethanol/mole hexose, or 1.67 mole ethanol/mole pentose degraded. These yields are, however, never obtained because part of the substrate is converted to biomass or other end products.

There are various factors that are known to influence ethanol production in thermophiles, including the influence of substrate, partial pressure of hydrogen, the initial substrate concentration, pH and temperature, and the presence of inhibitory compounds. Increased partial pressure of hydrogen is known to decrease the production of oxidized end products (acetate and butyrate) and increase reduced end product formation (ethanol and lactate). The main reason for this is due to the thermodynamics involved and the inhibitory effect on the key enzymes that are responsible for hydrogen production. This results in the production of ethanol and lactate instead of acetate and hydrogen, especially at lower temperatures. Several investigations on the effect of pH_2 on end-product formation have been performed to show this (Brynjarsdottir et al. 2012; Jessen and Orlygsson 2012), showing that by using either a biological scavenger (hydrogenotrophic methanogen) or chemical (such as thiosulfate or other alternative electron acceptor), most of the substrate ends up in acetate and hydrogen equivalents but increasing the partial pressure forces the microorganisms to produce more reduced end products like ethanol and lactate.

Another factor of great importance is the initial substrate concentration used. It is well known that thermophilic bacteria tolerate relatively low initial substrate concentration, often between 10 and 30 mM leading to incomplete substrate utilization. This inhibition may be caused either by increased partial pressure of hydrogen in a closed batch system or by the accumulation of acids lowering the pH in the culture broth. Recent investigation showed that the low tolerance of initial substrate concentration is more likely to be caused by increased pH_2, rather than the substrate or lowering of the pH (Vipotnik et al. 2016). Yeast on the other hand do not produce hydrogen and tolerate much higher initial substrate concentration, often greater than 1 M. Ideally, a good ethanol producer should tolerate at least 4% (v/v) ethanol for an economical ethanol recovery to occur (Taylor et al. 2009). Unfortunately, thermophiles often tolerate much lower ethanol concentrations as compared with yeasts

(Scully and Orlygsson 2015) with most wild-type thermophilic bacteria tolerating less than 3% (v/v), mainly due to their fatty acid membrane structure. Several efforts have been made to increase the ethanol tolerance of thermophiles, some of which have resulted in more tolerant strains but usually possessing lower ethanol yields (Lovitt et al. 1984).

4.2 Feedstocks

The main interest in using thermophilic bacteria is not because of their natural tolerance towards initial partial pressure of hydrogen, initial substrate concentration or ethanol tolerance and yields. Yeast are usually preferred for ethanol production when using simple substrates, namely, glucose, maltose, or sucrose. However, when using lignocellulose, a major part of biomass available today, yeast may not be the best choice of candidate. This is mainly due to the lack of wild-type yeasts to degrade pentoses, which is a major component in lignocellulose. Thus, in recent decades, using thermophilic bacteria degrading the wide variety of carbohydrates present in lignocellulose has been focused upon using species belonging to the genera of *Clostridium*, *Thermoanaerobacter*, and *Thermoanaerobacterium* (Taylor et al. 2009; Scully and Orlygsson 2015; Zheng et al. 2015). More recently, there has been a shift to using macro-algae (seaweeds) and their constituents as a raw material for fermentation although this is still in a very early stage of investigation (Chades et al. 2018; Moenaert et al. 2023).

One of the major challenges with utilizing second- and third-generation biomass is the costs associated with biomass pre-treatment and enzymatic hydrolysis of components into fermentable units. From a bioprocessing perspective, an integrated approach is done in which multiple process steps occur. To this end, there has been interest in microorganisms that can operate under conditions where biomass hydrolysis and fermentation occurs in tandem, as compared to the "traditional" approach in which biomass is processed in discrete steps, which is referred to as separate hydrolysis and fermentation (SHF). Simultaneous saccharification and fermentation (SSF) involves the degradation of the biomass via the action of exogenously added enzymes while the fermentative organism converts liberated carbohydrates to end products. The SSF approach is typically concerned with the hydrolysis of amylose or cellulose, liberating glucose or oligosaccharides thereof while other carbohydrates, such as xylose, are left unfermented necessitating a second fermentation step. To this end, situations in which multiple substrates (such as glucose and xylose) are used simultaneously are referred to as simultaneous saccharification and co-fermentation (SSCF). When the production of enzyme occurs in the same vessel as hydrolysis and fermentation, the term consolidated bioprocessing (CBP) is used. Of particular interest, *Clostridium thermocellum* and *Caldicellulosiruptor* are candidate CBP organisms for the production of bioethanol and biohydrogen, respectively. For more on processing approaches that combine steps, refer to the recent review by Scully and Orlygsson (2015).

4.3 Fermentation of Complex Biomass

The fermentation of biomass into bioethanol by thermophilic anaerobes has been widely reported since the late 1970s. Of particular interest is more recent work on the fermentation of lignocellulosic biomass, especially those with intrinsic cellulolytic capabilities such as *Clostridium thermocellum*, and those that are particularly selective for ethanol as the dominant fermentation product. Given the large variation in the biomass used, pre-treatment regimens, the fermentation conditions, and the organisms used, there is a wide range of ethanol yields and productivities have been reported in the literature with yields seldom approaching the theoretical yield of 0.51 g of ethanol per g glucose equivalent.

The highest yield of ethanol from glucose obtained is 1.9 mol ethanol/mol glucose by *Thermoanaerobacter ethanolicus* (Wiegel and Ljungdahl 1981). Other notable high yields (>1.3 mol ethanol/mol glucose) have been obtained by *Thermoanaerobacter* and *Thermoanaerobacterium* species such as *Thermoanaerobacterium* AK17 (Almarsdottir et al. 2012), *Thermoanaerobacter ethanolicus* (Avci and Donmez 2006), and other *Thermoanaerobacter* strains (Brynjarsdottir et al. 2012; Jessen and Orlygsson 2012). *Thermoanaerobacter ethanolicus* and *Thermoanerobacter finnii* produce 1.43 and 1.76 mol ethanol/mol xylose, respectively (Lacis and Lawford 1988; Fardeau et al. 1996). However, as stated above, these yields can be manipulated by various environmental factors (Scully and Orlygsson 2015). Thermophilic bacteria on lignocellulose never obtain the maximum yields, mainly because of the complex structure of lignocellulose. There have been recent investigations of yields of ethanol from various complex biomass using thermophilic bacteria. The highest yields obtained are 9.2 mM/g corn stover or wheat straw hydrolysates—pre-treated with acid or wet oxidation—by *Thermoanaerobacter* strain BG1L1 (Georgieva and Ahring 2007; Georgieva et al. 2008). *Thermoanaerobacterium* strain AK17 showed ethanol yields of 2.0 (paper) mM/g, 2.9 (grass) mM/g, and 5.8 (cellulose) mM/g biomass (Sveinsdottir et al. 2009). Optimization experiments showed an increase in ethanol yields on grass and cellulose up to 4.0 mM g^{-1} and 8.6 mM g^{-1}, respectively. The main culture factor increasing ethanol yields was obtained by lowering the substrate concentration from 7.5 g/L to 2.5 g/L (Almarsdottir et al. 2012). Recent investigations on two *Thermoanaerobacter* strains, AK5 and J1, showed promising results from various types of hydrolysates made from chemically and enzymatically pretreated lignocellulosic biomass (Brynjarsdottir et al. 2012; Jessen and Orlygsson 2012).

4.4 Genetic Engineering and Evolutionary Adaptation

For the production of ethanol from lignocellulosic biomass, several key processes and characteristics of microorganisms are needed (Taylor et al. 2009; Scully and Orlygsson 2015). The ideal microorganism for ethanol production should be homoethanologenic, have broad substrate spectra, high productivity, high ethanol tolerance, and tolerance to high initial substrate concentrations. Other factors of importance are for instance tolerance towards inhibitory compounds, cellulolytic properties, simple nutritional needs, low biomass production, and ease of genetic manipulation of the strains. No wild-type organism possesses all these properties. There are two main strategies of improving ethanol production by wild-type microorganisms, evolutionary adaptation, and genetic engineering (GE).

The use of evolutionary adaptation methods to enhance ethanol production has been applied to thermophilic bacteria, although on limited basis. In a study of increasing the ethanol concentration tolerance of *Thermoanaerobacter ethanolicus* strains, a selection based on pyruvate and iron deprivation was used (He et al. 2009). This led to an increased ethanol tolerance (10% v/v) at 10 g/L (55 mM) glucose concentrations. Other attempts have been made for enhancing ethanol production that have been done by increasing ethanol tolerance of *Clostridium thermocellum* by stepwise increasing and transferring cultures to increased ethanol concentrations (Shao et al. 2011), and of *Thermoanaerobacter pentosaceus*, by gradually increasing substrate concentration (Sittijunda et al. 2013).

As mentioned above, the main obstacle of using thermophilic bacteria for ethanol production is their natural spectra of end product produced resulting in lower ethanol yields and due to the fact that few thermophiles are cellulolytic. Genetic engineering techniques have thus been focused upon either increasing ethanol yields of a natural cellulolytic microorganism or by inserting cellulases to a high ethanol producing microorganism (Shaw et al. 2009). The first attempt to use GE in a thermophilic bacterium to increase ethanol production was performed on *Thermoanaerobacterium saccharolyticum* (Desai et al. 2004). Deletion of genes responsible for the production of other end products may be knocked out and has been performed on *Thermoanaerobacterium saccharolyticum* (Shaw et al. 2008, 2010), *Thermoanaerobacter mathranii* (Yao and Mikkelsen 2010a, b), *Clostridium thermocellum* (Argyros et al. 2011), *Geobacillus thermoglucosidasius* (Cripps et al. 2009), and *Caldicellulosiruptor bescii* (Chung et al. 2013).

Clostridium thermocellum has been intensively investigated for ethanol production from both cellulosic biomass as well as sugars. The end products formed from carbohydrates are ethanol, acetate, lactate, carbon dioxide, and hydrogen (Xu and Tschirner 2014). The strain was first genetically engineered in 2006 (Tyurin et al. 2006) leading to a development of genetic systems to knock out the pta gene to abolish acetate formation (Tyurin et al. 2006). This resulted in a strain with abnormal growth although the cellulase activity was intact. Later work resulted in increased ethanol yields in adapted strain (hpt, ldh, and pta) lacking both acetate and lactate formation pathways (Tyurin et al. 2006).

Mai and coworkers developed electroporation and shuttle vectors to engineer *Thermoanaerobacterium saccharolyticum* (Mai et al. 1997) later leading to a modified strain by inserting a cellobiohydrolase gene from *Clostridium thermocellum* into its genome (Biswas et al. 2014). Later, the *ldh* gene was knocked out from the bacterium (Desai et al. 2004) as well as knocking out both *ldh* and *ak* gene (Shaw et al. 2008). By knocking out the energy yielding acetate formation, less biomass was formed and ethanol yields increased. Finally, 44% increase in ethanol yields were obtained from the strain with GE work focused on the electron transfer (Shaw et al. 2009) where both *hfs* and *ldh* genes were knocked out that were responsible for hydrogen and lactate formation.

Thermoanaerobacter mathranii has been intensively genetically engineered to improve ethanol yields. Early work focused on knocking out the ldh gene resulting in a strain BG1L1 that produced twice as much ethanol as compared with the wild-type strain (Sommer et al. 2004). Further work on this strain involving overexpression of NAD(P)H-dependent alcohol dehydrogenase resulted in the formation of strain BG1E1 and ethanol yields were further improved (Yao and Mikkelsen 2010a, b).

Thermoanaerobacter BG1 "Pentacrobe" was genetically engineered by knocking out lactate dehydrogenase, phosphotransacetylase, and acetate kinase (Andersen et al. 2015) leading to near the maximum theoretical yields from both hexoses and pentoses. Finally, *Caldicellulosiruptor bescii* was genetically engineered recently of which the wild-type produces a mixture of lactic and acetic acid but almost no ethanol. Substantial genetic engineering work on this strain, however, increased ethanol to a yield of 0.66 mol ethanol/mol glucose (Chung et al. 2013).

4.5 Conclusion

Our understanding of the physiology of thermophilic anaerobic bacteria that are capable of producing high titers of ethanol has increased immensely in the past two decades. In general, anaerobes growing at temperatures between 50 and 80 °C produce a wide variety of end products from sugar metabolism, ranging from highly oxidized products like acetate and carbon dioxide or to highly reduced end products like ethanol and lactate. Reduced end products are more likely to be produced at the lower temperature range, but acetate and carbon dioxide at high temperatures are produced because of the thermodynamic nature of the reaction involved. Other environmental factors are of great importance also, like the ratio of liquid and gas phase in the culture mixture, as well as the concentration of the initial substrate. One of the main drawbacks of using thermophiles for ethanol production is because of their mixed end product formation. However, several attempts have been made by the use of genetic engineering to cut out end product formation routes of other products than ethanol. Thermophiles, however, have advantage over wild-type *Saccharomyces cerevisae* when dealing with hydrolysates made from lignocellulosic biomass because of their broad substrate range. On the other hand, wild-type thermophiles have lower ethanol tolerance as compared with industrial yeasts.

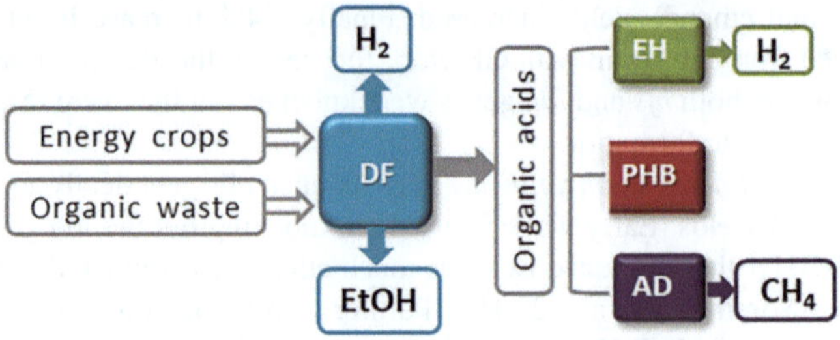

Fig. 3 Proposed biorefinery configurations to maximize biofuel and bioplastic production using thermophilic bacteria. Abbreviations: *DF* dark fermentation, *EtOH* ethanol, *EH* electrohydrogenesis, *PHB* poly-hydroxybutyrate, *AD* anaerobic digestion

5 Future Outlook

Most wastes contain a mixture of sugars, including pentoses and hexoses, which are excellent wastes for thermophilic processes due to their capacity to convert many types of sugars.

Caldicellulosiruptor species do not possess catabolite repression and have shown to prefer to take up pentoses. Applying these hydrogenic thermophiles in synthetic co-cultures with glucose-consuming ethanologens provides the opportunity to produce H_2 and ethanol simultaneously in one bioreactor operating (Fig. 3). Operating at 70–80 °C both H_2 and ethanol transfer into the gas phase, possibly promoted by a maximized gas-liquid interphase and a reduced pressure inside the bioreactor. The byproducts, mainly organic acids, can be converted in a second reactor into other products, for which there are several choices, such as H_2, CH_4, or poly-hydroxybutyrate. This biorefinery setup will, in general, maximize volumetric productivities and yields, making the hydrogen production economical feasible. The main obstacles for thermophilic ethanol production to be feasible in the near future are to increase their tolerance of higher initial substrate loadings and minimize their end-product formation spectrum. The former challenge may be solved by using fed-batch or continuous cultures but the latter with further use of genetic engineering.

Acknowledgments Part of this work was financially supported by the Swedish Research Council (Formas) (project no. 2017–00795).

References

Ahn Y, Park E-J, Oh Y-K et al (2005) Biofilm microbial community of a thermophilic biofilter used for continuous biohydrogen production. FEMS Microbiol Lett 249:31–38. https://doi.org/10.1016/j.femsle.2005.05.050

Almarsdottir AR, Sigurbjornsdottir MA, Orlygsson J (2012) Effect of various factors on ethanol yields from lignocellulosic biomass by *Thermoanaerobacterium* AK 17. Biotechnol Bioeng 109(3):686–694. https://doi.org/10.1002/bit.24346

Andersen RL, Jensen KM, Mikkelsen MJ (2015) Continuous ethanol fermentation of pretreated lignocellulosic biomasses, waste biomasses, molasses and syrup using the anaerobic, thermophilic bacterium *Thermoanarobacter italicus* Pentocrobe 411. PLoS One 10:8. https://doi.org/10.1371/journal.pone.0136060

Argyros DA, Tripathi SA, Barrett TF et al (2011) High ethanol titers from cellulose by using metabolically engineered thermophilic, anaerobic microbes. Appl Environ Microbiol 77(23): 8288–8294. https://doi.org/10.1128/AEM.00646-11

Avci A, Donmez S (2006) Effect of zinc on ethanol production by two *Thermoanaerobacter* strains. Process Biochem 41:984–989. https://doi.org/10.1016/j.procbio.2005.11.007

Bakonyi P, Kumar G, Koók L et al (2018) Microbial electrohydrogenesis linked to dark fermentation as integrated application for enhanced biohydrogen production: a review on process characteristics, experiences and lessons. Bioresour Technol 251:381–389. https://doi.org/10.1016/j.biortech.2017.12.064

Biswas R, Prabhu S, Lynd LR et al (2014) Increase in ethanol yield via elimination of lactate production in an ethanol-tolerant mutant of *Clostridium thermocellum*. PLoS One 9(2):86389. https://doi.org/10.1371/journal.pone.0086389

Brynjarsdottir H, Wawiernia B, Orlygsson J (2012) Ethanol production from sugars and complex biomass by *Thermoanaerobacter* AK5: the effect of electron-scavenging systems on end-product formation. Energy Fuels 26(7):4568–4574. https://doi.org/10.1021/ef300754q

Byrne E, Kovacs K, van Niel EWJ et al (2018) Reduced use of phosphorus and water in sequential dark fermentation and anaerobic digestion of wheat straw and the application of ensiled steam-pretreated lucerne as a macronutrient provider in anaerobic digestion. Biotechnol Biofuels 11:1–16. https://doi.org/10.1186/s13068-018-1280-z

Byrne E, Björkmalm J, Bostick JP et al (2021) Characterization and adaptation of *Caldicellulosiruptor* strains to higher sugar concentrations, targeting enhanced hydrogen production from lignocellulosic hydrolysates. Biotechnol Biofuels 14:210. https://doi.org/10.1186/s13068-021-02058-x

Call B, Logan DE (2008) Hydrogen production in a single chamber microbial electrolysis cell lacking a membrane. Environ Sci Technol 42:3401–3406. https://doi.org/10.1021/es8001822

Cao G-L, Zhao L, Wang A-J et al (2014) Single-step bioconversion of lignocellulose to hydrogen using novel moderately thermophilic bacteria. Biotechnol Biofuels 7:82. https://doi.org/10.1186/1754-6834-7-82

Cha M, Chung D, Elkins JG et al (2013) Metabolic engineering of *Caldicellulosiruptor bescii* yields increased hydrogen production from lignocellulosic biomass. Biotechnol Biofuels 6:85. https://doi.org/10.1186/1754-6834-6-85

Chades T, Scully SM, Ingvadottir EM et al (2018) Fermentation of mannitol extracts from Brown macro algae by Thermophilic Clostridia. Front Microbiol 9:1–13. https://doi.org/10.3389/fmicb.2018.01931

Cheng S, Logan BE (2007) Sustainable and efficient biohydrogen production via electrohydrogenesis. Proc Natl Acad Sci USA 104:18871–18873. https://doi.org/10.1073/pnas.0706379104

Christopher FC, Kumar PS, Vo D-VN et al (2021) A review on critical assessment of advanced bioreactor options for sustainable hydrogen production. Int J Hydrog Energy 46:7113–7136. https://doi.org/10.1016/j.ijhydene.2020.11.244

Chung D, Cha M, Snyder EN et al (2013) Cellulosic ethanol production via consolidated bioprocessing at 75°C by engineered *Caldicellulosiruptor bescii*. Biotechnol Biofuels 8:163. https://doi.org/10.1186/s13068-015-0346-4

Ciranna A, Larjo A, Kivistö A et al (2013) Draft genome sequence of the hydrogen- and ethanol producing anaerobic alkalithermophilic bacterium *Caloramater celer*. Genome Announc 1: e00471–e00413

Ciranna A, Pawar SS, Santala V et al (2014) Assessment of metabolic flux distribution in the thermophilic hydrogen producer *Caloramator celer* as affected by external pH and hydrogen partial pressure. Microb Cell Fact 13:48. https://doi.org/10.1186/1475-2859-13-48

Counts JA, Zeldes BM, Lee LL et al (2017) Physiological, metabolic and biotechnological features of extremely thermophilic microorganisms. WIREs Syst Biol Med 9:e1377. https://doi.org/10.1002/wsbm.1377

Cripps ER, Eley K, Leak DJ et al (2009) Metabolic engineering of *Geobacillus thermoglucosidasius* for high yield ethanol production. Metab Eng 11(6):398–408. https://doi.org/10.1016/j.ymben.2009.08.005

Desai SG, Guerinot ML, Lynd LR (2004) Cloning of L-lactate dehydrogenase and elimination of lactic acid production via gene knockout in *Thermoanaerobacterium saccharolyticum* JW/SL-YS485. Appl Microbiol Biotechnol 65(5):600–605. https://doi.org/10.1007/s00253-004-1575-9

Fardeau ML, Faudon C, Cayol JL et al (1996) Effect of thiosulphate as electron acceptor on glucose and xylose oxidation by *Thermoanaerobacter finnii* and a *Thermoanaerobacter* sp isolated from oil field water. Res Microbiol 147:159–165. https://doi.org/10.1016/0923-2508(96)80215-4

Georgieva TI, Ahring BK (2007) Evaluation of continuous ethanol fermentation of dilute-acid corn Stover hydrolysate using thermophilic anaerobic bacterium *Thermoanaerobacter* BG1L1. Appl Microbiol Biotechnol 77(1):61–68. https://doi.org/10.1007/s00253-007-1149-8

Georgieva TI, Mikkelsen MJ, Ahring BK (2008) Ethanol production from wet-exploded wheat straw hydrolysate by thermophilic anaerobic bacterium *Thermoanaerobacter* BG1L1 in a continuous immobilized reactor. Appl Biochem Biotechnol 145(1–3):99–110. https://doi.org/10.1007/s12010-007-8014-1

Han D, Norris SM, Xu Z (2012) Construction and transformation of a *Thermotoga-E. coli* shuttle vector. BMC Biotechnol 12:2. https://doi.org/10.1186/1472-6750-12-2

He Q, Lokken PM, Chen S et al (2009) Characterization of the impact of acetate and lactate on ethanolic fermentation by *Thermoanaerobacter ethanolicus*. Bioresour Technol 100(23):5955–5965. https://doi.org/10.1016/j.biortech.2009.06.084

Hiligsmann S, Beckers L, Masset J et al (2014) Improvement of fermentative biohydrogen production by *Clostridium butyricum* CWBI1009 in sequenced-batch, horizontal fixed bed and biodisc-like anaerobic reactors with biomass retention. Int J Hydrog Energy 39:6899–6911. https://doi.org/10.1016/j.ijhydene.2014.02.139

Holst O, Manelius Å, Krahe M et al (1997) Thermophiles and fermentation technology. Comp Biochem Physiol Part A Physiol 118:415–422. https://doi.org/10.1016/S0300-9629(97)00002-9

Islam AKMK, Dunlop PSM, Hewitt NJ et al (2021) Bio-hydrogen production from wastewater: a comparative study of low energy intensive production processes. Clean Technol 3:156–182. https://doi.org/10.3390/cleantechnol3010010

Jessen JE, Orlygsson J (2012) Production of ethanol from sugars and lignocellulosic biomass by *Thermoanaerobacter* J1 isolated from a hot spring in Iceland. J Biomed Biotechnol 186982. https://doi.org/10.1155/2012/186982

Jiang Y, Lu J, Lv Y et al (2019) Efficient hydrogen production from lignocellulosic feedstocks by a newly isolated thermophilic *Thermoanaerobacterium* sp. strain F6. Int J Hydrog Energy 44:14380–14386. https://doi.org/10.1016/j.ijhydene.2019.01.226

Kengen SWM, Goorissen HP, Verhaart M et al (2009) Biological hydrogen production by anaerobic microorganisms. In: Biofuels. Wiley, Oxford, pp 197–221

Kim M-S, Lee D-Y, Kim D-H (2011) Continuous hydrogen production from tofu processing waste using anaerobic mixed microflora under thermophilic conditions. Int J Hydrog Energy 36:8712–8718. https://doi.org/10.1016/j.ijhydene.2010.06.040

Koskinen PEP, Lay C-H, Puhakka JA et al (2008) High-efficiency hydrogen production by an anaerobic, thermophilic enrichment culture from an Icelandic hot spring. Biotechnol Bioeng 33:1168–1178. https://doi.org/10.1002/bit.21948

Kyazze G, Popov A, Dinsdale R et al (2010) Influence of catholyte pH and temperature on hydrogen production from acetate using a two chamber concentric tubular microbial electrolysis cell. Int J Hydrog Energy 35:7716–7722. https://doi.org/10.1016/j.ijhydene.2010.05.036

Lacis LS, Lawford HG (1988) Effect of growth-rate on ethanol-production by *Thermoanaerobacter ethanolicus* in glucose-limited or xylose-limited continuous culture. Biotechnol Lett 10:60608. https://doi.org/10.1007/BF01027138

Lal S, Levin DB (2016) Comparative genomics of core metabolism genes of cellulolytic and non-cellulolytic *Clostridium* species. In: Hatti-Kaul R, Mamo R, Mattiasson B (eds) Anaerobes in biotechnology. Springer, Cham, pp 79–112. https://doi.org/10.1007/10_2015_5007

Levin DB, Islam R, Cicek N et al (2006) Hydrogen production by *Clostridium thermocellum* 27405 from cellulosic biomass substrates. Int J Hydrog Energy 31:1496–1503. https://doi.org/10.1016/j.ijhydene.2006.06.015

Ljunggren M, Willquist K, Zacchi G et al (2011) A kinetic model for quantitative evaluation of the effect of hydrogen and osmolarity on hydrogen production by *Caldicellulosiruptor saccharolyticus*. Biotechnol Biofuels 4:1–15. https://doi.org/10.1186/1754-6834-4-31

Lovitt RW, Longin R, Zeikus JG (1984) Ethanol production by thermophilic bacteria: physiological comparison of solvent effects on parent and alcohol-tolerant strains of *Clostridium thermohydrosulfuricum*. Appl Environ Microbiol 48:171–177. https://doi.org/10.1128/AEM.48.1.171-177.1984

Mai V, Lorenz WW, Wiegel J (1997) Transformation of *Thermoanaerobacterium* sp. strain JW/SL-YS485 with plasmid pIKM1 conferring kanamycin resistance. FEMS Microbiol Lett 148:163–167. https://doi.org/10.1111/j.1574-6968.1997.tb10283.x

Maru BT, Bielen AAM, Constanti M et al (2013) Glycerol fermentation to hydrogen by *Thermotoga maritima*: proposed pathway and bioenergetics considerations. Int J Hydrog Energy 38:5563–5572. https://doi.org/10.1016/j.ijhydene.2013.02.130

Moenaert A, Lopez-Contreras AM, Budde M et al (2023) Evaluation of *Laminaria digitata* hydrolysate for the production of bioethanol and butanol by fermentation. Fermentation 9:59. https://doi.org/10.3390/fermentation9010059

Nagarajan D, Lee D-J, Chang J-S (2019) Recent insights into consolidated bioprocessing for lignocellulosic biohydrogen production. Int J Hydrog Energy 44:14362–14379. https://doi.org/10.1016/j.ijhydene.2019.03.066

Ngo TA, Kim MS, Sim SJ (2011) High-yield biohydrogen production from biodiesel manufacturing waste by *Thermotoga neapolitana*. Int J Hydrog Energy 36:5836–5842. https://doi.org/10.1016/j.ijhydene.2010.11.057

Oh S-E, Logan BE (2005) Hydrogen and electricity production from a food processing wastewater using fermentation and microbial fuel cell technologies. Water Res 39:4673–4682. https://doi.org/10.1016/j.watres.2005.09.019

Oh S-E, Iyer P, Bruns MA et al (2004) Biological hydrogen production using a membrane bioreactor. Biotechnol Bioeng 87:119–127. https://doi.org/10.1002/bit.20127

O-Thong S, Prasertsan P, Karakashev D et al (2008) High-rate continuous hydrogen production by *Thermoanaerobacterium thermosaccharolyticum* PSU-2 immobilized on heat-pretreated methanogenic granules. Int J Hydrog Energy 33:6498–6508. https://doi.org/10.1016/j.ijhydene.2008.07.060

Pachapur VL, Sarma SJ, Brar SK (2015) Co-culture strategies for increased biohydrogen production. Int J Energy Res 39:1479–1504. https://doi.org/10.1002/er.3364

Pauss A, Andre G, Perrier M et al (1990) Liquid-to-gas mass transfer in anaerobic processes: inevitable transfer limitations of methane and hydrogen in the biomethanation process. Appl Environ Microbiol 56:1636–1644. https://doi.org/10.1128/AEM.56.6.1636-1644.1990

Pawar SS, van Niel EWJ (2013) Thermophilic hydrogen production: how far are we? Appl Microbiol Biotechnol 97:7999–8009. https://doi.org/10.1007/s00253-013-5141-1

Pawar SS, Vongkumpeang T, Gray C et al (2015) Biofilm formation by designed co-cultures of *Caldicellulosiruptor* species as a means to improve hydrogen productivity. Biotechnol Biofuels 8:19. https://doi.org/10.1186/s13068-015-0201-7

Peintner C, Zeidan AA, Schnitzhofer W (2010) Bioreactor systems for thermophilic fermentative hydrogen production: evaluation and comparison of appropriate systems. J Clean Prod 18:S15–S22. https://doi.org/10.1016/j.jclepro.2010.06.013

Prasertsan P, O-Thong S, Birkeland N-K (2009) Optimization and microbial community analysis for production of biohydrogen from palm oil mill effluent by thermophilic fermentative process. Int J Hydrog Energy 34:7448–7459. https://doi.org/10.1016/j.ijhydene.2009.04.075

Rathinam NK, Bibra M, Salem DR et al (2019) Thermophiles for biohydrogen production in microbial electrolysis cells. Bioresour Technol 277:171–178. https://doi.org/10.1016/j.biortech.2019.01.020

Ren N-Q, Zhao L, Chen C et al (2016) A review on bioconversion of lignocellulosic biomass to H2: key challenges and new insights. Bioresource Technol 215:92–99. https://doi.org/10.1016/j.biortech.2016.03.124

Rinker KD, Kelly RM (2000) Effect of carbon and nitrogen sources on growth dynamics and exopolysaccharide production for the hyperthermophilic archaeon *Thermococcus litoralis* and bacterium *Thermotoga maritima*. Biotechnol Bioeng 69:537–547. https://doi.org/10.1002/1097-0290(20000905)69:5<537::AID-BIT8>3.0.CO;2-7

Rousseau R, Etcheverry L, Roubaud E et al (2020) Microbial electrolysis cell (MEC): strengths, weaknesses and research needs from electrochemical engineering standpoint. Appl Energy 257:113938. https://doi.org/10.1016/j.apenergy.2019.113938

Salimi F, Zhuang K, Mahadevan R (2010) Genome-scale metabolic modeling of a clostridial co-culture for consolidated bioprocessing. Biotechnol J 5:726–738. https://doi.org/10.1002/biot.201000159

Sander KB, Chung D, Klingeman DM et al (2020) Gene targets for engineering osmotolerance in *Caldicellulosiruptor bescii*. Biotechnol Biofuels 13:50. https://doi.org/10.1186/s13068-020-01690-3

Santibanez C, Varnero MT, Bustamanta M (2011) Residual glycerol from biodiesel manufacturing, waste or potential source of bioenergy: a review. Chil J Agric Res 71:469–475

Sarma SJ, Brar SK, Sydney EB et al (2012) Microbial hydrogen production by bioconversion of crude glycerol: a review. Int J Hydrog Energy 37:6473–6490. https://doi.org/10.1016/j.ijhydene.2012.01.050

Scully SM, Orlygsson J (2015) Recent advances in second generation ethanol production by thermophilic bacteria. Energies 8(1). https://doi.org/10.3390/en8010001

Selig M, Xavier KB, Santos H et al (1997) Comparative analysis of Embden-Meyerhof and Entner-Doudoroff glycolytic pathways in hyperthermophilic archaea and the bacterium *Thermotoga*. Arch Microbiol 167:2017–2232. https://doi.org/10.1007/BF03356097

Shao XJ, Raman B, Zhu MJ et al (2011) Mutant selection and phenotypic and genetic characterization of ethanol-tolerant strains of *Clostridium thermocellum*. Appl Microb Biotechnol 92(3):641–652. https://doi.org/10.1007/s00253-011-3492-z

Shaw AJ, Podkaminer KK, Desai SG et al (2008) Metabolic engineering of a thermophilic bacterium to produce ethanol at high yield. Proc Natl Acad Sci U S A 105(37):13769–13774. https://doi.org/10.1073/pnas.0801266105

Shaw AJ, Hogsett DA, Lynd LR (2009) Identification of the [FeFe]-hydrogenase responsible for hydrogen generation in *Thermoanaerobacterium saccharolyticum* and demonstration of increased ethanol yield via hydrogenase knockout. J Bacteriol 191(20):6457–6464. https://doi.org/10.1128/JB.00497-09

Shaw AJ, Hogsett DA, Lynd LR (2010) Natural competence in *Thermoanaerobacter* and *Thermoanaerobacterium* species. Appl Environ Microbiol 76(14):4713–4719. https://doi.org/10.1128/AEM.00402-10

Sheng T, Gao L, Zhao L et al (2015) Direct hydrogen production from lignocellulose by the newly isolated *Thermoanaerobacterium thermosaccharolyticum* strain DD32. RSC Adv 5:99781. https://doi.org/10.1039/c5ra20000h

Sittijunda S, Tomas AF, Reungsang A et al (2013) Ethanol production from glucose and xylose by immobilized *Thermoanaerobacter pentosaceus* at 70°C in an up-flow anaerobic sludge blanket

(UASB) reactor. Bioresource Technol 143:598–607. https://doi.org/10.1016/j.biortech.2013.06.056
Sommer P, Georgieva T, Ahring BK (2004) Potential for using thermophilic anaerobic bacteria for bioethanol production from hemicellulose. Bioch Soc Trans 32:283–289. https://doi.org/10.1042/BST0320283
Sveinsdottir M, Beck SR, Orlygsson J (2009) Ethanol production from monosugars and lignocellulosic biomass by thermophilic bacteria isolated from Icelandic hot springs. Icel Agric Sci 22: 45–58
Talluri S, Raj SM, Christopher LP (2013) Consolidated bioprocessing of untreated switchgrass to hydrogen by the extreme thermophile *Caldicellulosiruptor saccharolyticus* DSM 8903. Bioresour Technol 139:272–279. https://doi.org/10.1016/j.biortech.2013.04.005
Taylor MP, Eley KL, Martin S et al (2009) Thermophilic ethanologenesis: future prospects for second-generation bioethanol production. Trends Biotechnol 27(7):398–405. https://doi.org/10.1016/j.tibtech.2009.03.006
Thauer RK, Jungermann K, Decker K (1977) Energy conservation in chemotrophic anaerobic bacteria. Bacteriol Rev 41:100–180
Tyurin MV, Lynd LR, Wiegel J (2006) 13 Gene transfer systems for obligately anaerobic thermophilic bacteria. Meth Microbiol 35:309–330. https://doi.org/10.1016/S0580-9517(05)35013-6
Van Groenestijn JW, Geelhoed JS, Goorissen HP et al (2009) Performance and population analysis of a non-sterile trickle bed reactor inoculated with *Caldicellulosiruptor saccharolyticus*, a thermophilic hydrogen producer. Biotechnol Bioeng 33:1168–1178. https://doi.org/10.1002/bit.22185
Van Niel EWJ (2016) Biological processes for hydrogen production. In: Anaerobes in biotechnology. Springer, Cham, pp 155–193
Van Niel EWJ, Claassen PAM, Stams AJM (2003) Substrate and product inhibition of hydrogen production by the extreme thermophile, *Caldicellulosiruptor saccharolyticus*. Biotechnol Bioeng 81:255–262. https://doi.org/10.1002/bit.10463
Van Niel EWJ, Willquist K, Zeidan AA et al (2011) Hydrogen production by thermophilic fermentation. In: Azbar N, Levin DB (eds) State of the art and progress in production of biohydrogen. Bentham Science, Sharja, pp 137–159
Vipotnik Z, Jessen JE, Scully SM et al (2016) Effect of culture conditions on hydrogen production by *Thermoanaerobacter* strain AK68. Int J Hydrog Energy 41(1):181–189. https://doi.org/10.1016/j.ijhydene.2015.10.124
Vongkampang T (2021) Exploring strategies to improve volumetric hydrogen productivities of *Caldicellulosiruptor* species. PhD thesis, Lund University, Lund
Wang J, Yin Y (2021) *Clostridium* species for fermentative hydrogen production: an overview. Int J Hydrog Energy 46:34599–34625. doi: https://doi.org/10.1016/j.ijhydene.2021.08.052
Wang Y-Z, Zhang L, Xu T et al (2017) Influence of initial anolyte pH and temperature on hydrogen production through simultaneous saccharification and fermentation of lignocellulose in microbial electrolysis cell. Int J Hydrog Energy 42:22663–22670. https://doi.org/10.1016/j.ijhydene.2017.07.214
Wiegel J, Ljungdahl LG (1981) *Thermoanaerobacter ethanolicus* gen. nov., spec. nov., a new, extreme thermophilic, anaerobic bacterium. Arch Microbiol 128:343–348. doi: https://doi.org/10.1007/BF00405910
Xu L, Tschirner U (2014) Immobilized anaerobic fermentation for bio-fuel production by *Clostridium* co-culture. Bioprocess Biosyst Eng 37:1551–1559. https://doi.org/10.1007/s00449-014-1127-3
Yao S, Mikkelsen MJ (2010a) Identification and overexpression of a bifunctional aldehyde/alcohol dehydrogenase responsible for ethanol production in *Thermoanaerobacter mathranii*. J Mol Microb Biotechnol 19:123–133. https://doi.org/10.1159/000321498
Yao S, Mikkelsen MJ (2010b) Metabolic engineering to improve ethanol production in *Thermoanaerobacter mathranii*. Appl Microb Biotechnol 88(1):199–208. https://doi.org/10.1007/s00253-010-2703-3

Zeidan AA, van Niel EWJ (2009) Developing a thermophilic hydrogen-producing co-cultures for efficient utilization of mixed sugars. Int J Hydrog Energy 34:4524–4528. https://doi.org/10.1016/j.ijhydene.2008.07.092

Zeidan AA, Rådström P, van Niel EWJ (2010) Stable coexistence of two *Caldicellulosiruptor* species in a de novo constructed hydrogen-producing co-culture. Microb Cell Factories 9:102. https://doi.org/10.1186/1475-2859-9-102

Zheng J, Tashiro Y, Wang QH et al (2015) Recent advances to improve fermentative butanol production: genetic engineering and fermentation technology. J Biosci Bioeng 119:1–9. https://doi.org/10.1016/j.jbiosc.2014.05.023

Production of Fine Chemicals by Thermophilic, Anaerobic Bacteria

Sean Michael Scully and Johann Orlygsson

Abstract Thermophilic anaerobic bacteria possess many desirable metabolisms to produce various fine chemicals. The demand for more and more sustainable production of various chemicals has risen in the past decades, and some of these are produced by strict anaerobic, thermophilic bacteria. The present investigation covers the main chemical pathways used to produce 1,2-propanediol, 1,3-propanediol, branched-chain alcohols and the main microbes involved, namely *Thermoanaerobacter, Thermoanaerobacterium, Caloramator, Caldanaerobacterium, and Clostridium.*

1 Introduction

Thermophilic anaerobes are well known for their capacity to produce biofuels such as biohydrogen and bioethanol (Taylor et al. 2009; Scully and Orlygsson 2015a). While both of these molecules are important in the context of moving towards a circular economy, they are low-value molecules and do not necessarily provide an easy pathway to producing other bio-based molecules required to produce the other materials required for our modern societies. While there are dozens of papers on the production of both biohydrogen and bioethanol with thermophilic anaerobes, there is less knowledge regarding their capacity to produce higher-value compounds such as higher-order alcohols and diols. This chapter focuses on the latest development in the production of high-value compounds like 1,2-propanediol and branched-chain alcohols using thermophilic anaerobes. Other fine chemicals that are produced by fermentation are some specific pharmaceuticals (Revuelta et al. 2018), organic acids such as lactic and succinic acids (Prado-Rubio et al. 2020; Prabhu et al. 2020), and some other specialty chemicals that are produced by mesophiles and thus not mentioned further here.

S. M. Scully (✉) · J. Orlygsson
Faculty of Natural Resource Sciences, School of Health, Business and Natural Sciences, University of Akureyri, Akureyri, Iceland
e-mail: scully@unak.is

S. M. Scully, J. Orlygsson (eds.), *Thermophilic Anaerobes*, Emerging Issues and Trends, https://doi.org/10.1007/978-3-031-41720-7_8

As explored by Adalsteinsson and Hreggvidsson (2023), microorganisms capable of growth at high temperatures are classified as thermophiles and hyperthermophiles, where the latter grow at temperatures at and above 80 °C. The majority of hyperthermophiles are Archaea whereas Bacteria comprise most of the thermophilic group. Several thermophilic anaerobic bacteria are capable of the production of fine chemicals at high temperatures by species like *Clostridium, Thermoanaerobacter,* and *Caloramator* but also from hyperthermophiles such as strains within the genus of *Caldicellulosiruptor.*

The conversion of biomass to biofuel has been intensively investigated for the past 40 years but declined with the return of cheap oil as the main energy resource in the 1980s. That said, the production of other biomolecules that could offer alternatives to the use of petroleum-derived feedstocks for the production of other essential chemical building blocks was largely overlooked. However, due to new techniques and ever increasing awareness of environmental issues and geopolitical instability, biofuel production has once again found itself in the limelight with the sustainable production of other molecules that received much more attention as we have transition towards a circular economy (Scully and Orlygsson 2015a, b).

The term biorefining deals with the use of renewable resources, like agricultural crops or wastes, that are utilized for the extraction of intermediates that can either be further converted to biofuels or fine chemicals. The goal of the biorefinery is to produce both high-value, low-volume and low-value, high-volume products. Biomass can be used either directly as raw material for bioprocessing or be used as inexpensive substrates for fermentation processes from which products can be extracted (Solaiman et al. 2006). The types of biomass that can be used as a raw material are very diverse, including corn (Gaspar et al. 2005), wheat (Koutinas et al. 2004), sugar cane (Pye 2005), cassava (Enze 2006), and lignocellulose (Pan et al. 2006). There are two main types of biorefinery systems. One with one type of biomass that is converted to one main product or a mixture of biomass types that are converted to various end products. To obtain high yields, various types of pretreatment are usually needed: mechanical, biocatalytic, and chemical. It is worth mentioning that another very active area of research is the use of enzymes from extremophiles and thermophiles in particular for the production of highly specialized molecules through biocatalysis and biotransformation (Arbab et al. 2022; Ajeje et al. 2022) although this is the focus of a previous chapter (Scully and Orlygsson 2023).

According to the US Department of Energy, the main chemical building blocks produced from biomass used to produce various types of chemicals are compounds like 1,4 diacids, 2,5-furan dicarboxylic acid, 3-hydroxypropionic acid, aspartic acid, glucaric acid, glutamic acid, itaconic acid, levulinic acid, 3-hydroxybutyrolactone, glycerol, sorbitol and xylitol which can be further converted to other various products (Turner et al. 2007). However, most of these compounds are used by aerobic mesophiles but not thermophiles. Apart from well-known high-volume, low-value end products like ethanol, there are not many fine chemicals produced by the fermentation of thermophilic bacteria. The main focus of this chapter is on 1,2-propanediol, 1,3-propanediol and branched-chain alcohols.

2 Production of 1,2-Propanediol

The fine chemical propane-1,2-diol, also commonly referred to as α-propylene glycol or 1,2-propanediol (1,2-PD), is a three-carbon molecule that has a high boiling point and is highly hydrophilic. It is also a major commodity chemical and is highly valued because of its application in biodegradable plastics and polymer resins as well as a solvent in the anti-freezing industry, cosmetics, nutrition products, food industry, and hydraulic breaking system industry (Samad et al. 2015; Mhd Sawal et al. 2019; Siebert and Wendisch 2015; Zhao et al. 2018; Hatti-Kaul et al. 2018). Additionally, 1,2-PD can be found in nature as two enantiomeric forms: (*S*)-1,2-propandiol and (*R*)-1,2-propanediol. Each of these enantiomers are useful chiral building blocks like the conversion of 1,2-propanediol to D-2-hydroxypropionic acid (Gao et al. 2006). The annual sale of 1,2-PD was estimated at 1.36 million tons per year world wide (Tao et al. 2021).

While 1,2-PD can be produced chemically by hydrogenolysis of carbohydrates at high temperatures and pressure in presence of a metal catalyst, this method is though outdated and its production resulted in a racemic mixture of 1,2-PD and other polyols such as 1,3-propanediol and higher molecular weight polyols (Lenth and Puis 1945). While other routes from glycerol are known, they are also problematic due to their low chemoselectivity and further complicated by the challenges of separating 1,2-PD from 1,3-propanediol. At present, 1,2-PD is mainly produced by the hydration of propylene oxide, a hazardous molecule derived from petroleum sources being not only highly flammable but also acutely toxic and carcinogenic (Martin and Murphy 1994). Thus, in recent years, several attractive biological processes have been developed using renewable energy sources. The focus in this review will be on the production of 1,2-PD by fermentation.

2.1 Pathways for 1,2-Propanediol Production

Production of 1,2-PD occurs mainly through two pathways. Firstly, by degradation of deoxysugars, also known as methylpentoses, (like L-fucose and L-rhamnose) to yield the *S-enantiomer* as the end product (Turner and Robertson 1979) (Fig. 1a). Secondly, by a route known as the methylglyoxal bypass which proceeds through the glycolytic intermediate dihydroxyacetone phosphate (DHAP) via the formation of methylglyoxal and its subsequent reduction of the *R*-enantiomer of 1,2-PD (Fig. 1b) (Saxena et al. 2010a, b). It should be noted that a third pathway has been observed via lactic acid degradation with hypothetical routes to both enantiomers being possible (Elferink et al. 2001).

When L-rhamnose is used as a substrate, it is first phosphorylated to L-rhamnulose-1-phosphate which is cleaved to give DHAP and L-lactaldehyde (Badia et al. 1985). Similarly, the degradation of L-fucose renders the same end products and the latter is reduced to 1,2-PD (Fig. 1a). However, using the expensive

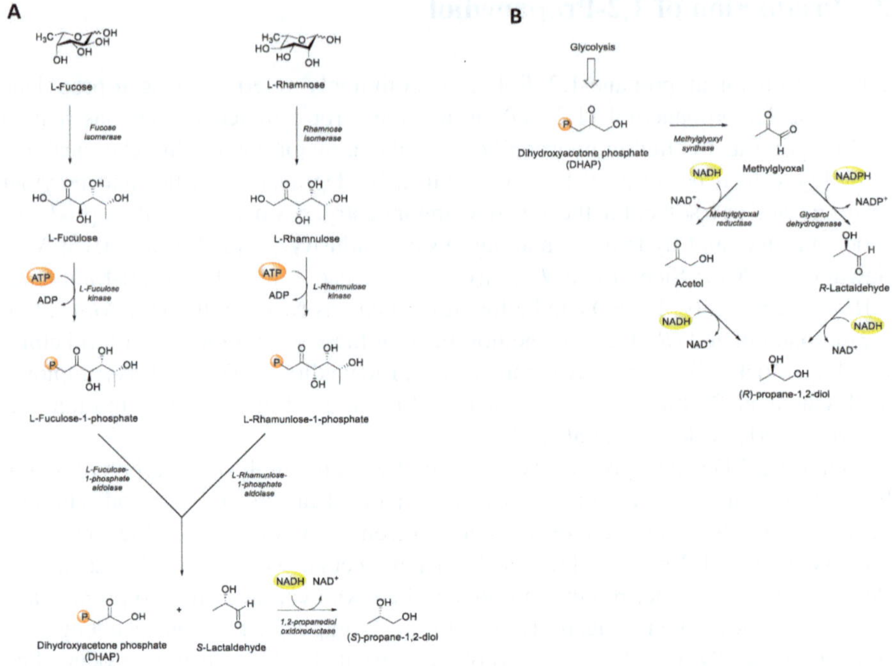

Fig. 1 Metabolic pathways for 1,2-PD production from deoxysugars (**a**) and from DHAP via the methylglyoxal bypase (**b**)

deoxysugars as a substrate for 1,2-PD is not economically feasible due to the high cost of these carbohydrates although it should be noted that both L-fucose and L-rhamnose can be sourced from various seaweeds.

The other route to produce 1,2-PD is via the methylglyoxal pathway (Cooper 1975). In this pathway, degradation of fructose-1,6-biphosphate is possible under phosphate limitation (Tran-Din and Gottschalk 1985) and the intermediate DHAP is converted to methylglyoxal and provides inorganic phosphate for the glyceraldehyde dehydrogenase reaction (Fig. 1b). Thereafter, the methylglyoxal is metabolized further to 1,2-PD and D (−)- lactate using the glyoxal bypass. The reduction of methylglyoxal to 1,2-PD can then continue via the dihydroxyacetone (acetol) (DeLey and Kersters 1964; Tanaka et al. 1975) or via lactaldehyde (Lin 1980). Finally, lactic acid bacteria have been shown to degrade lactic acid to a mixture of acetic acid and 1,2-PD (Elferink et al. 2001).

2.2 *Microorganisms Producing 1,2-Propanediol*

There are many microorganisms, both yeasts and bacteria, that can produce 1,2-PD but most of them are mesophilic. The mesophilic bacteria producing 1,2-PD are

Prevotella (Turner and Robertson 1979), *Salmonella* (Badia et al. 1985), *Klebsiella* (Badia et al. 1985), *Clostridium* (Sánchez-Riera et al. 1987; Ingvadottir et al. 2018; Cameron and Cooney 1986) and *Lentilactobacillus* (Elferink et al. 2001). Several fungal strains can also produce the compound (Suzuki and Onishi 1968; Dowd et al. 1994). There is however a great variety in both substrates and organisms in what pathways are used. One major difference is between bacteria and fungi concerning 1,2-PD is that the former is anaerobic but the latter aerobic. The first notification of 1,2-PD production was however by the thermophile *Clostridium thermobutyricum* in the 1950s (Enebo 1954) from cellulose and various sugars. No information is obtained whether the bacterium produces the R or S enantiomer of 1,2-PD. The mesophiles that can degrade deoxysugars to 1,2-PD are *Clostridium sphenoides* (Tran-Din and Gottschalk 1985) and *Salmonella typhimurium* (Badia et al. 1985) and produce 1,2-PD. In case of the former, degradation of glucose, fructose, mannose, and cellobiose also resulted in the production of 1,2-PD but only during phosphate limiting conditions since the methylglyoxal synthetase is strongly inhibited by phosphate. A mutant of *Thermoanaerobacterium thermosaccharolyticum* (formerly *Clostridium thermosaccharolyticum*) was found to degrade common sugars to 1,2-PD with maximum yields of 0.27 g 1,2-PD/g glucose in a batch culture using 45.0 g/L (Cameron and Cooney 1986; Sánchez-Riera et al. 1987) with lactate being the major fermentation end product. Later, the strain was shown to produce (*R*)-1,2-propanediol from both pentoses (arabinose, xylose), and hexoses (glucose, galactose) and lactose (Cameroon et al. 1998). It was reported to produce 2.8 g/L of 1,2-PD from hydrolyzed whey permeate but also other end products like lactate, acetate, and acetol. A recent investigation of *Clostridium* strain AK1 strain, a moderate thermophile isolated from a hot spring in Iceland, shows the ability of the production of (*S*)-1,2-propanediol from L-rhamnose (Ingvadottir et al. 2018). Initial studies showed that the strain produced 1,2-PD only from rhamnose but not fucose. The effect of various environmental culture parameters such as initial substrate concentrations, pH, temperature, the influence of the partial pressure of hydrogen as well as different initial phosphate concentrations was tested. The strain produces maximum 6.6 mM of 1,2-PD from 10 mM of rhamnose, or 61% of theoretical yields. At higher substrate concentrations, lower yields were observed. The optimum temperature and pH of growth were at 55 °C and 6.7, respectively. The 1,2-PD production was affected neither by partial pressure of hydrogen nor different initial phosphate concentrations (Ingvadottir et al. 2018).

Bielen and co-workers reported unpublished data of 1,2-PD formation by *Caldicellulosiruptor saccharolyticus* during growth on rhamnose (Bielen et al. 2013). Later, investigation of all nine species of the genus was tested for growth and end-product formation from both L-fucose and L-rhamnose (Ingvadottir et al. 2017). This study revealed that six of the strains degraded L-rhamnose (20 mM) to 1,2-PD ranging from 3.05 to 7.65 mM with other end products mainly being acetate and hydrogen but only traces of ethanol and lactate. Three strains could degrade fucose to mainly 1,2-PD together with acetate and hydrogen.

3 Production of 1,3-Propanediol

Propane-1,3-diol commonly referred to as β-propylene glycol or 1,3-propanediol (1,3-PD) is a bulk chemical with applications in polymers, cosmetics, adhesives, lubricants, foods, laminates, solvents, antifreeze, and medicine (Homann et al. 1990; Colin et al. 2000; Zhu et al. 2002; Cheng et al. 2007). Prior to 1990, 1,3-PD was only used in small quantities as compared with other bulk chemicals because of impurities and difficulties to produce it. Therefore, in the past three decades, some attractive biological processes have been developed for the production of high purity of the compound from low-cost renewable materials (Saxena et al. 2009).

Industrially, 1,3-PD has usually been produced from acrolein by Degussa DuPont and from ethylene oxide by Shell (Saxena et al. 2009). These chemical processes however involve high pressure in the hydroformylation and hydrogenation steps using high temperatures, expensive catalysts, and the release of toxic intermediates. Additionally, the yields are relatively low (40–80%) (Saxena et al. 2009). Thus, the interest in the past three decades has been more and more on the biological production of 1,3-PD mainly from glycerol (a waste product in the production of biodiesel) and glucose.

3.1 Pathways Involved in the Production of 1,3-Propanediol

The only known substrate to be converted to 1,3-PD is glycerol and the biochemistry of the production has been elucidated in detail (Seyfried et al. 2002; Biebl et al. 1999). The enzymes involved are glycerol dehydratase (GHD), 1,3-propandiol oxidoreductase (PDOR), glycerol dehydratase (GDH), 1,3-propanediol oxidoreductase (PDOR), glycerol dehydrogenase (GHD) and dihydroxyacetone phosphate kinase (DHAK). The dissimilation of glycerol involves either reductive or oxidative pathways. The former involves vitamin B12-dependent GHD that removes water from glycerol to form 3-hydroxypropionaldehyde (3-HPA) which is further reduced to 1,3-PD by NADH linked PDOR. In the oxidative pathway, glycerol is dehydrogenated to dihydroxyacetone (DHA) by NAD+ linked GHD and then converted to dihydroxyacetone phosphate (DHAP) by an ATP-dependent DHAK (Saxena et al. 2009) (Fig. 2).

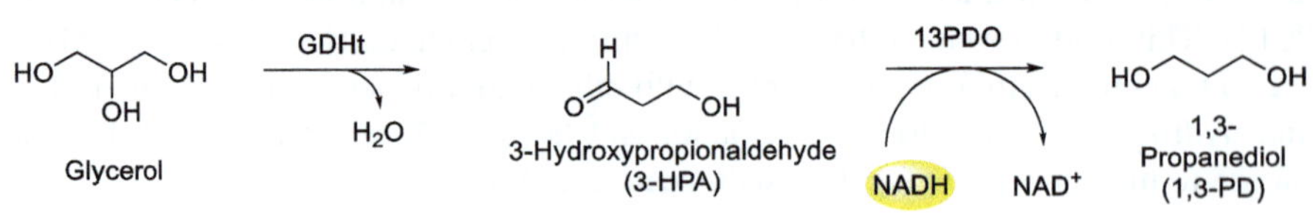

Fig. 2 Pathways with 1,3-propanol production

The biggest problems with the current fermentative techniques of 1,3-PD production are the relatively low yields and productivities. This can however overcome by using fed-batch or continuous culture methods or by mutagenesis of wild strains.

3.2 Microorganisms Producing 1,3-Propanediol

The only microorganisms that can produce 1,3-PD are bacteria but none can ferment sugars directly to 1,3-PD although some can use sugars to produce glycerol that is converted to 1,3-PD by other bacteria. Most bacteria that can produce 1,3-PD are mesophilic and have been well studied in the past. The well-known 1,3-PD producers are *Klebsiella* (Forage and Foster 1982; Yang et al. 2007), *Clostridia* (Forsberg 1987; Raynaud et al. 2003), *Citrobacter* (Homann et al. 1990; Seifert et al. 2001), *Enterobacter* (Barbirato et al. 1996) and *Lactobacilli* (Schutz and Radler 1984). The only thermophiles reported are two species within the genus *Caloramator, C. viterbiensis* and *C. boliviensis* (Seyfried et al. 2002; Crespo et al. 2012). Also, a study from 2001 showed that a strain most likely to belong to *Clostridium* was capable of 1,3-PD formation (Wittlich et al. 2001). Some data are from mixed cultures at thermophilic range in anaerobic reactors focusing on the production of hydrogen and 1,3-PD from crude glycerol (Sittijunda and Reungsang 2019; Simoes et al. 2021). *Caloramator viterbensis* was shown to produce 1,3-PD from glycerol in 2002 (Seyfried et al. 2002) producing approximately 2/3 of the substrate to 1,3-PD and on 1/3 to acetate. To the authors' knowledge, no experiments have been on the physiology to investigate enzymes or to optimize 1,2-PD with either of the *Caloramator* strains.

4 Production of Branched-Chain Alcohols

One of the major limitations of current biorefinery schemes using renewable biomass is the accumulation of protein without a strategy in place to convert protein products into biofuels. It has been well established that yeasts and some bacteria produce higher-order alcohols (sometimes referred to as "fusel alcohols") during fermentation of amino acids. Fusel alcohols are formed during fermentation under certain environmental conditions including high temperature, low pH, or during nitrogen limitation (Hazelwood et al. 2008; Smit et al. 2009).

It is well known that yeasts and some bacteria (*Staphylococcus, Enterococcus*, lactic acid bacteria) produce branched-chain alcohols (BCOH) from branched-chain amino acids (BCAA), but in low concentrations, via the Ehrlich pathway (Ehrlich 1907; Hazelwood et al. 2008; Beck et al. 2004; Ward et al. 2000, 1999). Using cheaper substrates, like sugars and carbon dioxide, to produce BCOH is only possible through genetic engineering. The highest concentration of iso-butanol

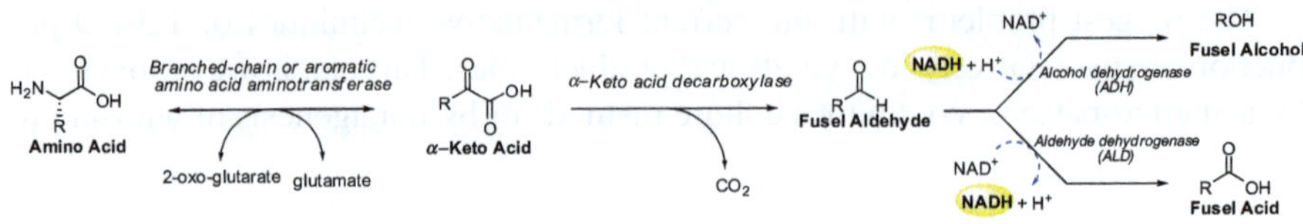

Fig. 3 Proposed scheme for the degradation of branched chain amino acids to a mixture of branched chain fatty acids and branched chain alcohols

reported in the literature is 50.8 g/L by genetically engineered *E. coli* from glucose, but usually, iso-butanol concentrations are less than 10 g/L (Huo et al. 2011).

Many thermophilic *Clostridia* can degrade amino acids to various end products. In most cases, the amino acids are first deaminated to their corresponding α-keto acids which are thereafter decarboxylated to a one-carbon shorter fatty acid (Orlygsson et al. 1995; Fardeau et al. 1996). The importance of interspecies hydrogen transfer has been well documented, especially for reduced amino acids like the BCAA. These amino acids degrade only when the electrons produced in the initial oxidative deamination step are scavenged due to the unfavorable thermodynamics involved (Orlygsson et al. 1995; Fardeau et al. 1996). The $\Delta G^{\circ\prime}$ for the degradation of the BCAAs to their corresponding branched-chain fatty acids (BCFA) is between +4.2 and + 9.7 kJ/mol (Fardeau et al. 1997). In cases where the amino acid degrading microorganism can use thiosulfate as an electron acceptor, degradation of these amino acids can take place (thiosulfate is reduced to H_2S or S^0). Another mechanism to degrade BCAA is co-cultivating the amino acid degrading microorganism with hydrogenotrophic methanogen where the electrons released are scavenged to the formation of methane. The catabolism of BCAAs by *Thermoanaerobacter brockii* has been studied in the presence of a hydrogen scavenging methanogen and in the presence of thiosulfate which led to the production of the corresponding fatty acid (Fardeau et al. 1997). Later, work on *T. brockii* and *Caldanaerobacter subterraneus* subsp. *yonseiensis* showed that the fermentation of BCAAs using thiosulfate as an electron sink resulted in the formation of a mixture of the corresponding BFCA and BCOH (Scully and Orlygsson 2014 (Fig. 3)). Later studies showed that this capacity of producing longer chain alcohols from the BCAA seems to be common among species within the genera of *Thermoanaerobacter* and *Caldanaerobacter* (Scully et al. 2015). Usually, the production of the BCFA was dominant over the alcohol, although some strains showed that more than 30% of the substrate ended up in the alcohol.

Several other studies by the same authors on other thermophilic bacteria to investigate the effect of various environmental factors on BCOH formation were performed. In a study with *Thermoanaerobacter* strain AK90, it was shown that the BCAA were converted to their corresponding BCFA in co-culture of an hydrogenotrophic methanogen but to a mixture of BCFA and BCOH in the presence of thiosulfate (Scully and Orlygsson 2015b). Not surprisingly, no degradation of none of the three BCAA occurred when used as a single substrate.

The influence of culture parameters on the degradation of BCAA was performed on *Thermoanaerobacter* strain AK85 (Scully and Orlygsson 2019). The main outcome from this investigation was that the concentration of the electron donor (thiosulfate) during the degradation of BCAA was of great importance for the ratio of BCFA/BCOH formed. At higher thiosulfate concentrations, a higher portion of the BCAA was converted to BCFA when compared with BCOH formation. It was however not clear from the data obtained whether the strain converts the BCAA to a mixture of the corresponding BCFA/BCOH or if the acid is the initial product and is further converted later on to BCOH. Finally, by using isotopically labelled BCFA, it was clear that the strain indeed produced both BCFA and BCOH from BCAA.

Finally, in a recent study on *Thermoanaerobacter pseudethanolicus,* it was showed that by varying the partial pressure of hydrogen in the cultivation system could change the ratio of BCFA and BCOH produced (Scully and Orlygsson 2020a, b). The amount of the BCOH was however significantly lower as compared with strain AK85. The issue of whether the BCFA is first produced in culture media and later during the stationary phase converted to its corresponding alcohol is interesting. Recent investigation on the conversion of fatty acids to their corresponding alcohols showed that this phenomenon is indeed active in several members within the genus of *Thermoanaerobacter.*

5 Conversion of Volatile Fatty Acids to Alcohols

Alcohol dehydrogenases (ADHs) of thermophilic bacteria have been of special interest in the past, and for instance, *Thermoanaerobacter pseudethanolicus* possess several ADSs with varying substrate specificity and cofactor preferences (Scully and Orlygsson 2020b). Production of higher alcohol carbon alcohol production has been known for a long time by using auto- or lithotrophic Clostridia such as CO and CO_2-utilizing bacteria. Recent work has also shown that some *Thermoanaerobacter* strains can covert carboxylic acids to their corresponding alcohols in the presence of organic substances like glucose (Hitschler et al. 2018; Scully et al. 2021) (Fig. 4). The potential use of cheap fatty acids as a substrate for more expensive alcohols is appealing and has a potential of high energy formation of higher alcohol fuels (Scully and Orlygsson 2014). This has been further investigated on *Thermoanaerobacter* strain AK152 where 1-propanol was produced from 1-propionic acid. This highly ethanologenic strain converted more than 40% of the fatty acid to the alcohol in initial investigations but could be maximized to 57.3% yields by varying various culture parameters (Scully and Orlygsson 2020c). Kinetic studies revealed that propionate conversion rate to propanol was rapid and indeed the substrate for the alcohol formation was from propionate.

Further studies on other type strains of *Thermoanaerobacter*, *T. pseudethanolicus* showed that the strain converts C2–C6 fatty acids to their corresponding alcohols. The conversion yields vary between 21.0 and 57.9% with pentanoate being the fatty acid with the highest conversion yields. The reduction of the higher carbon fatty

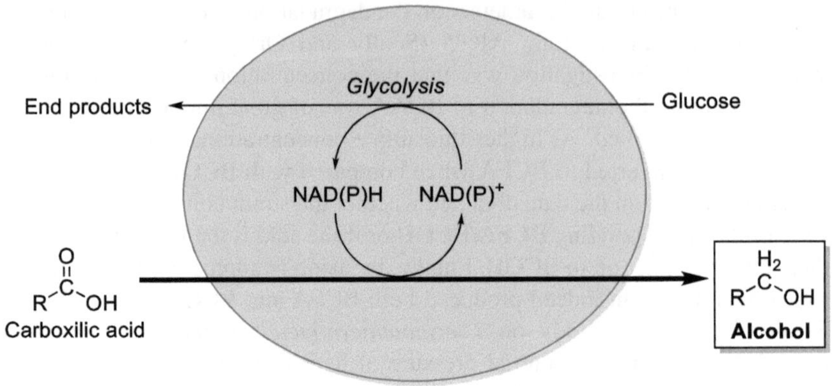

Fig. 4 Conversion of volatile fatty acids to alcohols

acids to their corresponding alcohols was further demonstrated by using ^{13}C labeled fatty acids and the conversion followed kinetically (Scully and Orlygsson 2020a).

Finally, a strain of *Thermoanaerobacter* strain AK91 was also shown to produce higher carbon alcohols from their corresponding fatty acids during a study of ethanol production from rhubarb leaf hydrolysate (Orlygsson and Scully 2021). Indeed, this was the first indication of higher chain alcohol formation from fatty acids and was observed by using fatty acids as inhibitors for ethanol production and surprisingly showed that the strain was not inhibited by the fatty acids used but used as electron acceptors to produce their corresponding alcohol. Given the ubiquity of carboxylic acids, the applications of a facile means of upgrading them to their corresponding alcohol may be of utility.

6 Conclusion

Production of fine chemicals by using thermophilic anaerobic bacteria is an exciting field of study. Several fine chemicals that are from non-renewable resources are produced today, but its production is often related with the use of toxic chemicals. Using more environmentally friendly processes are more and more sought up today, especially microbiological methods that often produce racemic pure compounds. Of the compounds produced today by thermophilic anaerobes, 1,2- and 1,3- PD are produced by several genera, namely *Clostridium*, *Thermoanaerobacterium,* and *Caldicellulosiruptor*. Producing 1,2-propandiol from renewable lignocellulosic waste is the most promising alternative for its production. Although promising as cell factories for these chemicals, there are no full-scale factories using fermentation techniques available today. Concerning the production of BCOH from BCFA, the main issue is the high cost of BCAA and low concentration of these amino acids in protein waste. Perhaps, the most promising alternative of using thermophilic, anaerobic bacteria is their capacity of using volatile fatty acids as electron donors, rendering the corresponding alcohol. This alternative opens the possibility to convert lignocellulose to fatty acids and further convert them to alcohols. The most likely process to be behind this would be a two-step fermentation process where pure culture of bacteria produces volatile fatty acids which are further converted to high energy fuel such as 1-propanol and 1-butanol.

References

Adalsteinsson BT, Hreggvidsson GO (2023) Geothermal habitats and adaptations of thermophilic microbes. In: Scully SM, Orlygsson J (eds) Thermophilic anaerobes – phylogeny, physiology and biotechnological applications. Springer, Physiology and biotechnological applications, p 2023

Ajeje SB, Hu Y, Song GJ et al (2022) Thermostable cellulases/xylanases from thermophilic and hyperthermophilic microorganisms: current perspective. Front Bioeng Biotechnol 9:794304. https://doi.org/10.3389/fbioe.2021.794304

Arbab S, Ullah H, Khan MIU et al (2022) Diversity and distribution of thermophilic microorganisms and their application in biotechnology. J Basic Microbiol 62:95–108. https://doi.org/10.1002/jobm.202100529

Badia J, Ros J, Aguilar J (1985) Fermentation mechanism of fucose and rhamnose in *Salmonella typhimurium* and *Klebsiella pneumoniae*. J Bacteriol 161:435–437

Barbirato F, Soucaille P, Bories A (1996) Physiologic mechanisms involved in accumulation of 3-hydroxypropionaldehyde during fermentation of glycerol by *Enterobacter agglomerans*. Appl Environ Microbiol 62:4405–4409. https://doi.org/10.1128/AEM.62.12.4405-4409.1996

Beck HC, Hansen AM, Lauritsen FR (2004) Catabolism of leucine to branched-chain fatty acids in *Staphylococcus xylosus*. J Appl Microbiol 96:1185–1193. https://doi.org/10.1111/j.1365-2672.2004.02253.x

Biebl H, Menzel K, Zeng A-P et al (1999) Microbial production of 1,2-propanediol. Appl Microbiol Biotechnol 52:289–297. https://doi.org/10.1007/s002530051523

Bielen AAM, Verhaart MRA, van der Oost J et al (2013) Biohydrogen production by the thermophilic bacterium *Caldicellulosiruptor saccharolyticus*: current status and perspectives. Life 3(1):52–85

Cameron DC, Cooney CL (1986) A novel fermentation: the production of R(−)1,2-propanediol and acetol by *Clostridium thermosaccharolyticum*. Nat Biotechnol 4:719–725. https://doi.org/10.1038/nbt0786-651

Cheng KK, Zhang JA, Liu DH et al (2007) Pilot-scale production of 1,3-propanediol using *Klebsiella pneumoniae*. Process Biochem 42:740–744. https://doi.org/10.1016/j.procbio.2007.01.001

Colin T, Bories A, Moulin G (2000) Inhibition of *Clostridium butyricum* by 1,3-propanediol and diols during glycerol fermentation. Appl Microbiol Biotechnol 54:201–205. https://doi.org/10.1007/s002530000365

Cooper RA (1975) The separation of 2,4-dinitrophenylhydrazones by thin layer chromatography. J Chromatogr 20:528–540

Crespo C, Pozzo T, Karlsson EN et al (2012) *Caloramator boliviensis* sp nov., a thermophilic, ethanol producing bacterium isolated from a hot spring. Int J Syst Evol Microbiol 62:1679–1686. doi: https://doi.org/10.1099/ijs.0.032664-0

DeLey J, Kersters K (1964) Oxidation of aliphatic glycols by acetic acid bacteria. Bacteriol Rev 28: 164–180

Dowd MK, Johansen SL, Cantarella L et al (1994) Low molecular weight organic composition of ethanol stillage from sugarcane molasses, citrus waste, and sweet whey. J Agric Food Chem 42: 283–288. https://doi.org/10.1021/jf00038a011

Ehrlich F (1907) Uver die bedingungen der fuselolbindung und uber ihren zusammenhang mit dem eiweissaufbau der hefe. Berichte der Deutschen Chemisten Gesellschaft 40:1027–1047

Elferink SJWHO, Krooneman J, Gottschal FC, Spoelstra SF, Faber F, Driehuis F (2001) Anaerobic conversion of lactic acid to acetic acid and 1,2-propanediol by *lactobacillus buchneri*. Appl Environ Microbiol 67:125–132. https://doi.org/10.1128/AEM.67.1.125-132.2001

Enebo L (1954) Studies in cellulose decomposition by an anaerobic thermophilic bacterium and two associated non-cellulolytic species. Viktor Pattersons Bokindustrie Akuebolag, Stockholm

Enze M (2006) Developing biorefinery by utilizing agriculture and forestry resources: striding forward the "carbohydrate" era. Progr Chem 18:132–141

Fardeau M-L, Faudon C, Cayol J-L et al (1996) Peptide and amino acid oxidation in the presence of thiosulfate by members of the genus *Thermoanaerobacter*. Curr Microbiol 31:152–157. https://doi.org/10.1007/BF00293546

Fardeau M-L, Patel BKC, Magot M et al (1997) Utilization of serine, leucine, isoleucine, and valine by *Thermoanaerobacter brockii* in the presence of thiosulfate or *Methanobacterium* sp. As electron acceptors. Anaerobe 3:405–410. https://doi.org/10.1006/anae.1997.0126

Forage RG, Foster AM (1982) Glycerol fermentation in *Klebsiella pneumoniae*: functions of the coenzyme B12-dependent glycerol and diol dehydratases. J Bacteriol 149:413–419. https://doi.org/10.1128/JB.149.2.413-419.1982

Forsberg CW (1987) Production of 1,3-propanediol from glycerol by *Clostridium acetobutylicum* and other *clostridium* species. Appl Environ Microbiol 53:639–634. https://doi.org/10.1128/AEM.53.4.639-643.1987

Gao KL, Song QX, Wei DZ (2006) Coupling of enantioselective biooxidation of DL-1,2-propanediol and bioreduction of pinacolone via regeneration cycle of coenzyme. Appl Microbiol Biotechnol 71:819–823. https://doi.org/10.1007/s00253-005-0231-3

Gaspar M, Juhasz T, Szengel Z et al (2005) Fractionation and utilization of corn fibre carbohydrates. Process Biochem 40:1183–1188. https://doi.org/10.1016/j.procbio.2004.04.004

Hatti-Kaul R, Chen L, Dishisha T, El EH (2018) Lactic acid bacteria: from starter cultures to producers of chemicals. FEMS Microbiol Lett 365:1–20. https://doi.org/10.1093/femsle/fny213

Hazelwood LA, Daran J-M, van Maris AJA et al (2008) The Ehrlich pathway for fusel alcohol production: a century of research on *Saccharomyces cerevisiae* metabolism. Appl Environ Microbiol 74:2259–2266. https://doi.org/10.1128/AEM.02625-07

Hitschler L, Kuntz M, Langschied F et al (2018) *Thermoanaerobacter* species differ in their potential to reduce organic acids to their corresponding alcohols. Appl Microbiol Biotechnol 102:8465–8476. doi: https://doi.org/10.1007/s00253-018-9210-3

Homann T, Tag C, Biebl H et al (1990) Fermentation of glycerol to 1,3-propanediol by *Klebsiella* and *Citrobacter* strains. Appl Microbiol Biotechnol 33:121–126

Huo Y-X, Cho KM, Rivera JGL et al (2011) Conversion of proteins into biofuels by engineering nitrogen flux. Nat Biotechnol 29:346–U160. https://doi.org/10.1038/nbt.1789

Ingvadottir EM, Scully SM, Orlygsson J (2017) Evaluation of the genus *Caldicellulosiruptor* for production of 1,2-propanediol from methylpentoses. Anaerobe 47:86–88. https://doi.org/10.1016/j.anaerobe.2017.04.015

Ingvadottir EM, Scully SM, Orlygsson J (2018) Production of (S)-1,2-propanediol from L-rhamnose using the moderately thermophilic Clostridium strain AK1. Anaerobe 54:26–30. https://doi.org/10.1016/j.anaerobe.2018.07.003

Koutinas AA, Wang R, Webb C (2004) Restructuring upstream bioprocessing: technological aspects for production of generic microbial feedstock from wheat. Biotechnol Bioeng 85:524–525. https://doi.org/10.1002/bit.10888

Lenth CW, Puis RND (1945) Polyhydric alcohol production by hydrogenolysis of sugars in the presence of copper-aluminum oxide. Ind Eng Chem Res 37:152–157

Lin ECC (1980) Conversion of reductases to dehydrogenase by regulatory mutations. In: Hollander A (ed) Trends in the biology of fermentation for fuels and chemicals. Plenum, New York, pp 305–313

Martin AE, Murphy FH (1994) Propylene glycols. In: Kroschwitz JI (ed) Kirk-Othmer encyclopedia of chemcial technology, vol 17, 4th edn. Wiley, New York, pp 715–726

Mhd Sawal MIF, Oslan AI, Hatta MHM et al (2019) Glycerol degradation in the absence of external hydrogen gas by using waste eggshell as heterogeneous catalyst. Malays J Anal Sci 23:880–891

Orlygsson J, Scully SM (2021) Influence of inhibitory compounds on biofuel production from oxalate-rich rhubarb leaf hydrolysate using *Thermoanaerobacter thermohydrosulfuricus* strain AK91. Fuels 2:71–86

Orlygsson J, Houwen FP, Svensson BH (1995) Influence of hydrogenotrophic methane formation on the thermophilic anaerobic degradation of protein and amino acids. FEMS Microbiol Ecol 13:327–334. https://doi.org/10.1016/0168-6496(94)90070-1

Pan X, Gilkes N, Saddler JN (2006) Effect of acetyl groups on enzymatic hydrolysis of cellulosic substrates. Holzforschung 60:398–401. https://doi.org/10.1515/HF.2006.062

Prabhu AA, Ledesma-Amaro R, Lin CSK et al (2020) Bioproduction of succinic acid from xylose by engineered *Yarrowia lipolytica* without pH control. Biotechnol Biofuels 13:1–15. https://doi.org/10.1186/s13068-020-01747-3

Prado-Rubio OA, Gasca-Gonzalez R, Fontalvo J et al (2020) Design and evaluation of intensified downstream technologies towards feasible lactic acid production. Chem Eng Process 158: 108174. https://doi.org/10.1016/j.cep.2020.108174

Pye EK (2005) Biorefing: a major opportunity for the sugar cane industry. Int Sugar J 107:222

Raynaud C, Sarcabal P, Meynial-Salles I et al (2003) Molecular characterization of the 1,3-propanediol operon of *clostridium butyricum*. Proc Natl Acad Sci U S A 100:5010–5015. https://doi.org/10.1073/pnas.0734105100

Revuelta JL, Serrano-Amatriain C, Ledesma-Amaro R et al (2018) Formation of folates by microorganisms: towards the biotechnological production of this vitamin. Appl Microb Biotechnol 102:8613–8620. https://doi.org/10.1007/s00253-018-9266-0

Samad WZ, Isahak WNRW, Nordin N et al (2015) Glycerol conversion over novel fuorine-doped tin oxide supported catalyst: effect of metal loadings and glycerol concentration. Malays J Anal Sci 19:55–64

Sánchez-Riera F, Cameron DC, Cooney CL (1987) Influence of environmental factors in the production of R(−)-1, 2-propanediol by *Clostridium thermosaccharolyticum*. Biotechnol Lett 9:449–454. https://doi.org/10.1007/BF01027450

Saxena RK et al (2009) Microbial production of 1,3-propanediol: Recent developments and emerging opportunities. Biotechnol Adv 27(6):895–913. https://doi.org/10.1016/j.biotechadv.2009.07.003

Saxena RK, Anand P, Saran S et al (2010a) Microbial production of 1,3-propanediol: recent developments and emerging opportunities. Biotechnol Adv 27:895–913. https://doi.org/10.1016/j.biotechadv.2009.07.003

Saxena RK, Anand P, Saran S et al (2010b) Microbial production and applications of 1,2-propanediol. Indian J Microbiol 50:2–11

Schutz H, Radler F (1984) Anaerobic reduction of glycerol to 1,3-propanediol by *Lactobacillus brevis* and *Lactobacillus buchneri*. Syst Appl Microbiol 5:169–178. https://doi.org/10.1016/S0723-2020(84)80018-1

Scully SM, Orlygsson J (2014) Branched-chain alcohol formation from branched-chain amino acids by *Thermoanaerobacter brockii* and *Thermoanaerobacter yonsenensis*. Anaerobe 30:82–84. https://doi.org/10.1016/j.anaerobe.2014.09.003

Scully SM, Orlygsson J (2015a) Recent advances in second generation ethanol production by thermophilic bacteria. Energies 8:1–30. https://doi.org/10.3390/en8010001

Scully SM, Orlygsson J (2015b) Amino acid metabolism of *Thermoanaerobacter* strain AK90: the role of electron scavenging systems in end product formation. J Amino Acids 410492

Scully SM, Orlygsson J (2019) Branched-chain amino acid catabolism of *Thermoanaerobacter* AK85 and the influence of culture conditions on branched-chain alcohol formation. Amino Acids 51:1039–1054. https://doi.org/10.1007/s00726-019-02744-z

Scully SM, Orlygsson J (2020a) Branched-chain amino acid catabolism of *Thermoanaerobacter pseudethanolicus* reveals potential route to branched-chain alcohol formation. Extremophiles 24:121–133. https://doi.org/10.1007/s00792-019-01140-5

Scully SM, Orlygsson J (2020b) Thermostable *Thermoanaerobacter* alcohol dehydrogenases and their use in organic synthesis. In: Physiological and biotechnological aspects of extremophiles. Elsevier, pp 183–193

Scully SM, Orlygsson J (2020c) Biotransformation of carboxylic acids to alcohols: characterization of *Thermoanaerobacter* strain AK152 and 1-propanol production via propionate reduction. Microorganisms 8:6

Scully SM, Orlygsson J (2023) Thermostable enzymes and their applications. In: Scully MS, Orlygsson J (eds) Thermophilic anaerobes – phylogeny, physiology and biotechnological applications. Springer, Berlin, p 2023

Scully SM, Iloranta P, Myllymaki P et al (2015) Branched-chain alcohol formation by thermophilic bacteria within the genera of *Thermoanaerobacter* and *Caldanaerobacter*. Extremophiles 19: 809–818. https://doi.org/10.1007/s00792-015-0756-z

Scully SM, Brown AE, Mueller-Hilger Y et al (2021) Influence of culture conditions on the bioreduction of organic acids to alcohols by *Thermoanaerobacter pseudethanolicus*. Microorganisms 9:1. https://doi.org/10.3390/microorganisms8060945

Seifert C, Bowien S, Gottschalk G et al (2001) Identification and expression of the genes and purification and characterization of the genes products involved in reactivation of coenzyme B12-dependent glycerol dehydratase of *Citrobacter freundii*. Eur J Biochem 68:2369–2378. https://doi.org/10.1046/j.1432-1327.2001.02123.x

Seyfried M, Lyon D, Rainey FA et al (2002) *Caloramator viterbensis* sp nov., a novel thermophilic, glycerol-fermenting bacterium isolated from a hot spring in Italy. Int J Syst Evol Microbiol 52: 1177–1184. doi: https://doi.org/10.1099/ijs.0.01472-0

Siebert D, Wendisch VF (2015) Metabolic pathway engineering for production of 1,2-propanediol and 1-propanol by *Corynebacterium glutamicum*. Biotechnol Biofuels 8:1–13. https://doi.org/10.1186/s13068-015-0269-0

Simoes AN, da Costa TB, de Menezes CA et al (2021) One waste and two products: choosing the best operational temperature and hydraulic retention time to recover hydrogen or 1,3-propanediol from glycerol fermentation. Bioprocess Biosyst Eng 44:2491–2502. https://doi.org/10.1007/s00449-021-02620-9

Sittijunda S, Reungsang A (2019) Bioconversion of crude glycerol into 1,3-propanediol by an anaerobic mixed microbial consortium. Chiang Mai J Sci 46:1–14

Smit BA, Engels WJM, Smit G (2009) Branched chain aldehydes: production and breakdown pathways and relevance of flavor in foods. Appl Microbiol Biotechnol 81:987–999. https://doi.org/10.1007/s00253-008-1758-x

Solaiman DKY, Ashby RD, Foglia TA et al (2006) Conversion of agricultural feedstock and coproducts into poly(hydroxyalkanoates). Appl Microbiol Biotechnol 71:783–789. https://doi.org/10.1007/s00253-006-0451-1

Suzuki T, Onishi H (1968) Aerobic dissimilation of l-rhamnose and the production of l-rhamnonic acid and 1,2-propanediol by yeasts. Agric Biol Chem 32:888–893

Tanaka Y, Fujii K, Tanaka A et al (1975) Metabolism of 1,2-propanediol in soil bacterium. J Ferment Technol 53:354–362

Tao YM, Bu CY, Zou LH et al (2021) A comprehensive review on microbial production of 1,2-propanediol: micro-organisms, metabolic pathways, and metabolic engineering. Biotechnol Biofuels 14:216. https://doi.org/10.1186/s13068-021-02067-w

Taylor MP, Eley KL, Martin S et al (2009) Thermophilic ethanologenesis: future prospects for second-generation bioethanol production. Trends Biotechnol 27:398–405. https://doi.org/10.1016/j.tibtech.2009.03.006

Tran-Din K, Gottschalk G (1985) Formation of D(−)-1,2-propanediol and D(−)-lactate from glucose by *Clostridium sphenoides* under phosphate limitation. Arch Microbiol 142:87–92. https://doi.org/10.1007/BF00409243

Turner KW, Robertson AM (1979) Xylose, arabinose, and rhamnose fermentation by *Bacteroides ruminicola*. Appl Environ Microbiol 38:7–12

Turner P, Mamo G, Karlsson EN (2007) Potential and utilization of thermophiles and thermostable enzymes in biorefining. Microb Cell Factories 6:9. https://doi.org/10.1128/AEM.38.1.7-12.1979

Ward DE, Ross RP, van der Weijden CC et al (1999) Catabolism of branched-chain a-keto acids in *enterococcus faecalis*: the bkd gene cluster, enzymes and metabolic route. J Bacteriol 181:5433–5442. https://doi.org/10.1128/JB.181.17.5433-5442.1999

Ward DE, van der Weijden CC, van der Merve MJ et al (2000) Branched-chain a-keto acid catabolism via the gene products of the bkd operon in *Enterococcus faecalis*: a new secreted metabolite serving as a temporary redox sink. J Bacteriol 182:3239–3246. https://doi.org/10.1128/JB.182.11.3239-3246.2000

Wittlich P, Themann A, Vorlop KD (2001) Conversion of glycerol to 1,3-propanediol by a newly isolated thermophilic strain. Biotechnol Lett 23:463–466. https://doi.org/10.1023/A:1010329321185

Yang G, Tian J, Li J (2007) Fermentation of 1,3-propanediol by a lactate deficient mutant of *Klebsiella oxytoca* under microaerobic conditions. Appl Microbiol Biotechnol 73:1017–1024. https://doi.org/10.1007/s00253-006-0563-7

Zhao H, Zheng L, Li X et al (2018) Hydrogenolysis of glycerol to 1,2-propanediol over cu-based catalysts: a short review. Catal Today 355:84–95. https://doi.org/10.1016/j.cattod.2019.03.011

Zhu MM, Lawman PD, Cameron DC (2002) Improving 1,3-propanediol production from glycerol in a metabolically engineered *Escherichia coli* by reducing accumulation of sn-glycerol-3-phosphate. Biotechnol Prog 18:694–699. https://doi.org/10.1021/bp020281+

Part IV
Future Aspects

Potential of Anaerobic Thermophiles and Future Prospects

Johann Orlygsson

Abstract Since the discovery of thermophilic and extremophilic bacteria, their potential in various biotechnological properties has been well studied. The present investigation dwells on the main areas of research where thermophilic bacteria are of great importance as well as discussion on possible future aspects of their utilization. The main use of thermophiles and extremophiles is within the production of biofuels (ethanol and hydrogen), their use in biorefineries (production of diols and branched-chain alcohols), production of thermostable enzymes, and industrial use in general. The main future prospect of the use of thermophiles is to upscale the production of ethanol (and hydrogen) to large-scale factories, using complex lignocellulosic biomass as substrate. Additionally, the production of most fine chemicals is not yet financially profitable as compared with the production from fossil fuels. The opportunities in this sector are most likely further genetic engineering work on present microorganism. Finally, although genetic engineering seems to be of more and more use in anaerobic thermophiles, the need for genetic tools is still lacking in their process.

1 Introduction

Thermophilic and extremophilic bacteria are metabolically active in extreme thermal environments. These harsh environments have resulted in the evolution of microorganisms capable of growth at extremes in salt (>1.0 mol/L), low and high pH (<5.0 or >8.0), high temperatures (>45 °C), atmospheric pressure, and many other (Rastogi et al. 2010; Urbieta et al. 2014a, b; Futterer et al. 2004; Giaveno et al. 2013; Ruepp et al. 2000). Microorganisms living in such extreme environments have many diverse cell structures, and a variety of different metabolisms and use ultimate survival strategies that allow them to withstand these harsh conditions (Urbieta et al. 2015).

J. Orlygsson (✉)
Faculty of Natural Resource Science, University of Akureyri, Akureyri, Iceland
e-mail: jorlygs@unak.is

S. M. Scully, J. Orlygsson (eds.), *Thermophilic Anaerobes*, Emerging Issues and Trends, https://doi.org/10.1007/978-3-031-41720-7_9

Thermophiles in general have been exploited for various processes but the main emphasis in the early days was on aerobic and chemolithotrophic thermophiles. Bacteria that do not use oxygen for growth at high temperatures are called thermophilic anaerobes. These bacteria have been of interest to both basic and applied science. The areas of the basic nature of thermophilic anaerobes are the various unique forms of microorganisms and the fact that they are very likely to possess similar metabolism as those of early life forms on Earth. The first life forms on Earth evolved 3.8 billion years ago when there was very little oxygen present, indicating that early life forms were of anaerobic nature. Thermophilic anaerobes are even more interesting due to the fact that during the early evolution of life on Earth, the temperature was much higher compared with the present day (Wagner and Wiegel 2008). It is however not possible to prove that the first microorganisms on Earth were anaerobic, but many believe that the evolution from low-temperature origin microorganisms evolving to adapt at higher temperatures is unlikely to happen. From an applied point of view, thermophilic anaerobic bacteria have been investigated intensively for the past few decades. Early investigations were on second-generation ethanol production and later hydrogen production. More recent investigations have since been on the production of low-volume, high-value compounds like 1,2-propanediol and other fine chemicals as well as thermostable enzymes. This chapter focuses on thermophilic anaerobic bacteria producing both high-volume, low-value and low-volume, high-value compounds, and future aspects are issued on these microbes.

2 Main Metabolic Pathways of Thermoanaerobic Bacteria

Heterotrophic anaerobic bacteria growing on various organic compounds may be categorized into glycolytic, cellulolytic, lipolytic, and proteolytic metabolism. The main glycolytic metabolic pathways used by these microbes are the same as for mesophilic microbes; the Embden-Meyerhof and Entner-Doudoroff pathways. However, thermophiles, especially of the archaea origin, have many modifications of these two mainstream metabolisms in cells (Bielen et al. 2010; Straub et al. 2020).

The degradation of glucose to pyruvate through the Embden-Meyerhof pathway (EMP) is the most common pathway used by thermophilic anaerobes. Degradation of glucose with EMP generates two NADHs, two pyruvates, the key intermediate in most organisms, together with the formation of two ATP by substrate-level phosphorylation. Pyruvate is the end product of glycolysis and can be converted to fermentation products like hydrogen, ethanol, acetate, lactate, alanine, and carbon dioxide.

The carbon flow depends on the microorganisms involved and the environmental conditions. Pyruvate can for instance be reduced to lactate by lactate dehydrogenase (LDH) but the most favorable pathway for anaerobic bacteria is to oxidize pyruvate to acetyl-CoA, H_2, and CO_2 by using pyruvate: ferredoxin oxidoreductase (PFOR) which can be converted to acetate with concomitant ATP synthesis from the acetyl-phosphate intermediate. Acetate is thus the oxidized product but the main advantage for the microorganisms is the extra ATP produced. The electrons are transported to reduced ferredoxin which acts as an electron donor for hydrogenases and H_2 is produced as the reduced product. Strict anaerobes can produce H_2 from two major breakpoints during the degradation of glucose; firstly, from a NAD(P)H by GAPDH and from pyruvate ferredoxin oxidoreductase (PFOR) (Jones 2008). The principal H_2 pathway is through PFOR because of the thermodynamics hindrance of reoxidizing NADH (Jones 2008). It is a well-known phenomenon that the low H_2 yields observed by mesophilic and moderate thermophilic bacteria are due to the fact that H_2 production from either ferredoxin or NAD(P)H are thermodynamically unfavorable (Jones 2008; Hallenbeck 2009). The redox potential of Fd_{red}/Fe_{ox} couple depends on the microorganism and temperature involved. In nature, high partial pressures of H_2 are relatively uncommon because of the activity of H2-scavenging microbes, e.g., methanogens or sulfate-reducing bacteria (Cord-Ruwisch et al. 1988). This results in a low partial pressure of H_2 which is favourable for a complete oxidation of glucose to acetate and CO_2. At high temperatures, the influence of the partial pressure of H_2 is less on the key enzymes responsible for H_2 production. This is the main reason why extremophilic bacteria have been reported to produce up to 4 moles of H_2 together with 2 moles of acetate in pure cultures and also for the fact that microorganisms growing at lower temperatures direct their end product formation to other reduced products (Pawar et al. 2013, 2015). At lower temperatures, the NADH ferredoxin oxidoreductase (NOR) that converts NADH to Fd_{red} is strongly inhibited. The E° is – 400 mV for Fd_{red}/Fd_{ox} couple but −320 mV for the $NADH/NAD^+$ couple (Jones 2008; Hallenbeck 2009). Therefore, at low temperatures, elevated H_2 concentrations inhibit H_2 evolution at much lower concentrations as compared to extreme temperatures. Mesophilic and moderate thermophilic bacteria respond to this by directing their reducing equivalents to other more favorable electron acceptors and consequently produce reduced products like ethanol, lactate, butyrate, and alanine.

Carbohydrate degradation by thermophilic bacteria is well known today. Most mesophiles degrade sugars to a mixture of acetate, ethanol, lactate, and butyrate, together with the formation of hydrogen and carbon dioxide. Moderate thermophiles mostly produce the same end-products although butyrate is less common but not unknown (Scully and Orlygsson 2015). Thermophiles usually never produce butyrate as the volatile end product, but more generally produce ethanol and lactate (and hydrogen) as reduced products and acetate, and carbon dioxide as oxidized end products.

Apart from the traditional fermentation glycolytic pathways, thermophilic anaerobes can degrade a wide variety of other organic compounds by fermentation. However, much less studied is amino acid degradation. The main difference in amino acid degradation as compared with carbohydrates is the fact that many amino acids are reduced and cannot be degraded unless with the help of interspecies hydrogen transfer or by the addition of external electron acceptors like thiosulfate (Fardeau et al. 1997; Scully et al. 2015). This is however not true for all amino acids because some of them can be degraded as a single substrate by a single microorganism, serine, and threonine being the best examples. Amino acids are usually first deaminated to their corresponding keto acid which is further decarboxylated to its corresponding fatty acid. However, there is much more variety for amino acid metabolism in general depending on the thermodynamics and the microbes involved (Orlygsson 1994). As an example of the degradation of reduced amino acids, the degradation of the branched-chain amino acids is discussed. Leucine, isoleucine, and valine are branched-chain amino acids with 5-carbon (valine) and 6-carbon structures (leucine and valine). The ΔG^{o1} for the deamination step is positive (4.2 to 9.7 kJ/mol), and these amino acids cannot be degraded as a single substrate under anaerobic conditions (Fardeau et al. 1997; Orlygsson 1994). By co-cultivating the amino acid degrading bacterium with a hydrogenotrophic methanogen, the overall thermodynamics change dramatically, and these amino acids are first deaminated and then decarboxylated, resulting in a production of one carbon shorter branched-chain fatty acids (Scully et al. 2015). Recent investigations have also shown that by cultivating *Thermoanaerobacter* and *Caldanaerobacter* species in the presence of thiosulfate as an electron acceptor, the branched-chain amino acids are degraded to a mixture of their corresponding fatty acid and alcohol, most likely these bacteria first produce the branched-chain fatty acid which can be used as an electron acceptor in competition with thiosulfate and convert them to the alcohol (Scully et al. 2015).

Examples of production of other valuable compounds produced by thermophilic, anaerobic bacteria are for instance production of 1,2-propanediol and 1,3-propanediol. Production of 1,2-propanediol by fermentation occurs mainly by two pathways, firstly by using deoxy sugars like fucose and rhamnose, and secondly by using "normal" sugars via the glycolytic intermediate dihydroxyacetone phosphate via the formation of methylglyoxal (Saxena et al. 2010). Examples of the former pathway in thermophilic microorganisms are *Clostridium* and *Caldicellulosiruptor* (Ingvadottir et al. 2018, 2017) and *Thermoanaerobacterium* species for the latter (Sánchez-Riera et al. 1987). The only known pathway for 1,3-propanediol formation is by using glycerol as the substrate. The only anaerobic thermophiles known to produce 1,3-propanediol are some members of the genus *Caloramator* (Seyfried et al. 2002). *Caloramator viterbiensis* degrades glycerol to a mixture of acetate (oxidized product) and 1,2-propanediol (reduced product).

3 Main Use of Thermoanaerobic Bacteria in Biotechnology

3.1 Biofuel Production

Production of ethanol, hydrogen, and methane by thermophilic bacteria is an exciting area of research. In this chapter, methane is omitted but the main focus is on ethanol and hydrogen.

The stochiometry of ethanol formation from glucose is as follows:

$$1 \text{ Glucose} \rightarrow 2 \text{ Ethanol} + 2 \text{ CO}_2 \quad (1)$$

Ethanol production from simple biomass like sugar cane and starch has been well known for many decades now and is a huge industry in many countries. This type of biomass is the so-called first-generation biomass with only simple sugars like sucrose and glucose. The vast majority of this type of biomass is fermented to ethanol by yeasts but not bacteria (Taylor et al. 2009). The main reason for this is because of the high yields of ethanol obtained by yeast and their high tolerance towards both substrate concentrations and high ethanol titers (Scully and Orlygsson 2015). However, using this type of biomass for biofuel production has been criticized heavily because of its competition with food and feed production. Therefore, in the past few decades, the focus has been on the utilization of more complex biomass, e.g. lignocellulosic biomass (Taylor et al. 2009; Scully and Orlygsson 2015). This type of biomass is however much more difficult to degrade, especially because of the complex sugar structure of hemicelluloses bound in lignocellulose (Hahn-Hagerdahl et al. 2006). Many of these sugars, e.g. xylose, arabinose, and many hexoses cannot be utilized by wild-type yeasts and thus lowers the ethanol yields of the process. Another drawback of using yeasts for the degradation of sugars released during pretreatment of lignocellulose is the so-called glucose effect many yeasts possess resulting in synchronized sugar utilization which leads to a longer degradation time of the substrate. However, thermophilic bacteria have much broader substrate spectra and can often degrade many sugars simultaneously (Taylor et al. 2009; Scully and Orlygsson 2015). Thus, they may be more suitable for the degradation of complex biomass as compared to yeasts. On the other hand, thermophilic bacteria also possess some negative factors like low tolerance towards ethanol and the production of other end products apart from ethanol, like lactate and acetate, lowering ethanol yields in the process (Almarsdottir et al. 2012; Brynjarsdottir et al. 2012). This has been circumvented by genetic engineering, mainly by knocking out metabolic pathways leading to the undesired end-product formation or by increasing ethanol tolerance (Shaw et al. 2008).

There are several ways to convert lignocellulose to ethanol after pretreatment: (a) separate hydrolysis and fermentation (SHF), (b) simultaneous saccharification and fermentation (SSF), (c) simultaneous saccharification and co-fermentation (SSCF), and (d) consolidated bioprocessing (Scully and Orlygsson 2015). Most relevant for thermophilic ethanol production are the SHF, SSCF, and CBP type microorganisms because there is no need for a separate hexose and pentose fermentation as needed in the SSF process. The CBP is a process where enzymatic production, cellulose degradation, and fermentation occurs in a single step. Thus, the choice of microorganisms in such a system is of great importance; it needs to have the enzymatic machinery to produce both cellulases and hemicellulases as well as to produce high titers of ethanol. No wild-type microorganism is known today to have such properties. Thus, active ongoing research focuses on genetic engineering, either to modify microorganisms that have the native ability to degrade lignocellulose and are then further engineered to become a powerful ethanol producer or to genetically engineer microorganisms that are a good ethanol producer but do not have the ability to produce the enzymes needed for the biomass breakdown (Scully and Orlygsson 2015).

Apart from ethanol, many thermophilic and extremophilic bacteria show high yields of hydrogen production. Production of hydrogen by fermentation is a feasible option because of the issue of renewable energy sources. Hydrogen production by microbes has been known for many years, especially by studies of mesophilic facultative and strictly anaerobic bacteria (Pawar et al. 2015; Scully and Orlygsson 2015). In general, the degradation of glucose to acetate is the most favorable process of producing hydrogen with a maximum of four moles per mole of glucose degraded:

$$1\ \text{Glucose} \rightarrow 2\ \text{Acetate} + 2\ CO_2 + 4\ H_2 \tag{2}$$

Other fermentative pathways may also yield hydrogen as a product, e.g. production of butyrate. However, other fermentative pathways like ethanol and lactate production pathways direct the electron flow away from hydrogen but towards to these reduced end products (Scully and Orlygsson 2015). Reaction equations for these reactions are as follows (and Eq. 1):

$$1\ \text{Glucose} \rightarrow 1\ \text{Butyrate} + 2\ CO_2 + 2\ H_2 \tag{3}$$

$$1\ \text{Glucose} \rightarrow 2\ \text{Lactate} + 2\ CO_2 \tag{4}$$

Hydrogen may be the fuel of the future once hydrogen fuel cells for propelling cars are more developed. Although hydrogen production from thermophilic bacteria has been extensively studied in the past three decades, there is no full-scale facility where these bacteria are utilized for hydrogen production. As for ethanol production, both culture parameters and microorganism properties are of great importance. At lower temperatures, the reaction equilibrium tends towards the formation of more reduced end products (lactate and ethanol) instead of acetate and hydrogen which results in lower hydrogen yields. The use of complex biomass for hydrogen production has been focused heavily upon in recent decades. Many experiments have been done on the use of lignocellulosic waste material using thermophilic bacteri. A co-culture of *Clostridium thermocellum* with non-cellulolytic thermophilic bacterium has been used for CBP-based hydrogen production (Liu et al. 2008; Ivanova et al. 2009). Additionally, several species within the genus of *Thermoanaerobacter* and *Thermoanaerobacterium* have been performed to possess their hydrogen production capacity from various substrates (Sigurbjornsdottir and Orlygsson 2012; Brynjarsdottir et al. 2014; Vipotnik et al. 2016). Because of the thermodynamic nature of hydrogen production, the best yields of hydrogen are with extremophilic bacteria, mainly belonging to the genera of *Thermotoga* and *Caldicellulosiruptor* (Pawar et al. 2015; Nguyen et al. 2008).

3.2 Thermophiles and Biorefineries

The use of thermophilic anaerobic bacteria to produce fine chemicals in low volumes but high in value is increasing because of the world demand for using renewable biomass for such production. These chemicals can be produced in biorefineries using sustainable processes. The fine chemicals produced by thermophiles are 1,2-propanediol, 1,3-propanediol, and branched-chain alcohols (Altaras et al. 2001; Raynaud et al. 2003; Scully and Orlygsson 2020a, b). The use of high temperature minimizes the risk of contaminants, gives better solubility of substrates, facilitates mixing, and enhances biomass transfer rate and bioconversion rate (Mesbah 2022). Relatively few thermophilic, heterotrophic anaerobes have been used so far and merely on a laboratory scale.

There are mainly two pathways used to produce 1,2-PD. Firstly, deoxysugars like fucose and rhamnose are used as substrates (Turner and Robertson 1979) and secondly, the glycolytic intermediate, dihydroxyacetone phosphate (DHAP) is used via the formation of methylglyoxal (Saxena et al. 2010). An example of a thermophilic bacterium that uses the first pathway is *Thermoanaerobacterium thermosaccharolyticum* (formerly *Clostridium thermosaccharolyticum*) was found to degrade common sugars to 1,2-PD with maximum yields of 0.2 g 1,2-PD/g glucose in batch culture using 45 g/L (Cameron and Cooney 1986; Sánchez-Riera et al. 1987) with lactate being the major fermentation end product. Recently, a strain of *Clostridium,* strain AK1, a moderate thermophile isolated from a hot spring in Iceland, show the ability of the production of (S)-1,2-propanediol from L-rhamnose (Ingvadottir et al. 2018). Finally, six of the nine *Caldicellulosiruptor* were shown to produce 1,2-PD from rhamnose and three species converted fucose to 1,2-PD (Ingvadottir et al. 2017).

The only known substrate to be converted to 1,3-PD is glycerol and the biochemistry of the production has been elucidated in detail. There are many microorganisms that can produce the compound but only one thermophilic bacterium, *Caloramator viterbiensis* (Seyfried et al. 2002). This strain degrades glycerol to a mixture of 1,3-PD and acetate.

Recent investigations have shown the ability of bacteria within the genera of *Thermoanaerobacter* and *Caldanaerobacter* to utilize branched-chain amino acids to a mixture of their corresponding branched-chain fatty acid and branched-chain alcohol. Because of the thermodynamics behind the oxidative deamination step of the branched-chain amino acids, they cannot be degraded as a single substrate unless with an external or biological hydrogen scavenging system (Orlygsson 1994). Recent studies showed that some bacteria within these two genera produced only the branched-chain fatty acid in a co-culture with a hydrogenotrophic methanogen, but produced a mixture of fatty acids and alcohols with an external electron acceptor, thiosulfate (Scully et al. 2015; Scully and Orlygsson 2020a, b). The most likely route of branched-chain alcohol formation is that first, the bacterium produces the branched-chain fatty acids which act as an electron acceptor and is reduced to their corresponding alcohol.

Recent investigations on the ability of thermophilic bacteria within the genera of *Thermoanaerobacter* and *Caldanaerobacter* to convert volatile fatty acids during the degradation of carbohydrates to their corresponding alcohols are interesting shuttle to produce high carbon alcohol forming complex alcohol (Scully and Orlygsson 2020a, b; Scully et al. 2021). Recent studies have as discussed above show that long-chain alcohols can be produced from long-chain amino acids, but additionally, we have shown that these bacteria can also degrade carbohydrates and convert fatty acids to their corresponding alcohols. Thus, there seems to be a competition of electrons produced during carbohydrate metabolism for either adding them to pyruvate or to an external electron acceptor, like a fatty acid. This is a completely new way of producing a higher alcohol carbon molecule as known before (Scully et al. 2021).

3.3 Thermophilic Enzymes

Enzymes are categorized into six different classes based on enzyme action mechanisms: ligases, isomerases, oxidoreductases, lyases, transferases, and hydrolyses (Rigoldi et al. 2017). Currently, more than 75% of the enzymes that are used commercially are hydrolyses (Elleuche and Antranikian 2013). Proteases are generally regarded as the major sub-type of the hydrolyses family and are widely used as detergents, in the starch industry as well as in animal feed and in the dairy industry. The second largest subgroup of hydroylases are enzymes degrading various carbohydrates, mostly amylases and cellulases. These enzymes are widely used in productive industrial sectors such as starch, textile, detergent, and in the baking industry.

Thermophiles have evolved in extreme conditions such as temperatures ranging up to 120C, or extreme pH, salt, and high pressures. Thus, adapted to such extremes, the enzymes of thermophiles and extremophiles are of great importance. Many of these enzymes tolerate extremes in various harsh conditions and are thus regarded as a suitable source for various industrial processes (Mesbah 2022). That said, however, most enzymes currently used in industry are of mesophilic origin (Turner et al. 2007). Below is a description of the main use of thermophilic enzymes.

Lignocellulose is the main structural component of all plant material. Its structure is a carbohydrate-(cellulose and hemicellulose) and phenolic-based (lignin) biopolymers. Cellulose is the most abundant organic compound on Earth and is the major component of plant cell material. It is a linear homopolymer of glucose which are linked with B-1,4-glycosidic bonds, with repeating units of cellobiose. Hemicellulose is the second major component of lignocellulose. This is a family of branched- and heterogeneous polymers, and its chemical structure is different from plant tissues and species. The composition of hemicellulose is C5 carbon sugars (i.e., xylose and arabinose) and/or C6 sugars (i.e., glucose, mannose, and galactose). Lignin is however a nonlinear polymer, composed of randomly linked aromatic compounds, e.g., coniferyl, sinapyl, and coumarin alcohols which are linked to both cellulose and hemicellulose, and acts like a barrier, preventing the penetration of chemicals and enzmyes (Huang et al. 2022).

Degradation of lignocellulose is a complex process and is dependent on many enzymes. Firstly, cellulose is degraded to cellobiase by the action of endoglucanases and exoglucanases resulting in the formation of cellobiose which is further degraded to glucose by the use of β-glucosidase. The hemicellulose fraction of lignocellulose is however more complicated because of its heterogenic structure. The most important enzymes are xylanases because the majority of the hemicellulosic structure is xylan. These enzymes degrade xylan mainly to xylose but other enzymes like β-mannanases, arabionofuranosidases, and α-L-larabinases are also needed. Thermophilic bacteria possess a broad variety of enzymes that are of huge importance for the degradation of lignocellulose, like xylanases, laccases, β-mannases, and cellulases (Ergun and Calik 2016). Thermophilic, anaerobic bacteria that can degrade cellulose and hemicellulose mainly belong to two genera, *Clostridium* and *Caldicellulosiruptor*. The best known thermophilic bacteria to decompose cellulose is *Clostridium thermocellum*. This bacterium possesses the well-studied cellulosome and is capable of degrading both amorphous and crystallin cellulose (Ichikawa et al. 2017). This bacterium is a well-known producer of various end products, like acetate, ethanol, hydrogen, and carbon dioxide. There are 14 species that belong to the extremophile genus *Caldicellulosiruptor* (Blumer-Schuette 2020). Most of them are capable of cellulose degradation. The thermophilic microorganisms that produce thermostable enzymes for industrial use are *Pyrococcus* species, *Anaerocellum*, and *Thermotoga maritima* producing cellulases active at 95–97 °C (Hebal et al. 2022; Bhalla et al. 2013) The main types of hemicellulases are endoxylanases, β-xylosidase, arabinofuranosidase, and acetyl-xylan esterase and more. The endoxylanases (EC: 3.2.1.8) mainly cleave the β-glycosidic bonds of the xylan backbone and release Xos as a product. The β-xylosidases (EC: 3.2.1.37) degrade xylobiose and other xylooligosaccharides to yield xylose. Arabinofuranosidases (EC: 3.2.1.55) and acetyl xylan esterases (EC:3.1.1.72) attack on side chains of heterogenous xylan substrate and help xylanases and β-xylosidases to degrade xylan completely (Collins et al. 2005). Additionally, the synergistic action of these enzymes facilitate xylan and lignin removal from cellulose without affecting the cellulose structure. The main use of hemicelluloses is in the pulp and paper industry where wood is used for the production of the pulp but this is most often done at high temperatures. Thermostable xylanases have been isolated from various thermophilic bacteria like *Pyrococcus furiosus* and *Thermotoga* species which are active between 50 and 80 °C (Bhalla et al. 2013).

The other thermostable enzymes are esterases, keratineases, and lipases. Most thermophilic esterases known today are from aerobic bacteria like *Bacillus* and *Thermus* (Finore et al. 2023). However, several investigations have been on esterases from anaerobic hyperthermophiles like *Thermotoga maritima* and *Pyrobaculum caldifontis* (Levisson et al. 2007; Palm et al. 2011). Bacteria within the genus *Thermoanaerobacter* and *Caldanaerobacter* have been shown to produce thermostable lipases (Cai et al. 2011; Royter et al. 2009).

Keratin is a natural fiber that is a component of recalcitrant structure of hair, nails, hoofs, feathers, and horns. Keratinases are proteolytic enzymes that attack disulfide bridges of keratin and thus are of great importance in various areas like textile processing, protein supplement, biomedical industries, and more (Brandelli et al. 2010). Keratinases are usually produced microbially from slaughterhouses, hair saloons, and mundan (donating hair). There is a wide variety of microorganisms capable of keratin degradation, both fungi, and aerobic and anaerobic bacteria. The thermophilic fungi that is known to degrade keratin are Epidermophyton and Trichophyton (Ignatova et al. 1999). Among bacteria degrading keratin, most are aerobic and mesophilic. Thermophilic anaerobes known to degrade the chemical are mainly found in the genus *Thermoanaerobacter*, e.g., *T. keratinophilus* (Kublanov et al. 2009; Riessen and Antranikian 2001).

3.4 Thermophiles in Industry

Enzymes of thermophilic origin have been used for a long time as detergents and for various food production (cheese, sourdough, beer, and wine) as well as for the production of indigo, linen, and leather (Turner et al. 2007; Blumer-Schuette et al. 2008; Mesbah 2022). Usually, such processes are triggered by the addition of enzymes to the substrate, in situ production of enzymes during processing, or by enzymes that are present in natural products. Today, there are many enzymes that are used in various industrial processes such as detergent, textile, and starch that are produced in an industrial scale by selected thermophilic microbes. The first commercial enzyme that was produced industrially was protease by Novozymes. More recent developments of enzyme production have been mostly done by protein and genetic engineering of microbes to tailor-made enzymes for various areas of applications. Most enzymes used in industrial processes are from yeast (50%), the rest belonging to bacteria (one-third), animals (8%), and plant (4%) (https://www.alliedmarketresearch.com/enzymes-market). Microbes that are used in industrial processes for enzyme production need to be economically feasible, easy of growth, and fast generation times. In the year 2020, the global market for industrial enzymes passed 12.46 billion dollars and its growth rate for the next two to three decades is estimated to be around 8% (https://www.grandviewresearch.com/industry-analysis/enzymes-industry).

Because of the robustness of thermophiles, they are often with great advantage compared to mesophilic microorganisms in various industrial applications (Ebaid et al. 2019). They usually grow faster and are more resistant to environmental stresses and produce thermotolerant enzymes. Their enzymes are thermostable and thus suitable for industrial processes that take place at high temperatures (Ebaid et al. 2019). Higher temperatures increase the rates of biochemical reactions and at the same time, reduce risk for mesophilic microbial contamination (Kuhad et al. 2011; Cuecas et al. 2016). Additionally, high temperatures reduce energy output because of increased solubility and the efficiency of substrate mixing (Dai et al. 2014).

Finally, higher temperatures facilitate downstream product recovery because of cheaper distillation and permeation membrane separation costs (Dai et al. 2014). The main disadvantage of using thermophiles in the industry is the fact that they have been much less investigated and often lack tools form genetically engineer them (Crosby et al. 2019). The main infrastructure of current fermentation technology is based on mesophilic microorganisms. Additionally, proteins usually found in mesophilic microorganisms are unsuitable for many thermophilic bacteria, mainly due to the risk of protein denaturation or poor enzymatic performance at high temperatures. At present, the majority of metabolic engineering microbes are well-known standard, mesophilic microorganisms like *E. coli* and *S. cerevisae*. The main reason for this is the fact that thermophiles in general are more recalcitrant to genetic manipulation than mesophilic microorganisms (Crosby et al. 2019).

The discovery of the thermostable DNA polymerases 25 years back has revolutionized genetic engineering and molecular biology and was rewarded with the Nobel Prize to Mullis and Smith in 1993. Thermostable microbes such as *Thermus aquaticus* (Kuznetsova et al. 2022), *Pyrococcus furiosus* (Ishino 2020), and *Thermococcus litoralis* (Terpe 2013) have been used for the production of DNA polymerases that survive denaturation temperatures and are now used as a routine work model for molecular biology world over. Another good example of the use of thermophilic enzymes use in industry is in the starch field. Starch degradation usually occurs in a two-step procedure, firstly by liquefaction of starch granules, and secondly by saccharification. These two steps were for a long time done at two different temperatures (105 °C for 5 min and 95 °C for 1 h, at pH 6.0), and at 60 °C for 3 h at pH 4.5. Traditionally, the key enzymes used were amylases, glucoamylases, and pullulanases that do not tolerate high temperatures and low pH used in the process, and thus, cooling and pH adjustments were necessary for the saccharification of starch. In 1990, the first archaeal amylase was investigated with an optimum temperature of 100 °C found in *Pyrococcus furiosus* (Koch et al. 1990). More recent observations involving acid-stable amylase was cloned from *Bacillus acidicola* to *E. coli* with a half-life of 30 min at 80 °C and pullulanase from *Thermococcus kodakarensis* KOD1 with a temperature optimum at 100 °C (Sharma and Satyanarayana 2012; Han et al. 2013). Many pharmaceutically active compounds contain nitrogen and can be derived from amino acids (Drauz 1997). Thermophilic L-aminoacylase (esterase) was cloned and overexpressed from archaeon *Thermococcus litoralis* (Toogood et al. 2002). This esterase gene responsible codes for pyroglutamyl carboxyl peptidase which is a cystein protease that cleaves the pyroglutamyl group from the N-terminus of biologcially important peptides. The commercial use of this enzyme is to cleave the pyroglutamyl group from blocked peptides. Other examples of commercial enzymes that have been identified are carboxyl esterase from *Thermogutta terrifontis* (Sayer et al. 2015), μ-lactamase from *Sulfolobus solfataricus* (Taylor et al. 1993), an α-carbonic anhydrase from *Thermovibrio ammonificans* (James et al. 2014), and transaminase from *Sulfolobus sulfataricus* (Sayer et al. 2012).

Although extremophiles and extremozymes have shown great potential in various industrial applications, their commercial use is still limited. Some recent trends show that extremozymes may emerge as available industrial enzymes (Sarimiento et al. 2015) although their full potential is yet to be realized. As for mesophiles, the basic lack of knowledge concerning growth, long generation times, and low yields are major limiting factors for industrial applications to fulfill their promise. The major limiting factor is the large-scale cultivation of extremophiles because of the robust conditions often needed. High temperatures result in low solubility of gases that may be growth limiting for the thermophiles involved together with product inhibition and difficulties with downstream and recovery processes (Gabani and Singh 2013).

3.5 Other Products/Use of Thermophiles

There are several other fields where thermophilic, anaerobic bacteria are used as for example conversion of glycerol to lactate, biodegradation of petroleum hydrocarbons, recovery of heavy metals, and remediation of textile dyes but most of these are done by mesophilic and aerobic thermophiles (Mehta et al. 2016). Other aerobic processes involved in the production of fine chemicals like 1,4, diacids, 3-hydroxy propionic acid, aspartic acid, glucaric acid, gluconic acid, sorbitol, and xylitol are also well-known products via aerobic fermentations (Turner et al. 2007).

4 Future Outlook

Although second-generation production of ethanol and hydrogen at high temperatures has been investigated intensively, there seem to be very few full-scale facilities using thermohilic bacteria. The first plant-producing lignocellulosic biomass to produce ethanol started in Italy (BioLyfe) in 2013 but they are using mesophilic temperatures with yeast as the fermenters. Until 2020, more than 60 production plants have been started, most of them in the US but almost half of them are not operative now. The main reason for fully industrializing second-generation ethanol production is in biomass pretreatment (recalcitrance of lignocellulosic feedstock to chemicals or enzymes, complete delignification processes, generation of inhibititors, and low sugar yields), enzymatic hydrolysis (cost of enzymes, effect of solid loadings), and fermentation (co-fermentation, rates of sugar uptake, tolerance of ethanol producing bacteria for high initial substrate concentrations, inhibitors and products). Another factor of importance is the fact that the energy content of ethanol is 33% lower as compared with gasoline. Therefore, the fact that thermophilic bacteria within the genera of *Thermoanaerobacter* and *Caldanaerobacter* are capable of converting high-carbon fatty acids to their corresponding alcohol, in the presence of carbohydrates may be an interesting option in near future. Full-scale production plants for hydrogen do still today not exist and more interest seems to be for non-biological hydrogen production. Increased knowledge about the bottlenecks for using either photobiological or fermentative bacteria for the production of biohydrogen is needed. Biorefineries for the production of fine chemicals known to be produced by thermophilic anaerobic bacteria like 1,2-propanediol, 1,3-propanediol and branched chain alcohols are still not economically viable to compete with chemical production routes. Currently, none of these compounds are produced in a large-scale facility because of the expensive cost. The most prosing area of research seems to be using genetic engineering to enhance yields and lower the production cost. The use of extremophilic enzymes used in various processes of industries like paper and pulp, saccharolytic degradation, and in the field of biofuel production of complex biomass is an exciting field of research. In many of these cases, extreme temperatures are needed, and thus, the importance of thermophiles will be of great importance in near future.

The robustness and high growth rates of thermophiles give a great advantage compared to mesophilic microorganisms in various industries. Their enzymes are thermostable and thus suitable for industrial processes that take place at high temperatures. The main disadvantage of using thermophiles in the industry is the fact that they have been much less investigated and often lack tools for genetically engineer them. The main enzymes of interest that are produced by thermophilic bacteria are saccharolytic enzymes used in the sugar and detergent industries. Opportunities are clearly for identifying suitable microorganisms and enzymes because most enzymes used today are thermotolerant enzymes produced from mesophilic microorganisms. As for mesophiles, the basic lack of knowledge concerning growth, long generation times, and low yields are major limiting factors for industrial applications to fulfill their promise. The major limiting factor is the large-scale cultivation of extremophiles because of the robust conditions often needed. High temperatures result in low solubility of gases that may be growth limiting for the thermophiles involved together with product inhibition and difficulties with downstream and recovery processes.

References

Almarsdottir AR, Sigurbjornsdottir MA, Orlygsson J (2012) Effects of various factors on ethanol yields from lignocellulosic biomass by *Thermoanaerobacterium* AK17. Biotechnol Bioeng 109:686–694. https://doi.org/10.1002/bit.24346

Altaras NE, Etzel MR, Cameron DC (2001) Conversion of sugars to 1,2-propanediol by *Thermoanaerobacterium thermosaccharolyticum* HG-8. Biotechnol Prog 17:52–56. https://doi.org/10.1021/bp000130b

Bhalla A, Bansal N, Kumar S et al (2013) Improved lignocellulose conversion to biofuel with thermophilic bacteria and thermostable enzymes. Bioresour Technol 128:751–759. https://doi.org/10.1016/j.biortech.2012.10.145

Bielen AAM, Willquist K, Engman J et al (2010) Pyrophosphate as a central energy carrier in the hydrogen producing extremely thermophilic *Caldicellulosiruptor saccharolyticus*. FEMS Microbiol Lett 307:48–54. https://doi.org/10.1111/j.1574-6968.2010.01957.x

Blumer-Schuette SE (2020) Insight into thermophilic plant biomass hydrolysis from *Caldicellulosiruptor* systems biology. Microorganisms 8:385. https://doi.org/10.1016/j.copbio.2008.04.007

Blumer-Schuette SE, Kataeva I, Westpheling J et al (2008) Extremely thermophilic microorganisms for biomass conversion: status and prospects. Science. Direct 19:210–217

Brandelli A, Daroit DJ, Riffel A (2010) Biochemical features of microbial keratinases and the production and applications. Appl Microbiol Biotechnol 85:1735–1750. https://doi.org/10.1007/s00253-009-2398-5

Brynjarsdottir H, Wawiernia B, Orlygsson J (2012) Ethanol production from sugars and complex biomass by *Thermoanaerobacter* AK5: the effect of electron scavenging systems on end-product formation. Energy Fuel 26:4568–4574. https://doi.org/10.1021/ef300754q

Brynjarsdottir H, Scully SM, Orlygsson J (2014) Production of biohydrogen from sugars and lignocellulosic biomass using *Thermoanaerobacter* GHL15. Int J Hydrog Energy 38:14467–14475. https://doi.org/10.1016/j.ijhydene.2013.09.005

Cai J, Xie Y, Song B et al. (2011) *Fervidobacterium changbaicum* Lip1: identification, cloning, and characterization of the thermophilic lipase as a new member of the lipase family V. Appl Microbiol Biotechnol 89:1463–1473. doi: https://doi.org/10.1007/s00253-010-2971-y.

Cameron DC, Cooney CL (1986) A novel fermentation: the production of R(−)1,2-propanediol and acetol by *Clostridium thermosaccharolyticum*. Nat Biotechnol 4:719–725. https://doi.org/10.1038/nbt0786-651

Collins T, Gerday C, Feller G (2005) Xylanases, xylanase families and extreme xylanases. FEMS Microbiol Rev 29:3–23. https://doi.org/10.1016/j.femsre.2004.06.005

Cord-Ruwisch R, Seitz H, Conrad R (1988) The capacity of hydrogenotrophic anaerobic bacteria to compete for traces of hydrogen depends on the redox potential of the terminal electron acceptor. Arch Microbiol 149:350–357. https://doi.org/10.1007/BF00411655

Crosby JR, Laemthong T, Lewis AP et al (2019) Extreme thermophiles as emerging metabolic engineering platforms. Curr Opin Biotechnol 59:55–64. https://doi.org/10.1016/j.copbio.2019.02.006

Cuecas A, Cruces J, Galisteo-Lopes JF et al (2016) Cellular viscosity in prokaryotes and thermal stability of low molecular weight biomolecules. Biophys J 111:875–8882. https://doi.org/10.1016/j.bpj.2016.07.024

Dai YM, Chen KT, Chen CC (2014) Study of the microwave lipid extraction from macroalgae for biodiesel production. Chem Eng J 250:267–273. https://doi.org/10.1016/j.cej.2014.04.031

Drauz K (1997) Chiral amino acids: a versatile tool in the synthesis of pharmaceuticals and fine chemicals. Int J Chem Chimia 51:310–314

Ebaid R, Wang H, Sha C et al (2019) Recent trends in hyperthermophilic enzymes production and future perspectives for biofuel industry. J Clean Prod 238:117925. https://doi.org/10.1016/j.jclepro.2019.117925

Elleuche S, Antranikian G (2013) Starch-hydrolyzing enzymes from thermophiles. In: Satyanarayana T (ed) Thermophilic microbes in environmental and industrial biotechnology: viotechnology of thermophiles, 2nd edn. Springer, Hoboken, pp 509–533. https://doi.org/10.1007/978-94-007-5899-5_20

Ergun BG, Calik P (2016) Lignocellulose degrading extremozymes produced by *Pichia pastoris*: current status and future prospects. Bioprocess Biosyst Eng 39:1–36. https://doi.org/10.1007/s00449-015-1476-6

Fardeau MI, Patel BKC, Magot M et al (1997) Utilization of serine, leucine, isoleucine and valine by *Thermoanaerobacter brockii* in the presence of thiosulfate or *Methanobacterium* sp. as electron acceptors. Anaerobe 3:405–410. https://doi.org/10.1006/anae.1997.0126

Finore I, Feola A, Russo L et al (2023) Thermophilic bacteria and their thermozymes in composting processes: a review. Chem Biol Technol Agric 10:7. https://doi.org/10.1186/s40538-023-00381-z

Futterer O, Angelov A, Liesgang H et al (2004) Genome sequence of *Picrophilus torridus* and its implications for life around pH 0. Proc Natl Acad Sci USA 101:9091–9096. https://doi.org/10.1073/pnas.0401356101

Gabani P, Singh OV (2013) Radiation-resistant extremophiles and their potential in biotechnology and therapeutics. Appl Microbiol Biotechnol 97:993–1004. https://doi.org/10.1007/s00253-012-4642-7

Giaveno MA, MSa U, Ulloa JR et al (2013) Physiologic versatility and growth flexibility as the main characteristics of a novel thermoacidophilic *Acidianus* strain isolated from Copahue geothermal area in Argentina. Microb Ecol 65:336–346. https://doi.org/10.1007/s00248-012-0129-4

Hahn-Hagerdahl B, Galbe M, Gorwa-Grauslund MF et al (2006) Bioethanol the fuel of tomorrow from residues today. Trends Biotechnol 24:549–556. https://doi.org/10.1016/j.tibtech.2006.10.004

Hallenbeck PC (2009) Fermentative hydrogen production: Principles, progress and prognosis. Int J Hydrog Energy 27:1185–1193. https://doi.org/10.1016/j.ijhydene.2008.12.080

Han T, Zeng F, Li Z et al (2013) Biochemical characterization of a recombinant pullulanase from *Thermococcus kodakarensis* KOD1. Lett Appl Microbiol 57:336–343. https://doi.org/10.1111/lam.12118

Hebal H, Hamalainen J, Makkonen L et al (2022) Enhanced activity of hyperthermostable *Pyrococcus horikoshii* endoglucanase in superbase ionic liquids. Biotechnol Lett 44:961–974. https://doi.org/10.1007/s10529-022-03268-5

Huang CX, Jiang X, Shen XJ et al (2022) Lignin-enzyme interaction: A roadblock for efficient enzymatic hydrolysis of lignocellulosics. Renew Sustain Energy Rev 154:111–822. https://doi.org/10.1016/j.rser.2021.111822

Ichikawa S, Nishida A, Yasui S et al (2017) Characterization of lignocellulose particles during lignocellulose degradation by *Clostridium thermocellum*. Biosci Biotechnol Biochem 81:2028–2033. https://doi.org/10.1080/09168451.2017.1364619

Ignatova Z, Gousterova A, Spassov G et al (1999) Isolation and partial characterization of extracellular keratinase from wool degrading thermophilic actinocycete strain *Thermoactionmyces candidus*. Can J Microbiol 45:217–222. https://doi.org/10.1139/cjm-45-3-217

Ingvadottir EM, Scully SM, Orlygsson J (2017) Evaluation of the genus *Caldicellulosiruptor* for production of 1,2-propanediol from methylpentoses. Anaerobe 47:86–88. https://doi.org/10.1016/j.anaerobe.2017.04.015

Ingvadottir EM, Scully SM, Orlygsson J (2018) Production of (S)-1,2-propanediol from l-rhamnose using the moderately thermophilic *Clostridium* strain AK1. Anaerobe 54:26–30. https://doi.org/10.1016/j.anaerobe.2018.07.003

Ishino Y (2020) Studies on DNA-related enzymes to elucidate molecular mechanisms underlying genetic information processing and their application in genetic engineering. Biosci Biotechnol Biochem 84:1749–1766. https://doi.org/10.1080/09168451.2020.1778441

Ivanova G, Rakhely G, Kovacs KL (2009) Thermophilic biohydrogen production from energy plants by *Caldicellulosiruptor saccharolyticus* and comparison with related studies. Int J Hydrog Energy 34:3659–3670. https://doi.org/10.1016/j.ijhydene.2009.02.082

James P, Isupov MN, Sayer C et al (2014) The structure of a tetrameric alpha-carbonic anhydrase from *Thermovibrio ammonificans* reveals a core formed around intermolecular disulfides that contribute to its thermostability. ACTA Crystallogr D Biol Crystallogr 70:2607–2618. https://doi.org/10.1107/S1399004714016526

Jones P (2008) Improving fermentative biomass-derived hydrogen production by engineered microbial metabolism. Int J Hydrog Energy 33:5122–5130. https://doi.org/10.1016/j.ijhydene.2008.05.004

Koch P, Zablowski P, Spreinat A et al (1990) Extremely thermostable amylolytic enzyme from the archaebacterium *Pyrococcus furiosus*. FEMS Microbiol Lett 71:21–26

Kublanov IV, Tsiroulnikov KB, Kaliberda EM et al (2009) Keratinase of an anaerobic thermophilic bacterium *Thermoanaerobacter* sp. strain 1004–09 isolated from a hot spring in the Baikal rift zone. Microbiology 78:67–75. https://doi.org/10.1134/S0026261709010093

Kuhad RC, Gupta R, Khasa YP et al (2011) Bioethanol production from pentose sugars: current status and future prospects. Renew Sust Energ Rev 15:4950–4962

Kuznetsova AA, Fedorova OS, Kuznetsov NA (2022) Structural and molecular kinetic features of activities of DNA polymerase. Int J Mol Sci 23:12. https://doi.org/10.3390/ijms23126373

Levisson M, van der O, SWM K (2007) Characterization and structural modeling of a new type of thermostable esterase from *Thermotoga maritima*. FEBS J 274:2832–2842. https://doi.org/10.1111/j.1742-4658.2007.05817.x

Liu Y, Yu P, Song X et al (2008) Hydrogen production from cellulose by co-culture of *Clostridium thermocellum* JN4 and *Thermoanaerobacterium thermosaccharilyticum* GD17. Int J Hydrog Energy 33:2927–2933. https://doi.org/10.1016/j.ijhydene.2008.04.004

Mehta R, Singhal P, Singh H et al (2016) Insight into thermophiles and their wide-spectrum applications. 3 Biotech 6:81. https://doi.org/10.1007/s13205-016-0368-z

Mesbah NM (2022) Industrial biotechnology based on enzymes from extreme environments. Front Bioeng Biotech 10:870083. https://doi.org/10.3389/fbioe.2022.870083

Nguyen TAD, Kim JP, Kim MS et al (2008) Optimization of hydrogen production by hyperthermophilic eubacteria, *Thermotoga maritima* and *Thermotoga neapolitana* in batch fermentations. Int J Hydrog Energy 33:1483–1488. https://doi.org/10.1016/j.ijhydene.2007.09.033

Orlygsson J (1994) The role of interspecies hydrogen transfer on thermophilic protein and amino acid metabolism. Ph.D. thesis. Swedish University of Agricultural Sciences

Palm GJ, Fernandez-Alvaro E, Bogdanovic X et al (2011) The crystal structure of an esterase from the hyperthermophilic microorganism *Pyrobaculum calidifontis* VA1 explains its enantioselectivity. Appl Microbiol Biotechnol 91:1061–1072. https://doi.org/10.1007/s00253-011-3337-9

Pawar SS, Nkemka VN, Zeidan AA et al (2013) Biohydrogen production from wheat straw hydrolysate using *Caldicellulosiruptor saccharolyticus* following by biogas production in two-step uncoupled process. Int J Hydrog Energy 38:9121–9130. https://doi.org/10.1016/j.ijhydene.2013.05.075

Pawar SS, Vongkumpeang T, Grey C et al (2015) Biofilm formation by designed co-cultures of *Caldicellulosiruptor* species as a means to improve hydrogen productivity. Biotechnol Biofuels 8:19. https://doi.org/10.1186/s13068-015-0201-7

Rastogi G, Bhalla A, Adhikari A et al (2010) Characterization of thermostable cellulases produced by *Bacillus* and *Geobacillus* strains. Bioresour Technol 101:8798–8806. https://doi.org/10.1016/j.biortech.2010.06.001

Raynaud C, Sarcabal P, Meynial-Salles I et al (2003) Molecular characterization of the 1,3-propanediol operon of *Clostridium butyricum*. Proc Natl Acad Sci USA 100:5010–5015. https://doi.org/10.1073/pnas.0734105100

Riessen S, Antranikian G (2001) Isolation of *Thermoanaerobacter keratinophilus* sp. nov., a novel thermophilic, anaerobic bacterium with keratinase activity. Extremophiles 5:399–408. https://doi.org/10.1007/s007920100209

Rigoldi F, Donini S, Redaelli A et al (2017) Review: engineering of thermostable enzymes for industrial applications. APL Bioeng 2:011501. https://doi.org/10.1063/1.4997367

Royter M, Schmidt M, Elend C et al (2009) Thermostable lipases from the extremophilic anaerobic bacteria *Thermoanaerobacter thermohydrosulfuricus* SOL1 and *Caldanaerobacter subterraneus* subsp. *tengcongensis*. Extremophiles 13:769–783. https://doi.org/10.1007/s00792-009-0265-z

Ruepp A, Graml W, Santos-Martinez M-L et al (2000) The genome sequence of the thermoacidophilic scavenger *Thermoplasma acidophilum*. Nature 407:508–513. https://doi.org/10.1038/35035069

Sánchez-Riera F, Cameron DC, Cooney CL (1987) Influence of environmental factors in the production of R(−)-1, 2-propanediol by *Clostridium thermosaccharolyticum*. Biotechnol Lett 9:449–454. https://doi.org/10.1007/BF01027450

Sarimiento F, Peralta R, Blamey JM (2015) Cold and hot extremozymes: industrial relevance and current trends. Front Bioeng Biotechnol 3:148. https://doi.org/10.3389/fbioe.2015.00148

Saxena RK, Anand P, Saran S et al (2010) Microbial production and applications of 1,2-propanediol. Indian J Microbiol 50:2–11. https://doi.org/10.1007/s12088-010-0017-x

Sayer C, Bommer M, Ward JM et al (2012) Crystal structure and substrate specificity of the thermophilic serine: pyruvate aminotransferase from *Sulfolobus solfataricus*. Acta Crystallogr D Biol Crystallogr D68:763–772. https://doi.org/10.1107/S0907444912011274

Sayer C, Isupov MN, Bonch-Osmolovskaya E et al (2015) Structural studies of thermophilic esterase from a new *Planctomycetes* species *Thermogutta terrifontis*. FEBS J 282:2846–2857. https://doi.org/10.1111/febs.13326

Scully SM, Orlygsson J (2015) Recent advances in second generation ethanol production by thermophilic bacteria. Energies 8:1–30. https://doi.org/10.3390/en8010001

Scully SM, Orlygsson J (2020a) Branched-chain amino acid catabolism of *Thermoanaerobacter pseudethanolicus* reveals route to branched-chain alcohol formation. Extremophiles 24:121–133. https://doi.org/10.1007/s00792-019-01140-5

Scully SM, Orlygsson J (2020b) Biotransformation of carboxylic acids to alcohols: characterization of *Thermoanaerobacter* strain AK152 and 1-propanol production via propionate reduction. Microorganisms 8:6. https://doi.org/10.3390/microorganisms8060945

Scully SM, Iloranta P, Myllymaki P et al (2015) Branched-chain alcohol formation by thermophilic bacteria within the genera of *Thermoanaerobacter* and *Caldanaerobacter*. Extremophiles 19: 809–818. https://doi.org/10.1007/s00792-015-0756-z

Scully SM, Brown AE, Mueller-Hilger Y et al (2021) Influence of culture conditions on the bioreduction of organic acids to alcohols by *Thermoanaerobacter pseudethanolicus*. Microorganisms 9:1. https://doi.org/10.3390/microorganisms9010162

Seyfried M, Lyon D, Raney FA et al. (2002) *Caloramator viterbensis* sp. nov., a novel thermophilic, glycerol-fermenting bacterium isolated from a hot spring in Italy. Int J Syst Evol Microbiol 52:1177–1184. doi: https://doi.org/10.1046/j.1432-1327.2001.02123.x.

Sharma A, Satyanarayana T (2012) Cloning and expression of acid stable, high maltose forming Ca2+ independent alpha amylase from an acidophile *Bacillus acidicola* and its applicability in starch hydrolysis. Extremophiles 16:515–522. https://doi.org/10.1007/s00792-012-0451-2

Shaw AJ, Podkaminer KK, Desai SG et al (2008) Metabolic engineering of a thermophilic bacterium to produce ethanol at high yield. Proc Natl Acad Sci U S A 105:13769–13774. https://doi.org/10.1073/pnas.0801266105

Sigurbjornsdottir MA, Orlygsson J (2012) Combined hydrogen and ethanol production from sugars and lignocellulosic biomass by *Thermoanaerobacterium* AK54, isolated from hot spring. Appl Energy 97:785–791. https://doi.org/10.1016/j.apenergy.2011.11.035

Straub CT, Schut G, Otten JK et al (2020) Modification of the glycolytic pathway in *Pyrococcus furiosus* and the implications for metabolic engineering. Extremophiles 24:511–519. https://doi.org/10.1007/s00792-020-01172-2

Taylor SJC, McCague R, Wisdom R et al (1993) Development of the biocatalyctic resolution of 2-azabicyclo [2.2.1] hept-5-en-3-one as an entry to single enantiomer carbocyclic nucleosides. Tetrahedron Assymetry 4:1117–1128. https://doi.org/10.1016/S0957-4166(00)80218-9

Taylor MP, Eley KL, Martin S et al (2009) Thermophilic ethanologenesis: future prospects for second generation bioethanol production. Trends Biotechnol 27:398–405. https://doi.org/10.1016/j.tibtech.2009.03.006

Terpe K (2013) Overview of thermostable DNA polymerases for classical PCR applications: from molecular and biochemical fundamentals to commercial systems. Appl Microbiol Biotechnol 97(10243–10):254. https://doi.org/10.1007/s00253-013-5290-2

Toogood HS, Hollingsworth EJ, Brown RC et al (2002) Thermostable L-aminoacylase from *Thermococcus litoralis*: cloning, overexpression, characterization, and applications in biotransformations. Extremophiles 6:111–122. https://doi.org/10.1007/s007920100230

Turner KW, Robertson AM (1979) Xylose, arabinose, and rhamnose fermentation by *Bacteroides ruminicola*. Appl Environ Microbiol 38:7–12. https://doi.org/10.1128/AEM.38.1.7-12.1979

Turner P, Mamo G, Karlsson EN (2007) Potential and utilization of thermophilic and thermostable enzymes in biorefining. Microbiol Cell Fact 6:9. https://doi.org/10.1186/1475-2859-6-9

Urbieta MS, Rascovan N, Castro C et al (2014a) Draft genome sequence of the novel thermoacidophilic archaeon *Acidianus copahuensis* strain ALE1, isolated from the Copahue volcanic area in Neuquen, Argentina. Genome Announc 2:e00259–e00254

Urbieta MS, Toril EG, Giaveno MA et al (2014b) Archaeal and bacterial diversity in five different hydrothermal ponds in the Copahue region in Argentina. Syst Appl Microbiol 37:429–441. https://doi.org/10.1016/j.syapm.2014.05.012

Urbieta MS, Donati ER, Chan KG et al (2015) Thermophiles in genomic era: biodiversity, science, and applications. Biotechnol Adv 33:633–647. https://doi.org/10.1016/j.biotechadv.2015.04.007

Vipotnik Z, Jessen JE, Scully SM et al (2016) Effect of culture conditions on hydrogen production by *Thermoanaerobacter* strain AK68. Int J Hydrog Energy 41:181–189. https://doi.org/10.1016/j.ijhydene.2015.10.124

Wagner ID, Wiegel J (2008) Diversity of thermophilic anaerobes. Incredible anaerobes: from physiology to genomics to fuels. Ann NY Acad Sci. 1125:1–43

MIX
Papier aus verantwortungsvollen Quellen
Paper from responsible sources
FSC® C105338

FSC
www.fsc.org

If you have any concerns about our products,
you can contact us on
ProductSafety@springernature.com

In case Publisher is established outside the EU,
the EU authorized representative is:
Springer Nature Customer Service Center GmbH
Europaplatz 3, 69115 Heidelberg, Germany

Printed by Libri Plureos GmbH
in Hamburg, Germany